**Cambridge Studies in Ecology** presents balanced, comprehensive, up-to-date, and critical reviews of selected topics within ecology, both botanical and zoological. The Series is aimed at advanced final-year undergraduates, graduate students, researchers, and university teachers, as well as ecologists in industry and government research.

It encompasses a wide range of approaches and spatial, temporal, and taxonomic scales in ecology, including quantitative, theoretical, population, community, ecosystem, historical, experimental, behavioural and evolutionary studies. The emphasis throughout is on ecology related to the real world of plants and animals in the field rather than on purely theoretical abstractions and mathematical models. Some books in the Series attempt to challenge existing ecological paradigms and present new concepts, empirical or theoretical models, and testable hypotheses. Others attempt to explore new approaches and present syntheses on topics of considerable importance ecologically which cut across the conventional but artificial boundaries within the science of ecology.

CAMBRIDGE STUDIES IN ECOLOGY



# The ecology of bird communities

*Volume 2*

Processes and variations

*Also in the series*

H. G. Gauch, Jr *Multivariate Analysis in Community Ecology*

R. H. Peters *The Ecological Implications of Body Size*

C. S. Reynolds *The Ecology of Freshwater Phytoplankton*

K. A. Kershaw *Physiological Ecology of Lichens*

R. P. McIntosh *The Background of Ecology: Concept and Theory*

A. J. Beattie *The Evolutionary Ecology of Ant–Plant Mutualisms*

F. I. Woodward *Climate and Plant Distribution*

J. J. Burdon *Diseases and Plant Population Biology*

J. I. Sprent *The Ecology of the Nitrogen Cycle*

N. G. Hairston, Sr *Community Ecology and Salamander Guilds*

H. Stolp *Microbial Ecology: Organisms, Habitats and Activities*

R. N. Owen-Smith *Megaherbivores: the Influence of Large Body Size on Ecology*

J. A. Wiens *The Ecology of Bird Communities:*
*Volume 1 Foundations and Patterns*
*Volume 2 Processes and Variations*

N. G. Hairston, Sr *Ecological Experiments*

R. Hengeveld *Dynamic Biogeography*

C. Little *The Terrestrial Invasion: an Ecophysiological Approach to the Origins of Land Animals*

P. Adam *Saltmarsh Ecology*

M. F. Allen *The Ecology of Mycorrhizae*

D. J. von Willert *et al.* *Life Strategies of Succulents in Deserts: With Special Reference to the Namib Desert*

J. A. Matthews *The Ecology of Recently-deglaciated Terrain: a Geoecological Approach to Glacier Forelands*

# The Ecology of Bird Communities

*Volume 2*

## Processes and variations

JOHN A. WIENS

*Professor of Ecology*
*Department of Biology*
*Colorado State University*
*Fort Collins, Colorado*

Published by the Press Syndicate of the University of Cambridge
The Pitt Building, Trumpington Street, Cambridge CB2 1RP
40 West 20th Street, New York, NY 10011–4211, USA
10 Stamford Road, Oakleigh, Victoria 3166, Australia



First published 1989

First paperback edition 1992

Printed in Great Britain at the University Press, Cambridge

*British Library cataloguing in publication data*
Wiens, J A
The ecology of bird communities
Vol. 2: Processes and variations
1. Birds. Ecology
I. Title
598.2′5

*Library of Congress cataloguing in publication data*
Wiens, John A.
The ecology of bird communities / John A. Wiens
p. cm. – (Cambridge studies in ecology)
Includes bibliographies and indexes.
Contents: v. 1. Foundations and patterns – v. 2. Processes and variations.
ISBN 0 521 26030 2 (v. 1) (hb) – ISBN 0 521 36558 9 (v. 2) (hb)
ISBN 0 521 42634 0 (v. 1) (pb) – ISBN 0 521 42635 9 (v. 2) (pb)
1. Birds–Ecology. I. Title. II. Series.
QL673.W523 1989
598.2′5–dc19 88-38331 CIP

ISBN 0 521 26030 2 vol. 1 hardback
ISBN 0 521 42634 0 vol. 1 paperback
ISBN 0 521 36558 9 vol. 2 hardback
ISBN 0 521 42635 9 vol. 2 paperback
ISBN 0 521 42636 7 set of 2 vols. paperback

SE

*For Bea*

# Contents of volume 2

# Contents of volume 1

# Preface

This book and its companion volume represent a personal statement about the ecology of bird communities – what they are, what we know about them, and what we need to know. They are also about how avian community ecology has been practiced as a science – how we have gone about gaining our knowledge of bird comunities and how logical and methodological considerations affect the certainty we can attach to that knowledge. Because studies of birds have contributed a good deal to the foundation of contemporary community ecology and because concerns about logic, methodology, and epistemology are central to any science, I believe that the themes and viewpoints I develop are relevant to community ecology well beyond the somewhat artificial boundaries dictated by my focus on birds. My topic is really community ecology as it has been practiced on birds rather than bird communities *per se*.

Avian community ecology is a complex, multifaceted discipline that is enriched by controversy. I have written these volumes partly in an attempt to examine the complexity of communities and to probe the dimensions of the controversies, but partly also out of a simple enjoyment of the subject. I have directed my comments particularly toward advanced undergraduates and graduate students with interests in avian community ecology or, more broadly, in birds or in ecology, for I feel that they are in the best position to put my comments into practice or to challenge my views. I hope that my colleagues – practicing ecologists – will also find much to interest (or outrage!) them.

My objective when I began this project was to provide a critical assessment of the current state of affairs in avian community ecology. I rapidly discovered that this required that many studies and areas of investigation be developed in considerable detail. It also led to excursions into aspects of the philosophy of science and the history of community ecology that provide essential perspectives on current thinking in this discipline. When I finished, it was apparent that the manuscript would result in a book of such formidable size (and cost!) that only the most dedicated students would be likely to read it. Accordingly, it was decided to publish the work in two volumes. Although these volumes are a closely integrated set, each has a particular focus. In the first volume, I emphasize the historical, logical, and methodological background of studies of bird communities and the sorts of patterns that these studies have (or have not) revealed. In this volume, I consider how these patterns have been explained in terms of causal processes, how the operation of those processes has been determined, and how the patterns and our efforts to discern and understand them are influenced by the complexity and variability of natural environments.

It is clear that we know rather less about bird communities than we thought we knew. Our explanations of patterns have more often been founded on beliefs than on empirical tests, and our recognition of the temporal and spatial variability of natural systems has only complicated matters further. The erosion of our certainty about the patterns of communities and their interpretation, however, has its rewards: we can now see more clearly what needs to be done in order to pose and answer questions (or test hypotheses) in community ecology with greater rigor, so that when answers are obtained we may have greater confidence in their being correct. Community ecology is a difficult science, but it is not impossible. I hope that the views I develop will not discourage readers but will inspire them to design and conduct appropriate field studies and to generate useful theory for dealing with the complexity that, after all, *is* nature.

I should say what this book is not. It is not a review or synthesis of all the literature dealing with bird communities. Instead, I have used selected examples that, for one reason or another, seemed to me appropriate. Most importantly, it is not a book about community theory. There are lots of concepts presented, but very little formal theory *per se*. There are few equations in the book. Others (e.g. Roughgarden 1979, Pielou 1974, 1977, Vandermeer 1981, May 1976) do that well, and I am not a theorist (although I could be called a conceptualist).

I think that the time is appropriate for a clear, critical examination of our observations of nature, to determine where both our field studies and our development of theory should be directed. I hope that this book and its companion volume will provide the stimulus and guidance for these developments.

# Acknowledgments

I contemplated writing a book about bird communities for many years, but the opportunity and the impetus to stop contemplating and start writing came during a sabbatical leave from the University of New Mexico that I spent as a Fulbright Senior Scholar in Australia in 1984–85. Peter Sale, Tony Underwood, Rich Bradley, Charles Birch, and other members of the School of Biological Sciences at the University of Sydney provided support, good discussions, and southern hemispheric insights that broadened and sharpened my thinking about communities. At the University of New Mexico, Don Duszynski and Mary Alice Root provided assistance and understanding as I continued to write, and faculty, students, and administrators at Colorado State University tolerated my absentminded preoccupation as I completed the project. My graduate students, Gary, Luke, Hanni, Rachel, and Tasha, provided a refreshing diversity of viewpoints about communities for me to think about.

Several of my colleagues found themselves volunteering to read chapters in varying stages of development: Joel Cracraft (Chapters 1, 2, 3), John Emlen (1, 2, 3), Luke George (1, 2, 3), Yrjö Haila (5), Dennis Heinemann (1, 2), Dick Holmes (4), Janice Moore (3), Bill Rice (4), Terry Root (1, 2), Tom Sherry (1, 2, 3), Dean Urban (5), and Bea Van Horne (6). Their comments were unfailingly helpful and deeply appreciated, and of course I followed all of their suggestions to the letter. Jean Ferner read the entire manuscript in its initial draft, kindly pointing out my lapses in style and my affronts to the English language. John Birks did the same for the penultimate draft; thanks to Jean, his job was relatively easy. I owe a special debt of gratitude to each. Sylvia Sullivan handled the subediting at Cambridge University Press, and was extremely perceptive in finding small inconsistencies that no one else had noticed and quite gracious in pointing them out.

Authors usually thank members of their family for putting up with them during the writing of a book. I never understood what they were talking about. It seems, however, that there have been moments during the writing of this book when my behavior was not entirely normal, when I seemed detached from what was going on about me, or when I even was not a joy to be around. Thanks, Bea, for your understanding and support.

# PART I

## Explaining community patterns: process hypotheses

The study of bird communities begins with the search for patterns, but rarely does it end there. A pattern, by its very existence, begs for explanation. What processes have caused it to be the way it is? The search for patterns in bird communities has usually been conducted within the framework of the MacArthurian paradigm, which focuses on interspecific competition as the major (often the only) process determining these patterns. As a consequence, patterns consistent with that view have been emphasized, even when methodological flaws render them suspect or when they cannot be distinguished from patterns that might be generated by other processes. There is thus a bias to the sorts of patterns we document and discuss. The well-worn tenets of competition theory fit comfortably over those patterns, providing explanations that are satisfying to many ecologists. Processes are inferred from patterns and then treated as demonstrated facts rather than as hypotheses awaiting tests. Unfortunately, it is considerably easier to assert the operation of a process such as competition than to test that assertion, and this has reinforced the tendency to rely on inference and assertion in process explanations.

To break away from this doctrinaire approach to interpreting community patterns, it is necessary to recognize the folly of single-factor explanations and to evaluate alternative process hypotheses. In order to test these hypotheses, however, we must consider the criteria for documenting the operation of a process with a given degree of certainty and evaluate the evidence at hand. This is my objective in the following three chapters. Because of its historical prominence, I consider competition first and then examine various other processes that may affect community patterns.

# 1

# Competition

The most frequently offered explanation of the patterns of communities is that they are products of interspecific competition, acting either contemporaneously or in the past. Writing in 1975, Ricklefs stated that 'few ecologists doubt that competition is a potent ecological force or that it has guided the evolution of species relationship within communities' (1975: 581), and, in his more recent review, Giller (1984) attributed most of the community patterns he discussed to competition. The challenges that have brought controversy to community ecology have centered largely on whether or not competition produces the patterns we see and how applicable a good deal of competition theory is to nature.

It is important to distinguish between the process of competition and its application as an explanation of community patterns. When Brown (1981) wrote that he found competition theory 'a disappointment', he was concerned with the *theory* and its application, not the *process* itself. Competition theory applies the process of competition under specified conditions to predict community patterns. If the predictions fail to match observations, it is not because the process is flawed but because the theory is incorrect or has been misapplied. The challenges to the competition paradigm have questioned the simplicity of the theory and its overly enthusiastic or illogical applications, not the reality of the process itself. There is nothing about the process of competition, as an interaction among individuals, that requires equilibrium or optimization or precludes opportunism, nonequilibrium, or stochastic influences. The notion that equilibrium and optimization are associated with competition derives from conventional applications of the theory that predict population-level consequences of the interactions, usually in the framework of the Lotka–Volterra equations. These applications lead one to expect that the process of competition, acting under equilibrium conditions or until an equilibrium is attained, will produce optimal, adaptive patterns. If competition occurs intermittently, the optimal patterns may not be produced (Wiens 1977), but the process nonetheless does occur. Contrary to Martin's (1986) claim, most ecologists have not viewed competition as an 'all-or-none' phenomenon, although they have shown an excessive preoccupation with conditions close to equilibrium or carrying capacity. Even in the simplistic Lotka–Volterra formulations, competition occurs in populations well below carrying capacity.

The difference between the occurrence of competition among individuals and its effects on individual, population, or community attributes is expressed in the distinction between the *intensity* of competition (its proximate, physiological or behavioral effects on individuals) and its *importance* (the ecological or evolutionary consequences of those effects) (Welden and Slauson 1986). Competition may at times be quite intense but nonetheless be relatively unimportant if individual fitness or community attributes are determined largely by other factors. Because there may be variation in the degree to which different individuals in a population experience competition, some individuals may be influenced by competition without the effects of these interactions translating into population- or community-level consequences (Martin 1986). Conversely, competition may occur at relatively low intensity but be very important if it is the major process influencing individual fitness in populations or the niche space occupied by a species. Our failure to recognize this distinction has certainly contributed to some of the arguments about the frequency of competition in nature and its role in determining community patterns.

## What is competition?

Competition as a process has been defined in various ways. Miller (1967) defined it as 'the active demand by two or more individuals . . . for a common resource that is actually or potentially limiting', whereas Arthur (1982) considered two populations to be competing 'if each exerts an inhibitory effect on the growth or equilibrium size of the other'. Aarssen (1984) termed it 'an interaction between two (or more) organisms in which each reduces the availability of resource units to the other from a limited supply on which they both make demands'. Wilson (1975) said it was the 'active demand by two or more organisms (or two or more species) for a common resource'. Birch's (1957) definition is perhaps the broadest: competition occurs 'when a number of animals . . . utilize common resources, the supply of which is short; or if the resources are not in short supply, competition occurs when the animals seeking that resource nevertheless harm one another in the process'.

A careful reading of these and other definitions reveals subtle but important differences (Table 1.1). First, there are differences in how competition is held to operate. The most common distinction is between *exploitation* and *interference* competition (Miller 1967, 1969) (Schoener (1983b) has distinguished six types of competition, but only three of these are applicable to birds and these represent variations on the exploitation–interference theme). In exploitation competition, individuals have free access to resources and the use of those resources by some individuals diminishes their availability to other individuals. Interference competition usually incorporates a spatial component, some individuals being denied access to resources by the actions (often aggressive) of others. Most definitions include both, although not always explicitly. Exploitation and interference differ not only in their mechanism but in their possible effects. Interference may promote temporal separation among competing species, for example (Carothers and Jaksić 1984), or lead to the exclusion of one species by another when resource measures do not indicate obvious limitation. As a consequence, the response of individuals or populations to experimental manipulations of resources or species removals will differ depending on the form of competition that is present. Roughgarden (1979) and Case and Gilpin (1974) have suggested that interference is an evolutionary consequence of prior exploitation competition; this led Roughgarden to suggest that interspecific aggression should occur only under conditions of limiting resources, as 'there should not be conflict over nothing' (1979: 412). On the other hand, Maurer (1984) argued that, because interference competition incurs the additional costs of aggressive behavior, it should occur only when resources are abundant enough to offset those costs. According to this view, then, exploitation competition occurs when resources are so scarce that the costs of active interference cannot be met; interference competition may be more frequent when resources are abundant. Both Maurer's and Roughgarden's arguments rely on optimality assumptions, and other explanations of interspecific aggression can be offered (e.g. Murray 1971, Payne and Groschupf 1984, Martin 1986).

There are also differences in the level at which the process is presumed to operate (Table 1.1). Miller and Aarssen, for example, explicitly focus on individuals, whereas Arthur and others frame competition in terms of population interactions. Those who emphasize populations as the competing entities also tend to define the process in terms of its *effects* (e.g. depression of densities, population growth, alteration of age structure). This contrasts with other definitions, in which the *prior conditions* promoting the interaction (i.e. shared use of resources) are the focus. There are also differences of opinion regarding the necessary consequences of a competi-

Table 1.1. *Characteristics of definitions of competition, from a sampling of references*

| Reference | Level of action | | Defined by | | Nature of effects | | | Resource limitation | | | Form of competition | | |
|---|---|---|---|---|---|---|---|---|---|---|---|---|---|
| | Individual | Population | Effects | Shared resource use | -,- | -,0 | No effects stated | Stated | Implied | Not involved | Exploitation | Interference | Both implied |
| Birch (1957) | X | | | X | X | | | X | | X | X | X | |
| Milne (1961) | X | | | X | | | X | X | | X | | | X |
| Miller (1967) | X | | | X | | | X | X | | X | | | X |
| Odum (1971) | | X | X | | X | | | | | X | | | X |
| Reynoldson & Bellamy (1971) | X | | X | | | X | | | | X | X | X | |
| MacArthur (1972) | | X | X | | | X | | | | X | | | X |
| Pielou (1974) | X | X | X | | | X | | X | | | | | X |
| Wilson (1975) | X | X | | X | | | X | | | X | | | X |
| Ricklefs (1979) | X | | X | X | | X | | | | X | X | X | |
| Roughgarden (1979) | X | X | X | | X | | | | X | | | | X |
| McNaughton & Wolf (1979) | X | | | X | | | X | X | | | | | X |
| Pontin (1982) | | X | X | | X | | | | X | | | | X |
| Arthur (1982) | | X | X | | X | | | | X | | | | X |
| Mac Nally (1983) | | X | X | | | X | | | X | | | | X |
| Aarssen (1984) | X | | | X | X | | | X | | | | | X |
| Giller (1984) | X | X | X | X | X | | | X | | | | | X |
| Welden & Slauson (1986) | X | | X | X | | X | | | X | | | | X |

tive interaction – must both participants be affected negatively, only one, or perhaps neither? Traditionally, competition has been defined in terms of reciprocal negative effects, a −, − interaction (Odum 1971). Asymmetrical interactions, in which only one of the individuals or populations suffers (−,0; termed amensalism), have attracted increasing attention among ecologists (e.g. Lawton and Hassell 1981, Schoener 1983b, Underwood 1986b), and such interactions should be included in the definition of competition.

It is not appropriate, however, to consider interactions in which no effects on individuals or populations are evident to be competition. Neither Miller's phrase 'potentially limiting', nor Wilson's definition requires that the interaction have an effect. If one accepts strict optimality arguments, any interaction must have negative effects on one or more of the participants. Aggression between species, for example, may be an energetically costly activity, and it should therefore take place only when benefits accrue, as through the exclusion of a competitor from a limited resource. By the same token, the displaced species should be negatively affected by such interactions, even if the effects are apparently slight. If constraints on optimal behavior or energy balancing are not tight, however, aggression may not incur real costs or produce benefits (Wiens 1984b). This is especially likely if resources are not directly limiting to individuals. Thus, although aggression is clearly a form of interference, whether or not it produces competitive effects depends on its relationship to limiting resources.

Most definitions of competition also imply limitation of supplies of shared resources as a precondition of competition, although this is not always the case (Table 1.1). If resources are not limiting, the exploitation of the resources by individuals of species A is not likely to reduce the well-being of individuals of species B that share those resources. It makes no sense to consider such a situation as competitive. If individuals of species A interfere with species B by reducing their access to a resource pool that is otherwise sufficient for both, however, the interaction is competitive *if* B suffers as a consequence.

The necessary conditions of interspecific competition, then, are these: (1) the species must share resources (i.e. they are members of overlapping resource systems), and (2) the joint exploitation of those resources and/or interference interactions related to the resources must negatively affect the performance of individuals of either or both species; often these effects have population consequences as well. We may thus define interspecific competition as an interaction between members of two or more species that, as a consequence either of exploitation of a shared resource or of interference related to that resource, has a negative effect on fitness-related characteristics of at least one of the species.

Devoting so much attention to the definition of a term so commonplace as competition may seem unwarranted. Definitions are not trivial, however. A definition of a process such as competition specifies the conditions under which the process may be expected to operate and provides guidance in evaluating whether or not the results of observations constitute evidence of that process. The adoption of different definitions of competition by ecologists (Table 1.1) may have contributed to the uncertainty about some of the evidence that I discussed in Volume 1.

## The question of resource limitation

In order for competition to occur, the resources shared by the individuals or populations must be limiting. Documenting resource limitation in communities, however, is difficult, and in most studies it has simply been assumed. To some degree, this reflects the impact of David Lack's thinking about food limitation of populations (Lack 1954, 1966), but it also stems from optimization and equilibrium arguments and the predominant position of the Lotka–Volterra formulations in community theory. According to these views, populations should expand until they are limited by resources, and communities should be structured so as to exploit the overall resource base most efficiently. If this is so, resources *should* be limiting at almost any time. These arguments are based more on belief than evidence, however, and there are increasing indications that resource limitation may occur only intermittently (see Chapter 4).

### *Defining resource limitation*

Resource limitation occurs when demand exceeds supply (availability, not abundance) and the performance of individuals or populations is diminished as a result. Limitation may be viewed at either an individual or a population level (Wiens 1984b). A shortage of food, for example, may influence the behavior or physiology of individuals. Whether or not these effects are translated into population consequences depends on how many individuals in the population are affected, how directly the limitation affects natality or mortality, and how long the period of limitation continues. To some degree, this is a matter of the scale on which the consequences are viewed. Jones and Ward (1979), for example, elegantly documented the physiological effects of food limitation on individuals in a breeding colony of Red-billed Quelea (*Quelea quelea*) in Africa and the consequences of these effects on mortality and reproductive success. The population conse-

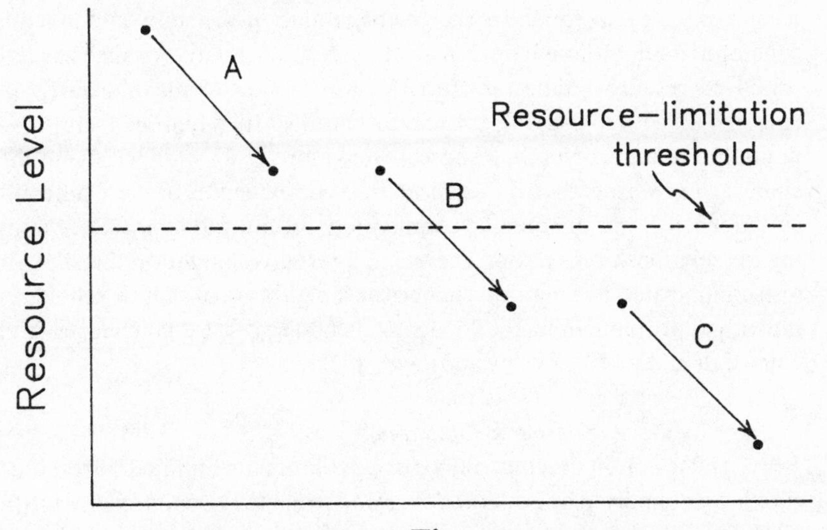


Fig. 1.1. Hypothetical changes in resource levels through time in three different situations in relation to a resource-limitation threshold of a species. In A, resources are always superabundant, in C, always limiting, and in B levels change from nonlimiting to limiting, even though the magnitude and rate of resource change are the same in all three situations. For simplicity, the resource-limitation threshold is assumed to remain constant, although in nature it will of course vary.

quences of this episode, however, were confined to a local area during a single breeding cycle.

Simply documenting the fact that two species share a resource says nothing about resource limitation, of course, and interpreting patterns of resource overlap in the context of resource limitation is unwarranted without additional information (Collins *et al.* 1982, Hastings 1987). Even if resource levels are also measured, limitation can only be inferred at best. Extending such measurements to document changes in resource levels over time or between locations may permit statements about the relative abundance or scarcity of the resource, but these levels cannot be related to limitation unless the supply-demand relationship or resource-limitation threshold is also known. A glance at Fig. 1.1 will show why this is so. Suppose one monitors resource levels in a location over time and records a substantial decrease. Customarily, this decrease has been interpreted to mean that resource limitation has become more severe and competition has intensified (e.g. Schoener 1982). Whether or not this is so, however, depends on how the changes relate to the resource-limitation threshold. In (A), resources were superabundant at the beginning of the study and are still superabundant (although less so) at the end, so the relative change has not produced resource limitation. In (B), on the other hand, the change in resource levels crossed the threshold, shifting the situation from non-limiting to limiting. In (C), resources were limiting to begin with, and the change simply exacerbated this effect. The consequences of these different situations, in terms of potential competition, are quite different. Determining the relationship of resource levels to a resource-limitation threshold is also complicated by temporal changes in the position of that threshold, as individual or population energy demands undergo seasonal changes (e.g. Kendeigh *et al.* 1977, Wiens and Dyer 1977).

***Evidence of resource limitation***

The notion that individuals or populations are limited by resource supplies is supported by evidence of varying strength (Newton 1980, Martin 1988a). Often limitation is simply inferred on the basis of little or no evidence. Puttick (1981), for example, used bioenergetic calculations to support his contention that populations of Curlew Sandpipers (*Calidris ferruginea*) were not limited by winter food supplies, but he then inferred that food must be limiting on the breeding grounds, largely on the basis of morphological differences between the sexes. Toft *et al.* (1982) suggested that the factor limiting the breeding waterfowl assemblages they studied was food for young ducklings, but no evidence was provided to support this contention. Dhondt (1977) demonstrated that the reproductive success of tits in Belgium was associated with conspecific and/or interspecific densities; from this, he argued that exploitation competition was involved and that food was therefore limiting. In general, differences in resource use between coexisting species have been used as evidence of resource limitation, on the mistaken premise that one would not expect such differences if resources were not limiting.

Other sorts of observations provide progressively stronger evidence of resource limitation. Often the evidence is correlative, variations in the density or reproductive output of populations being associated with variations in some measure related to resource availability. On the basis of a 21-yr study in West Germany, Wendland (1984) documented a concordance between cycles of voles (*Apodemus*) and breeding success (but not population density) of Tawny Owls (*Strix aluco*). Several studies have linked reproductive success, overwinter survival, and densities of Great Tits (*Parus major*) with variations in food abundance (Gibb 1960, Royama 1970, Minot 1981, Krebs 1971, van Balen 1980, Klomp 1980, Bejer and

Rudemo 1985). Densities of several seed-eating species in northern Europe vary in association with the cone crop of conifers (Newton 1972, 1980). In deserts and grasslands in the southwestern USA, variations in summer rainfall determine seed production the following autumn, and wintering densities of some (but by no means all) sparrow species vary accordingly (Pulliam and Parker 1979, Dunning and Brown 1982, Pulliam 1985). In the northern Great Basin, on the other hand, rainfall occurs primarily in winter and early spring. There, rainfall and sparrow densities the following winter are not correlated; seasonal climatic factors such as temperature and winter snow cover appear to be more important (Laurance and Yensen 1985).

Such correlative relationships are not necessarily found in all years, however, nor in all groups (e.g. frugivores; Lack 1971, Herrera 1985) or ecological settings (e.g. the Serengeti grasslands; Folse 1982). Moreover, some of the tests relate bird abundances to variables that only indirectly index food availability, such as rainfall or forest productivity (e.g. von Haartman 1971, Newton 1980). Even where food levels have been measured directly, the correlations provide only suggestive evidence of food limitation at best, for direct effects have not been shown and other environmental factors could also account for the observed patterns (Newton 1980).

Another source of evidence of food limitation relates to depletion of resources by the birds. During some years, granivores appear to consume most of the available seeds over the winter (Pulliam and Enders 1971, Pulliam and Parker 1979), and wintering insectivores may make substantial inroads on the abundance of arthropods or arachnids in woodlands (Norberg 1978, Solomon *et al.* 1976, Askenmo *et al.* 1977, Gunnarsson 1983, Stairs 1985). The capacity of shorebirds to deplete clam populations was noted in Volume 1, Chapter 10, and other studies have reported that shorebirds may remove substantial portions of their prey populations (Evans and Dugan 1984, Marsh 1986a, b, Piersma 1987), at least in some situations (Quammen 1982). By experimentally excluding birds from foliage during the breeding season, Holmes *et al.* (1979) demonstrated that they had a significant depressive effect on levels of lepidopteran larvae but not on other arthropod groups. Joern (1986) reported a 27% reduction in grasshopper density and a drop in species diversity in uncaged areas of grassland as compared with plots caged to exclude avian predators. When we conducted similar experiments in a shrubsteppe habitat, however, we could detect no differences in arthropod abundance or diversity between caged and uncaged shrubs (Wiens *et al.*, unpublished). Although these studies generally indicate that birds may sometimes have major effects on populations of their prey, they do not directly address the matter of resource limitation. Demonstrating that the birds have an effect on the prey does not necessarily demonstrate that prey availability is limiting to the birds, although it is suggestive (Wiens 1984b). Prey depletion is a necessary, but not a sufficient, criterion of food limitation (Maurer 1984; *contra* Martin 1986).

Other studies provide stronger circumstantial evidence of food limitation. These generally involve not only an assessment of the relative food levels but also indications of direct effects on the birds. Högstedt (1980a) reported that many nestlings of Jackdaws (*Corvus monedula*) and Magpies (*Pica pica*) in southern Sweden starve, which strongly suggests that food (or access to it) must be limiting. Some species, such as waterfowl or waders, meet their daily protein demands by drawing from protein reserves (chiefly pectoral muscles) when foraging conditions are poor and supply does not meet demand (Evans and Smith 1975, Ankney and MacInnes 1978). Davidson and Evans (1982) examined dead Redshanks (*Tringa totanus*) and Oystercatchers (*Haematopus ostralegus*) found during a prolonged period of severe winter weather in Britain. The dead birds had much lower fat and protein reserves than did healthy birds at the same time. The birds that died presumably suffered food limitation.

Perhaps the most compelling nonexperimental evidence of food limitation comes from the studies of Galápagos finches discussed in Volume 1. This work has shown that there are seasonal and yearly variations in seed abundance and that, on both between-island and within-island scales, finch biomass is positively correlated with seed biomass (especially small seeds) (Smith *et al.* 1978, Grant and Grant 1980, Schluter 1982b, Boag and Grant 1984). When this evidence is combined with the demonstrated mortality of birds during the 1977 drought and the attendant strong selection on morphology that was related to the shifts in seed availability (Boag and Grant 1981, Price *et al.* 1984, Grant 1986a), the case for food limitation seems about as strong as can be made without experiments. Newton (1980) has reviewed several other studies that provide strong circumstantial evidence of food limitation in birds (Table 1.2).

Experiments generally produce the least ambiguous evidence of resource limitation, and a fair number have been conducted. Some, such as fencing off areas of headland on Peruvian guano islands (which produced a dramatic increase in densities of nesting seabirds and indicated nest-site limitation; Duffy 1983), were not really designed to assess resource limitation. Others have assessed limitation by testing for its effects. In the Oregon shrubsteppe, for example, we experimentally increased the brood size of Brewer's Sparrows (*Spizella breweri*) from the norm of three to as many as

Table 1.2. *Examples of studies providing evidence of population limitation by food shortages*

| | |
|---|---|
| Rook *Corvus frugilegus* in northeast Scotland (Feare 1972, Feare *et al.* 1974) | Population at lowest density, and spread over largest area, in mid-summer. At this time food-stocks were minimal, because of a reduction in the numbers of fields in which birds could feed, an absence of grain, and the disappearance of large invertebrates (earthworms and tipulid larvae) from the soil surface. Associated with high mortality (especially of juveniles) and low weights, and with lower feeding rates (150 cals/day) and longer feeding periods (90% of daylight, or 15 h) than at any other season. Another potential food-shortage occurred during periods of deep snow in winter, when birds competed for space at localized feeding sites; however, the birds then spent only 30% (3 h) of the active day feeding and obtained 240 cals per day. |
| Red-billed Dioch *Quelea quelea* in Nigeria (Ward 1965). | Period of sudden population decline (due partly to emigration) coincided each year with temporary shortage of food, when grass seeds germinated at the onset of the rains. Starved and underfed birds were most prevalent then. |
| Coal Tit *Parus ater* in southeast England (Gibb 1960). | Over 5 yr, a close correlation was found between winter bird density and winter foodstocks (invertebrates on foliage): survival from October to March varied greatly from year to year in relation to measured food-stocks. During this period, the birds ate around 50% of several main prey species, and in mid-winter spent more than 90% of the day feeding. |
| Wood-pigeon *Columba palumbus* in southeast England (Murton *et al.* 1964, 1966). | Period of population decline in winter coincided with depletion of grain, then clover stocks. The lowest bird densities, low weights, low feeding rates, and most starving birds occurred in late winter, when clover stocks were minimal. Temporary food-shortages occurred during periods of deep snow, when the birds were concentrated on localized brassica sites. |
| Red Grouse *Lagopus l. scoticus* in northeast Scotland (Jenkins *et al.* 1967, Lance 1978, Miller *et al.* 1966, 1970, Moss 1969, Watson & Moss 1972). | Greater mean breeding densities and success on moors overlying base-rich rock than on moors on base-poor rock, corresponding with greater nutrient content of heather on base-rich areas. Within areas, an inverse correlation between the territory size of individuals and the nitrogen content (taken to reflect nutritive value) of the heather in each territory. Also, an increase in breeding density following the experimental fertilizing or burning of areas of heather (the food plant) when compared with densities on nearby control areas. |

*Source:* From Newton (1980).

six in order to determine whether or not the adults could rear that many offspring successfully (Wiens and Rotenberry unpublished). A reduction in nestling growth rates or survival would indicate food limitation, as gauged by the inability of the adults to deliver sufficient food to the nestlings. Even with the doubled brood sizes, we found no depressive effects on growth rates or fledging success, which reinforced our observational conclusions (Rotenberry 1980, Wiens and Rotenberry 1981a) that food is apparently not limiting in this system during most breeding seasons. Similar results of brood-size manipulations have been reported by Gaubert (1985), although other such manipulations have produced results more consistent with the food-limitation hypothesis (Martin 1988a).

In most experiments, the resources themselves have been manipulated, usually by supplementing the natural supply. The provisioning of nest boxes has resulted in dramatic increases in the densities of hole–nesting passerines in many areas of Europe (van Balen *et al.* 1982, Slagsvold 1975, Perrins 1979, Alatalo *et al.* 1985, Alerstam 1985). Jansson *et al.* (1981) provided supplemental food to populations of wintering Crested and Willow tits (*Parus cristatus* and *P. montanus*) in Sweden and recorded improved overwinter survival of marked individuals as well as considerable immigration into the supplemented area. Breeding populations were consequently doubled over those in control areas, although there was a subsequent density-dependent reduction of breeding success in *montanus*. The supplementation of food supplies of breeding populations of these species (Brömssen and Jansson 1980) produced changes in breeding phenology but not in population sizes, clutch sizes, or fledging success. In a similar experiment, Samson and Lewis (1979) supplemented the food supplies of wintering parids in eastern North America. Black-capped Chickadee (*Parus atricapillus*) flock size increased dramatically, but there was no response among Tufted Titmice (*Parus bicolor*). In another experiment in Sweden, Källander (1981) provided supplemental food to Great and Blue tits (*Parus caeruleus*) during two winters. Great Tit densities increased on the experimental plots while decreasing in the control area during the first winter, but they increased in both the experimental and control areas in the second winter. The first winter was severe, with most of the beech mast being unavailable under snow cover, whereas the following winter was normal and mast was readily available, apparently in nonlimiting quantities. Blue Tits did not respond to the experimental supplementation in either year. This contrasts with the results Krebs (1971) obtained in Britain, where Blue Tits increased in density following the addition of food to one wood during one winter but Great Tits did not.

Table 1.3. *The conclusions that are justified from the results of experiments involving a change in food supplies*

| | Response | |
|---|---|---|
| Experiment | Population increases | Population stays the same or decreases |
| Food increased | Food is a sufficient or necessary explanation | (a) Increased food is not sufficient to prevent a decline<br>(b) Reduced food is not necessary to cause a decline |
| Food decreased | (a) Increased food is not necessary to allow an increase<br>(b) Reduced food is not sufficient to prevent an increase | Food is a sufficient or necessary explanation |

*Source:* From Newton (1980).

These and other experiments provide compelling evidence of food limitation, although the interpretation of experimental results must be made with caution (Newton 1980; Table 1.3). Moreover, the design of these experiments has not always been tight. Rarely has the magnitude of food supplementation been carefully related to existing densities and per capita consumption rates so that a quantitative prediction of the magnitude of response could be made. Some experiments have suffered from a lack of replication (e.g. Krebs 1971) or from only a small control sample (one flock in the studies of Sampson and Lewis), and in others the supplemental food has been provided at the wrong time (e.g. Franzblau and Collins 1980).

It is not an easy matter to document resource limitation beyond a reasonable doubt, but this does not justify basing studies of potential competitive interactions on an untested assumption that limitation is continuously present. Diamond (1978: 330) has stated that 'doubters of the role of interspecific competition object that, until resource levels are actually measured, an essential link in the argument remains hypothetical and the "overwhelming" evidence for competition remains circumstantial'. He is right. There is sufficient evidence of variation in the existence or magnitude of resource limitation to make the assumption of continuous limitation unacceptably fragile and to require careful measurement and experimental manipulation of resource levels. Some of the best evidence of resource limitation has been obtained in studies of nectarivorous species, which are considered in detail in the next chapter.

**Criteria for establishing the operation of competition**

How does one determine what constitutes valid and compelling evidence for interspecific competition? Traditionally, competition has been established operationally by associating changes in the abundance or resource-use patterns of a species with changes in another species (Thomson 1980, Connell 1983). This is the sort of situation that characterizes island-mainland comparisons and that has led some to believe that the associated patterns of density compensation and niche shifts provide the most compelling evidence of competition (MacArthur and Wilson 1967, Diamond 1978).

This approach is not satisfactory. To demonstrate competition, one must show that the species overlap in their use of resources and that the exploitation of the resource by one of the species or the interference with one species' access to the resource by another has negative effects. To demonstrate this in practice involves meeting several criteria, which are often approached in a stepwise fashion (Reynoldson and Bellamy 1971, Mac Nally 1983). As one proceeds through the following list (Table 1.4), it becomes progressively more difficult to meet the criteria, but the evidence gathered is progressively stronger and more convincing.

1. The observed patterns must be consistent with expectations based on the operation of interspecific competition. If they are not, the competition hypothesis probably has been falsified (although there are reasons why this falsification might be invalid; see Volume 1). Unless a pattern has been documented accurately through sound methodology, of course, the pattern is not relevant to the testing of *any* hypothesis.
2. The species must overlap in their use of resources. If they do not, competition at some time in the past may have produced the divergence (a proposition not amenable to direct testing), but proximate competition is not likely. An assessment of this criterion requires a careful consideration of how resources are defined and measured.
3. Intraspecific competition must be present in the species' populations. Because members of the same species are normally much more similar to one another than they are to members of the other species, reductions in resource availability should first be felt among members of the same population. Intraspecific competition tends to produce different patterns than does interspecific competition (e.g. niche broadening rather than niche contraction;

Table 1.4. *Summary of criteria that establish the occurrence of interspecific competition with varying degrees of certainty.*

| Strength of evidence | Criteria |
|---|---|
| Weak | 1. Observed patterns consistent with predictions |
| ↓ | 2. Species overlap in resource use |
| | 3. Intraspecific competition occurs |
| Suggestive | 4. Resource use by one species reduces availability to another species |
| ↓ | 5. One or more species is negatively affected |
| Convincing | 6. Alternative process hypotheses are not consistent with patterns |

*Source:* Modified and extended from Reynoldson and Bellamy (1971).

Svärdson 1949), so the effects of the two processes should be distinguishable. In relatively few investigations of interspecific competition has the occurrence or magnitude of intraspecific competition been considered, but it is quite important (Abrams 1980, Connell 1983, Reynoldson and Bellamy 1971, Underwood 1986a,b). The outcome of interspecific interactions in both nature and experiments depends on the relative importance and intensities of intra- and interspecific competition. Martin's (1986) contention that intraspecific competition is not necessary for interspecific competition to occur may be true in some situations for individuals, but at the population level it is not. Martin's conclusion apparently stems from his view that competition theory assumes that intraspecific competition will occur only when a population is at or near carrying capacity.

4. The use of the resource by one species (exploitation) or nonconsumptive pre-emption of the resource (interference) must reduce resource availability to the other species. If this does not happen, there will be no negative effects from the interaction.
5. One or more of the species must be negatively affected. Demonstrating a negative effect generally requires a manipulative experiment in order to assure that the effect is related to the actions of the other species. Usually, such experiments involve alterations in the densities of one or the other species. Ideally, however, both the populations and the resources should be manipulated separately, in order to link resource dynamics with population effects (Reynoldson and Bellamy 1971, Martin 1986). The choice of

response variables is critical (Mac Nally 1983), especially if one intends to separate exploitation from interference competition.

6. Alternative hypotheses must be tested. Any observed pattern is likely to be consistent with several possible process explanations. Although well-designed manipulative experiments may narrow the possibilities considerably, alternatives may still exist.

None of these criteria provides sufficient evidence of competition by itself. A great many studies have inferred competition by satisfying criterion 1 alone or 1 and 2 together. This is extremely weak evidence. Demonstration of criterion 4 strengthens the case, but the evidence remains circumstantial. Criteria 1–4 together do not necessarily implicate competition as the process responsible for an observed pattern, as the required documentation of negative effects is still lacking. Likewise, a failure to document criteria 1–4 does not necessarily indicate an absence of competition in the past, although it is suggestive of its current absence. Satisfying criterion 5 by experimental manipulations substantially strengthens the case, but a failure to demonstrate negative effects may be due either to an absence of the process or to an inappropriate experimental design (Mac Nally 1983, Underwood 1986b). Neither Reynoldson and Bellamy (1971) nor Mac Nally (1983) included criterion 6 in their listings of desiderata for documenting competition, but it is critically important. This will become evident in Chapter 3. Some additional criteria proposed by others are inappropriate; falling into this category is Pontin's (1982) requirement that the system return to equilibrium after an experimental or natural disturbance of the relative proportions of the species.

**Circumstantial evidence of competition**

Most of the evidence that competition is an important process in producing community patterns is observational or correlative rather than experimental. Indeed, Diamond (1978: 322) concluded that 'ecologists using only the simplest observational methods now routinely document competition', and Giller (1984) reviewed a variety of observations of community patterns that he interpreted as strong evidence of competition. Observations gathered from comparative studies or correlative analyses, however, are almost always open to interpretations based on processes other than competition (including the null hypothesis that the observed pattern does not differ from that generated by random processes). To restrict an explanation of these observations to competition requires the acceptance of the *ceteris paribus* assumption, and many ecologists object to this as unjustified. Diamond (1978: 327), however, has argued that 'this objection strains one's credulity' and that the observations constitute compelling evidence of competition. This view is not compatible with a rigorous approach to testing hypotheses (see Volume 1).

Another complication has to do with the effects of competition in the past. Often, observations that are not consistent with predictions based on proximate competitive processes are explained in terms of competitive interactions in the evolutionary past (e.g. Noon 1981). Past competition (as well as other historical processes) *may* have contributed to contemporary patterns, of course, but, as Strong (1984a: 40) has observed, 'only peculiar circular logic underlies interpreting the absence of competition as proof that it once existed'.

Experiments usually provide the most unambiguous evidence of the action of competition (Mac Nally 1983, Connell 1980, 1983, Underwood and Denley 1984, Underwood 1986b), but often they are not feasible or are morally or legally forbidden. We must then rely on comparative observations, so-called 'natural experiments', to provide insights into processes. Before we consider some of the nonexperimental evidence relating to competition, however, three points should be emphasized. First, observations of patterns have no standing as evidence for or against competition if they are gathered with a flawed methodology. Second, such circumstantial evidence is usually not as strong as that obtained from well-designed experiments (although it may be superior to that from poor experiments). The strength of circumstantial evidence is a function of the care with which a study is conducted, the completeness with which environmental variables such as resource levels are measured, and the degree to which the evidence is incompatible with the predictions of other process hypotheses (see Chapter 3). Third, the adoption of a comparative, nonexperimental approach requires the acceptance as assumptions of the criteria listed in Table 1.4. This should be justified, either by measurement or by sound logical arguments.

We can now consider some of the kinds of evidence that bear on the occurrence of competition and its contributions to community structuring. Many of the most important patterns and some process interpretations of them were reviewed in Volume 1. Here, I will summarize what those studies suggest about competition and comment on some specific aspects of this evidence.

### Morphological evidence

Morphological patterns of communities have frequently been related to the operation of competition, and some ecologists (e.g. Ricklefs and Cox 1977, Schoener 1974a, 1982) have suggested that morphological data may be superior to information on resource use as evidence of competition because they contain a 'genetic memory' of competition and are not so sensitive to exactly when or where the data are obtained. This justification may enhance the standing of morphological data as inferential evidence of past competition, but it does not bear on competition as a proximate process.

Several general patterns in community morphology have been predicted from competition theory, but support for each is varied and equivocal. The notion that coexisting species should be separated in morphological space, perhaps in accord with some constant size ratio, has been challenged on logical and methodological grounds, and the actual observations exhibit varying degrees of adherence to the predictions. In some groups, such as the West Indies birds examined by Case *et al.* (1983) or the Galápagos finches studied by Grant and Schluter (1984), the species that occur together do appear to differ morphologically more than would be expected by chance. Interspecific competition might well have produced this pattern, but, in the absence of additional ecological information, the evidence remains only suggestive. The support for a relationship between competition and other ecomorphological patterns, such as character displacement or convergence, either diminishes when the supposed examples are subjected to close scrutiny or relates equally well to other noncompetitive processes (Grant 1972, Arthur 1982).

Morphological attributes do differ among coexisting species, however, and in some situations these differences are clearly consistent with what one might expect from the effects of competition. To consider these differences to be evidence of competition requires that the features be ecologically relevant to the resource(s) over which competition is presumed to occur. This is usually inferred or assumed. In the absence of more direct information, such evidence is strongest if the comparison is restricted to a well-defined guild or to closely related species. A comparison involving groups as disparate as finches, warblers, and flycatchers (Ricklefs and Cox 1977), for example, strains the assumption that morphological differences directly index niche differences much more than a comparison among small foliage insectivores (Alatalo 1982) or *Geospiza* finches (Grant 1986a). In the finches, both bill morphology and body-size differences between species do relate closely to their feeding habits, and the ecological relevance of morphology is clearly demonstrated by the stabilizing and directional selection accompanying shifts in resource distributions during the drought on Daphne (Boag and Grant 1981, 1984, Price *et al.* 1984, Grant and Schluter 1984, Grant 1985). This is strong circumstantial evidence of the importance of resource limitation and competition in determining morphological patterns.

Overall, however, ecomorphological community patterns that are consistent with competitive explanations do not appear to be general in their occurrence, and where they are found they usually provide only ambiguous evidence of competition on either proximate or evolutionary time scales. This is primarily because the associations between morphological features, their ecological functions, and niche relationships among species are rarely established. When documentations of within-guild or within-genus morphological patterns are combined with careful field studies, the evidence merits close attention, especially when studies of functional morphology (e.g. Norberg 1979, 1981, Leisler and Thaler 1982) provide additional perspective. Demonstrations that species within such groups that do not co-occur are especially close morphologically (e.g. Grant and Schluter 1984, Alatalo *et al.* 1986, Moulton 1985) are also compelling.

***Distributional evidence***

Patterns in the distributions of species over geographical locations or habitat types have also been interpreted as evidence of competition. 'Checkerboard distributions' of species on islands or among isolated habitat patches may be produced by total interspecific competitive exclusion, as Diamond (1975a, 1978) has claimed, but the complete explanation of such patterns is probably considerably more complex (Abbott 1981, Vuilleumier and Simberloff 1980).

Often species co-occur in some locations and are found in the absence of the putative competitor elsewhere. By applying association or contingency analyses to such presence–absence patterns, one can determine whether the species co-occur significantly more or less often than expected by chance (Pielou 1972, 1974, Schluter 1984). Similar tests can be used to determine whether the densities of two species vary independently, in parallel, or inversely. If competition is occurring between the species, one possible expression of its effects is a negative association or density relationship between the species. The example of the Willow Warbler (*Phylloscopus trochilus*) and Brambling (*Fringilla montifringilla*) discussed in Volume 1 (Chapter 10) is of this sort, although this pattern is not consistent over time or space. Thompson and Lawton (1983) used a similar approach to examine patterns of microspatial association of wintering birds feeding on experimental patches that differed in food supplies. They found significant negative associations that matched the patterns of interspecific aggression they observed. In England, Minot and Perrins (1986) determined that year-to-year changes in densities of Great and Blue tits were negatively correlated when nest-box densities were low but positively correlated at high nest-box densities; they attributed the negative correlation to successful interference competition by Great Tits when boxes were limiting. In Hungary, Great Tit clutch size was reduced in years of high Blue Tit density, and hatching success of both tit species was lower when Collared Flycatcher (*Ficedula albicollis*) densities were high (and vice versa) (Sasvári *et al.* 1987). When James and Boecklen (1984) examined the pairwise density correlations among a large number of species breeding in an upland forest in Maryland over a 7-yr period, they found that the distribution of correlations was symmetrical about a mean value of approximately 0. Here, the densities of most species varied independently of one another, although some significant negative correlations did emerge.

Such association analyses usually indicate some significant positive as well as negative correlations between species. In their analysis of density responses of wintering sparrows to rainfall variations, Dunning and Brown (1982) found that the abundances of the different species were positively correlated over time – a good (or bad) year for one species tended to be good (or bad) for all. Laurance and Yensen (1985) found a similar positive correlation of density variations among the wintering sparrows they studied in the northern Great Basin. Dunning and Brown suggested that the abundances of resource-limited, potentially competing species might be positively rather than negatively associated over time if the resource base varies substantially. Be that as it may, it is apparent that a negative (or positive) association between species in their occurrence or abundances may have little direct bearing on whether or not competition has occurred. The patterns may easily reflect different responses to environmental variations in time or space, a tracking of resources whose variations are negatively related, or the operation of processes other than species interactions (Birch 1979, Andrewartha and Birch 1984, Minot 1981, James and Boecklen 1984, Schluter 1984). It may also be difficult to detect real competitive effects through linear correlation or association analyses if competitive interactions are nonlinear (e.g. it takes a much greater density of individuals of species A to produce a given effect on species B when the latter is rare than when it is common; Schoener 1983b). Moreover, it may be difficult to detect the effects of competition in species co-occurrence patterns even when competition is intense without large sample sizes (Hastings 1987). In the absence of other information, negative relations between the distribution and abundance of species provide only weak evidence of competition.

A particularly detailed approach to the detection of competition through

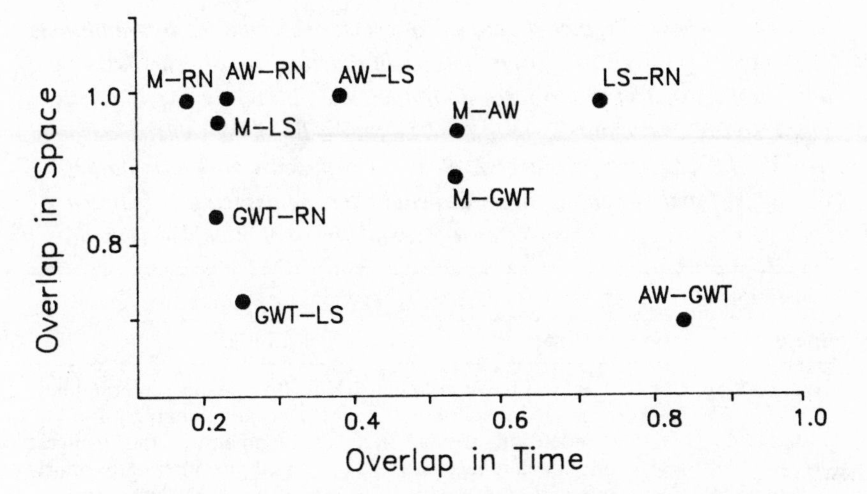


Fig. 1.2. Overlap in time and space for pairs of waterfowl species occupying small ponds in Canada. M = Mallard (*Anas platyrhynchos*), RN = Ring-necked Duck, AW = American Wigeon, LS = Lesser Scaup, GWT = Green-winged Teal. From Toft *et al.* (1982).

association analysis has been followed by Toft *et al.* (1982) in their analysis of the distributional patterns of breeding waterfowl among small ponds in northwestern Canada. Toft and her colleagues found that the species differed in their density response (measured by number of broods per pond) to pond size (measured by pond perimeter); this suggested that density was limited by some space-related factor and that the differences between the species might be consequences of competition for space. If ecological overlap between species is proportional to the intensity of competition when resources are limiting, one should expect species with high overlap in pond use to be negatively associated in distribution or abundance. Toft *et al.* calculated the overlap between species pairs in space (pond size) and time (hatching dates) (Fig. 1.2) and generated specific predictions of co-occurrence patterns for the alternative hypotheses of no resource limitation versus resource limitation and competition (Table 1.5). The results of both parametric and nonparametric association analyses (Table 1.5) were in general agreement with the predictions of the hypothesis of resource limitation and competition. Lesser Scaups (*Aythya affinis*) and Ring-necked Ducks (*Aythya collaris*) exhibited the greatest overlap in time and space and were strongly negatively related in their abundance-distribution patterns, whereas scaups and Green-winged Teal (*Anas crecca*), which overlapped least, showed the least distributional interaction. None of the species pairs with intermediate overlap exhibited a significant distributional association.

Table 1.5. *Above: Predicted patterns of species associations for conditions of low and high overlap between species in time and space. The null hypothesis is that of no resource limitation; the alternative hypothesis is that resource availability is limiting. The four cells of the table correspond with the four quarters of Fig. 1.2. Below: Coefficients of determination* ($r^2$, *above*) *and of contingency* (C, *below*) *for the association between each pair of waterfowl species over a sample of small ponds in Canada*

| Dimension: | Time | |
|---|---|---|
| Space | Low overlap | High overlap |
| High overlap | Null: 0 or positive interaction<br>Alt.: negative interaction is competition; at most an intermediate negative interaction is expected if resources are limiting | Null: 0 or positive interaction<br>Alt.: negative interaction is competition; the strongest negative interaction must occur here if resources are limiting |
| Low overlap | Null: 0 or negative interaction<br>Alt.: even if resources are limiting, little or no effect of competition should be seen here | Null: 0 or negative interaction<br>Alt.: a strong negative interaction is not necessarily competition if there is a negative interaction for species with low overlap in both dimensions |

| | | | |
|---|---|---|---|
| High overlap | M-RN: 0.001+<br>0.101+ | LS-AW: 0.005+<br>0.082+ | LS-RN: 0.016−<br>0.125− |
| | AW-RN: 0.004+<br>0.045− | AW-M: 0.003<br>0.059+ | |
| | LS-M: 0.006+<br>0.136+ | GWT-M: 0.037+<br>0.055+ | |
| | GWR-RN: 0.001+<br>0.045− | | |
| Low overlap | LS-GWT: 0.0001−<br>0.009− | | LS-AW: 0.015+<br>0.178+ |

*Source:* After Toft *et al.* (1982).

The strong positive association between American Wigeon (*Anas americana*) and Green-winged Teal, however, was unexpected. This value was the result of a strong association in only one year when resources may have been nonlimiting. Toft and her colleagues concluded from these patterns that the null hypothesis that resources normally are not limiting in this system could be rejected and that the results confirmed a close relationship between overlap and competition intensity, the negative associations indicating strong competition.

This study is unusual in that it follows a careful approach to association analysis and presents explicit predictions of alternative hypotheses. Some difficulties should be noted, however. Spatial overlap was calculated from brood densities, which also formed the basis for the association tests. Toft *et al.* acknowledged this circularity problem but suggested that it made the test for competition effects conservative. Space (more specifically, pond perimeter) and time (specifically, hatching date) were assumed to be the limiting resources. Quite apart from the fact that breeding time is probably not a resource, there is no empirical basis for believing these to be the critical resources other than their *post hoc* association with density patterns. Ponds were also assumed to be fully saturated with breeding ducks (i.e. resources are limiting to densities), but again the empirical justification for this assumption follows from the association tests rather than being independently derived. The patterns by and large are suggestive of competition, but of the criteria listed in Table 1.4, numbers 1 and 5 are demonstrated, numbers 2 and 4 are only indirectly satisfied or inferred, number 3 is not documented, and number 6 is only partially met (alternatives other than that of nonlimiting resources were not considered).

One of the uncertainties in this study has to do with whether or not the pond habitats really are similar. On the basis of floristic surveys and their general impressions, Toft and her coworkers assumed that they were similar and that any habitat differences were therefore unimportant in determining the distributional patterns of the ducks. If habitat effects can be evaluated quantitatively, the strength of inferences about competition derived from distributional evidence may be increased. Several ecologists (Schoener 1974b, Crowell and Pimm 1976, Hallett and Pimm 1979, Hallett 1982) have suggested one way to do this. Under equilibrium conditions in a homogeneous environment, two competitors will occur at densities that reflect their competitive effects on one another. If their densities are perturbed from that point by any factors, their return to equilibrium will follow a trajectory determined by their competitive interactions. The slope of a line drawn through a succession of points on that trajectory (or through points sampled from different locations in a homogeneous system) will indicate the competition coefficients between the species (Rosenzweig *et al.* 1984).

Natural systems are not homogeneous, of course, so for this method to be useful, habitat variation must be taken into account. If one records the densities of species and measures habitat variables over a large number of locations or samples, the values may be subjected to multiple regression analysis to determine the relationship of density variations to habitat features for each species. If densities of other species are included in the regression model, the results can be examined to determine whether or not the densities of the presumed competitors account for deviations from the habitat regression. If they do, the partial regression coefficients may be taken as estimates of the competitive effects of one species on the other, corrected for habitat influences (Crowell and Pimm 1976, Hallett and Pimm 1979, Rosenzweig *et al.* 1984). The point of the habitat analysis, then, is not to determine which habitat variables are ecologically important to the species but simply to permit their effects to be separated from those presumably due to competition.

This approach suffers from a number of damaging assumptions and constraints. As originally applied, the procedure employed linear analyses and therefore presumed that responses to habitat variation are linear. This is unlikely, and the use of nonlinear analysis (e.g. Rosenzweig *et al.* 1984) produces quite different estimates of competitive effects (Schoener 1985). The approach may also be sensitive to violations of the assumption of independence among variables, although Hallett and Pimm (1979) suggested that this bias is not serious. Like virtually all other methods of overlap analysis, the regression approach also assumes that interaction coefficients do not change as habitats vary (Rosenzweig *et al.* 1984). Dunning and Brown (1982) pointed out that if there are fluctuations in the availability of shared limiting resources, species may be positively rather than negatively correlated, and the 'interaction coefficients' derived from the regression model will be interpreted incorrectly. If the species differ demographically (e.g. in mean longevity), the meaning of numerical changes in populations may be obscure, as the species will differ in the rate of turnover of individuals and the capacity of the populations to track environmental changes (T. Sherry, personal communication). Most important, different locations in nature differ in a great many respects, and it is unlikely that all of these effects will be included in the habitat variables used in the model. To presume that only the effects of competition remain to be calculated after the removal of such habitat effects and that the calculated values are synonymous with competition coefficients (alphas) in Lotka–

Volterra equations is simplistic. Overall, there are formidable problems to calculating competition coefficients by this method, and it is unjustified to conclude anything about competition from temporal or spatial patterns of density fluctuations without a detailed knowledge of resource levels and the mechanisms of resource utilization (Bender *et al.* 1984, Dunning and Brown 1982, Abramsky *et al.* 1986).

The procedure described above has been applied primarily in investigations of mammal communities (e.g. Hallett 1982, Dueser and Hallett 1980, Hallett *et al.* 1983) and has received little attention from those working with birds. Mountainspring and Scott (1985), however, used a variation on this analytical theme to investigate potential interactions among forest birds on several of the Hawaiian Islands. Using information derived from unusually careful and extensive census surveys (Scott *et al.* 1986), they conducted multiple regression analyses of bird densities and habitat variables, followed by an examination of the resulting partial correlations between bird species abundances (indicators of species associations that they considered to be superior to the regression coefficients used in the Hallett-Pimm procedure). Correlation tests among bird species conducted before partialling out the influences of habitat indicated a substantial number of significant negative associations, which might be suggestive of competition. Most of these disappeared, however, when habitat effects were taken into account. Of the 37 significantly negative correlations found over all sites before removal of habitat effects, 24 became nonsignificant in the subsequent partial correlation analysis, 7 became significantly positive, and only 6 remained significantly negative. If nothing else, this indicates the magnitude of the effect that habitat variables can have on patterns of species associations.

Mountainspring and Scott did not interpret the partial correlations as measures of competitive intensity but used the significant negative correlations as indicators of species pairings deserving closer scrutiny. In fact, two species pairs, Japanese White-eye (*Zosterops japonicus*)/Iiwi (*Vestiaria coccinea*) and white-eye/Elepaio (*Chasiempis sandwichensis*), showed consistent negative associations. On the basis of this and other evidence, Mountainspring and Scott attributed these associations to competitive interactions. This approach still suffers from some of the assumptions contained in the Hallett–Pimm approach, although Mountainspring and Scott explicitly considered nonlinearity in habitat responses and avoided inferring competition directly from the correlation coefficients.

Mountainspring and Scott found that native-exotic species pairs had a significantly greater proportion of negative partial correlation coefficients

(37%) than either native–native pairs (8%) or exotic–exotic pairs (0%). This suggested that interspecific competition might be occurring rather broadly between members of a coevolved native fauna and recently introduced exotic species. The association analyses provide little indication of potential competition among the exotic species, but Moulton and Pimm (1983, 1986, Moulton 1985) found suggestive evidence of such interactions when they restricted attention to the lower-elevation zones in which only the introduced species occur. In addition to their morphological evidence, Moulton and Pimm documented that the per-species extinction rates of introduced species on the Hawaiian Islands increase in a nonlinear, accelerating fashion with the number of species present. Moulton and Pimm interpreted this pattern as indicating that the species mutually affect each other's chances of extinction, and they concluded that the pattern was strong evidence of interspecific competition. They considered the alternative hypotheses that the pattern might be due to habitat alterations or to intrinsic properties of the species, but these were dismissed (perhaps a bit too hastily). The pattern is suggestive of interactions, and in some instances anecdotal observations of species replacements seem to lend support to this notion (but see Jehl and Parkes 1983; Chapter 4). A great many factors may influence the persistence of populations (especially of exotics) on islands, and in this example only some of the criteria necessary to document competition (numbers 1, 2 (perhaps), 5, and 6 (partially)) have been satisfied.

One other sort of distributional pattern has been offered as strong evidence of the effects of competition. When two species (especially congeners) have abutting but nonoverlapping (parapatric) geographical or local distributions, competition may be involved (Diamond 1978, Terborgh and Weske 1975). This is more likely if the zone of contact does not correspond with a zone of habitat change, and both Diamond and Terborgh claim to have documented numerous such patterns among birds on mountain slopes in the tropics (see Volume 1, Chapter 8). In neither set of studies were habitats actually measured; rather, the investigators used subjective evaluations to record changes between a few major habitat zones. Both Mountainspring and Scott (1985; Scott *et al.* 1986) and Sherry and Holmes (1985) documented through detailed studies that bird species may respond strongly to subtle variations in particular habitat elements in a species-specific manner. This casts doubt on studies in which parapatric distributions are interpreted to be the result of competition, unless they include detailed analyses of habitat variation and species' responses.

***Niche overlap***

The idea that niche differences or overlap among species are closely related to the probability and intensity of interspecific competition has been especially attractive and persistent in avian community ecology. The basic premise is that species will be increasingly likely to compete when resources are limiting as they become increasingly similar in ecology. Unfortunately, one may argue equally logically either that resource scarcity will lead to increased overlap between species as they are restricted to common resources, promoting competition (e.g. Ford and Paton 1977) *or* that resource scarcity will promote diversification in resource use among species, circumventing competition (e.g. Herrera and Hiraldo 1976). Measures of niche overlap are also generally insensitive to the availability of resources and, in the absence of information on resource supplies and demands, imply nothing about the potential for competition (Mac Nally 1983, Abrams 1980). Moreover, the overlap that one calculates between species is sensitive to exactly how resources are defined or categorized (Harner and Whitmore 1977, Högstedt 1980b). As a consequence of these differing interpretations and operational problems, considerable confusion has developed among ecologists about the use and meaning of niche overlap (Wiens 1977, 1983a, b; Schoener 1982; Rosenzweig *et al.* 1984). Clearly, in the absence of additional information on resources and their limiting effects, patterns of niche overlap between species alone provide very weak and ambiguous evidence of competition. Documentation of overlap satisifes only criterion 2 of Table 1.4.

This does not mean, however, that niche-overlap patterns have no bearing at all on the question of whether or not competition occurs in specific situations. Schoener (1982) examined patterns of change in niche overlap through time in communities studied over several seasons or years. A change in resource availability from abundant to scarce, he argues, should foster intensified competition, and this should be reflected in a reduction of overlap during the lean periods. After reviewing some 30 studies (over half of which dealt with birds), Schoener concluded that, with few exceptions, overlap is less during the relatively lean period. Schoener tabulated the results of his survey, expanding on an earlier treatment by Smith *et al.* (1978). In Table 1.6 I have expanded in turn on Schoener's listing, including some studies of birds he omitted and adding some that have appeared since his survey. My interpretation of the studies Schoener considered agrees with his, and in most (but not all) instances the additional studies I include confirm his impression that overlap generally decreases in

Table 1.6. *Patterns of temporal change in resource overlap among various coexisting bird species.*

| Species/Area | Overlap dimension | Period of least overlap | Time scale of comparison | Evidence of limitation? | Authors' interpretation | Reference |
|---|---|---|---|---|---|---|
| Shorebirds/Florida & Canada | Foraging behavior/ microhabitat | Lean (winter) | Seasonal | No | Competition | Baker & Baker 1973 |
| Shorebirds/Canadian arctic | Diet | Lean (?) (early spring) | Within-season (summer) | No | Competition | Holmes & Pitelka 1968 |
| *Buteo* hawks/ shrubsteppe Idaho | Diet | Lean (low prey year) | Years | Yes (1 species) | No competition (see text) | Steenhof & Kochert 1985 |
| *Accipiter* hawks/ Netherlands | Diet | Lean (winter) | Seasonal | No | Competition? | Opdam 1975 |
| Hummingbirds/ Trinidad, Tobago | Nectar (flowers) | Lean | Seasonal | No | Competition | Feinsinger & Swarm 1982 |
| Hummingbirds/Costa Rica | Nectar (flowers) | Lean | Seasonal | No | Competiiton | Feinsinger 1976 |
| Hummingbirds/Mexico | Nectar (flowers) | Fat | Seasonal | No | Opportunism | DesGranges 1979 |
| Galliforms/Finland forests | Habitat | Lean (winter) | Seasonal | No | | Alatalo 1980 |
| Corvids/Swedish meadows | Foraging habitat | No change | Seasonal | No | Opportunism | Loman 1980 |
| Woodpeckers/Norway forests | Foraging strata | Lean (winter) | Seasonal | No | None | Hogstad 1971 |
| Foliage gleaners/ Finland forests | Habitat | Fat (summer) | Seasonal | No | | Alatalo 1981 |
| Foliage gleaners/ Finland | Microhabitat, foraging behavior | Lean (winter) | Seasonal | No | Competition | Alatalo 1980 |

| | | | | | | |
|---|---|---|---|---|---|---|
| Foliage gleaners/ Swedish pine woods | Foraging microhabitat | fat (summer; see text) | Seasonal | No | Competition | Ulfstrand 1977 |
| Foliage insectivores/ English broadleaved woods | Diet | Lean (winter) | Seasonal | Yes | Competition | Gibb 1960 |
| Foliage insectivores/ English pine woods | Diet | No change | Seasonal | Yes | Competition | Gibb 1960 |
| Thrashers/Texas woods | Diet composition | Fat (spring) | Seasonal | No | | Fischer 1981 |
| Thrashers/Texas woods | Prey size | No change | Seasonal | No | | Fischer 1981 |
| Waterthrushes/woods | Foraging substrate | Fat (summer) | Within-season | No | Opportunism | Craig 1984 |
| Finches/English woods | Diet | Lean (see text) | Seasonal | No | Opportunism | Newton 1967 |
| Finches/English farmlands | Diet | No change (see text) | Seasonal | No | Opportunism | Newton 1967 |
| Finches/Galápagos islands | Diet/ foraging behavior | Lean (dry) | Seasonal, yearly | Yes | Competition | Smith *et al.* 1978, Schluter 1982a, Grant 1985 |
| Passerines/Washington shrubsteppe | Diet | Little change | Seasonal | No | Opportunism | Rotenberry 1980 |
| Nestling passerines/ Saskatchewan prairie | Diet | Fat (mid summer) | Within-season | No | Competition | Maher 1979 |
| Sparrows/Arizona grassland | Habitat | Lean | Years | No | Opportunism | Pulliam 1986 |
| Foliage insectivores/ France | Foraging microhabitat | Lean (late winter) | Seasons/years | No | Competition | Laurent 1986 |
| Thornbills/Australian woodlands | Foraging behavior | Lean (?) (drought) | Seasons/years | No | | Bell & Ford MS |

the period of relative resource scarcity. Schoener (1982) suggested that habitat or microhabitat overlap may not be very useful in indicating competition, as habitats may be the 'arenas' rather than the objects of competition (Schoener 1974a).. Later (1983b), he argued that macrohabitat overlap may in fact be inversely related to competition; overlap should therefore be greatest during relatively lean times. In fact, among the studies I surveyed, overlap was least during the period of relative scarcity as often for habitat measures (6 out of 10 studies) as for diet measures (8 of 16 studies) (Table 1.6).

Some specific comments on some of the studies surveyed in this table are in order. Ulfstrand (1977) found overlap in foraging microhabitat use among foliage insectivores in Sweden to be less in summer than in winter. He attributed this to a relative scarcity of food at that time, following the dubious argument that the low overlap indicates competition and resources should therefore be relatively scarce then. Newton (1967) found little seasonal change in overlap in diet and feeding location among farmland finches, although seeds became scarce in early spring. This was especially so when there was heavy snow, making the seeds locally unavailable and forcing many of the species to move elsewhere (e.g. farmyards). This complicates the interpretation of niche-overlap changes, as does the tendency among among species to form flocks that increase in size in late winter.

The two *Buteo* hawks studied by Steenhof and Kochert (1985) demonstrated the expected decrease in overlap during a year of prey decline on a general scale. When overlap was calculated separately for individuals of the two species occupying overlapping foraging ranges and individuals occupying separate ranges, however, the reduction in diet overlap was greater in the individuals occupying unshared ranges. These individuals should have felt less competitive pressure than those occupying shared ranges and thus should have decreased their overlap less. The sparrows that Pulliam (1986) studied in Arizona grasslands do not fit Schoener's expectations. These birds exhibited a reduction in the range of habitats occupied by each species when food was scarce but expanded the variety of seeds they consumed. Arguing from the predictions of optimal foraging theory, Pulliam (1985) suggested that the birds should be expected to specialize on the seed types that they consume most efficiently when seeds are very abundant, increasing their diet variety as the favored seed types become scarcer.

In most of the studies, seasonal rather than yearly changes in overlap were considered, and actual documentation of any effects of resource limitation was rarely provided. Gibb (1960), however, reported that winter mortality in the foliage insectivores he studied was closely related to food supplies, and Grant and his colleagues have shown the importance of food limitation in *Geospiza* finches during both dry seasons and drought years.

There are three general problems with inferring the occurrence of competition from patterns of temporal change in overlap, even when the inference is accompanied by information on relative resource levels. First, one may interpret high overlap in 'fat' times and low overlap in 'lean' times either as a response to the lean conditions through specialization (whether produced by competition or by intrinsic properties of the species) or as a common opportunistic response of the species to the fat conditions. These alternatives are not distinguishable unless one also knows how the 'fat' or 'lean' conditions relate to resource availability and the position of the resource-limitation thresholds of the populations (Fig. 1.1). A reduction in overlap during relatively 'lean' times may be suggestive of possible competition, but the evidence is weak unless accompanied by information on the effects of resource limitation. The form of the reduction in resource levels may also influence how overlap changes. If resource availability decreases over the entire resource spectrum, potential competitors may diverge in resource use, whereas an uneven reduction that leads to scarcity of some resource types but leaves others relatively abundant may lead to convergence among species on the more common resources, producing greater overlap (Ford, personal communication, Fig. 1.3).

Second, unless changes in population densities are assessed along with the changes in niche overlap, the pattern may easily be misinterpreted (Llewellyn and Jenkins 1987). If the 'lean' time corresponds with a period in which the densities of the species in question are reduced, for whatever reasons, the breadth of habitat occupancy of each may be reduced as a consequence of lowered intraspecific competition. If the species respond to habitat conditions in species-specific ways, this will have the effect of reducing interspecific niche overlap. Interspecific interactions might or might not also be involved in producing the pattern.

The third difficulty is that the amount of overlap between species may not decrease monotonically with increasing resource scarcity even in a competitive system. Under conditions of resource superabundance, several species may jointly exploit the same highly profitable resources, and overlap will be high. As conditions worsen, competition may foster increasing niche separation. There may be a point, however, at which resources become so scarce that the individuals remaining in an area are compelled to use the same resources. This will once again increase niche overlap (Fig. 1.4). The latter situation is obviously unstable should the dire straits continue, but it may

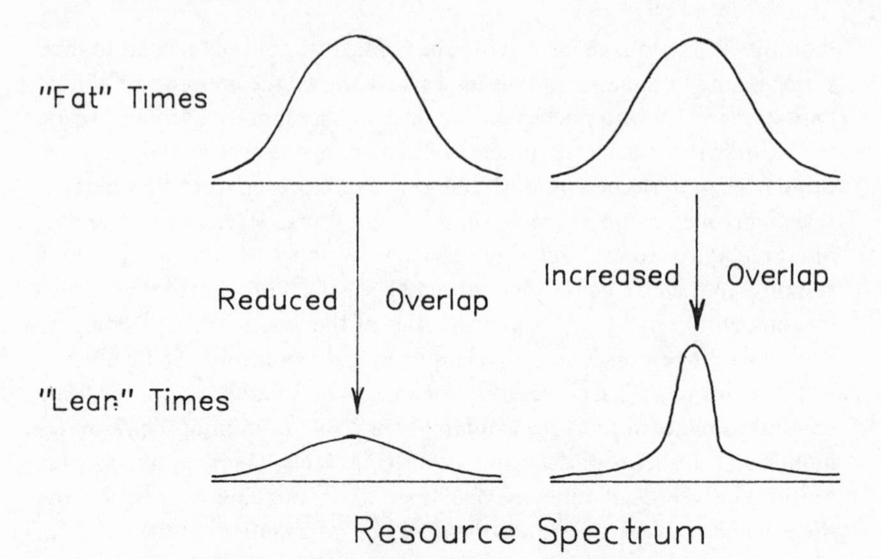


Fig. 1.3. The way in which resource levels are reduced during periods of relative scarcity can influence whether potential competitors reduce or increase overlap in resource use. In the left diagram, resource levels are reduced over the entire resource spectrum, and overlap among species may decrease as competition restricts their resource use. In the right diagram, some resources remain relatively abundant while others become very scarce; the species converge to use the abundant resources, and overlap increases. From H. Ford (personal communication).

occur during temporary resource shortages. Bell's (1983, Bell and Ford MS) studies of thornbills (*Acanthiza*) in eastern Australia clearly exemplify this sort of pattern. Bell's studies included a period of increasing drought, during which time plant growth decreased and arthropod abundance fell. Overlap among the three species was lowest when food was scarce, but it rose again to relatively high levels well before food abundance increased (Fig. 1.5). Bell thought that this reversal in overlap patterns might reflect the effects of heightened intraspecific competition, which would promote a broadening of the foraging niche, or the abandonment by the species of specialized foraging modes under conditions of extreme scarcity, which would produce an opportunistic convergence on what little food was available.

Studies of niche relationships have documented that coexisting species often differ in a variety of ways, although at times they may be remarkably similar in ecology. The patterns in most cases are relatively clear-cut; what is uncertain is how they should be interpreted and whether or not they provide firm evidence of competition. In most cases the evidence, although sugges-

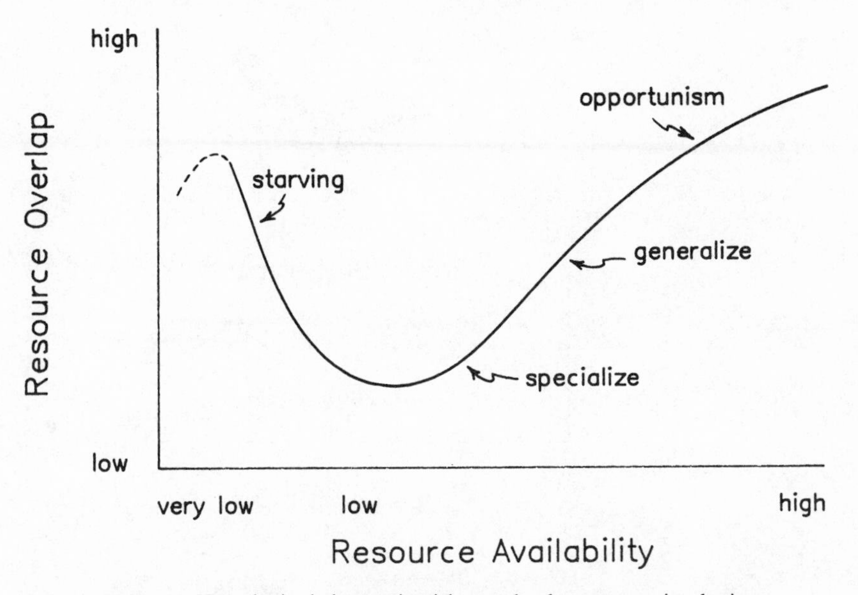


Fig. 1.4. Hypothesized changes in niche overlap between species sharing a resource as resource availability changes from very abundant to very scarce.

tive, is weak. There may be a great many reasons for species to differ in aspects of their niches, and relating any specific patterns or levels of overlap or dissimilarity to competition in preference to other processes is not realistic in the absence of additional information. Interpreting the patterns is all the more difficult if they are variable in time or space. This variability is the basis of investigations of niche shifts.

***Niche shifts and density compensation***

Comparisons of areas containing or not containing putative competitors are often held to be the best sorts of natural experiments, and when niche features or densities of a species change between such areas the evidence for competition is considered to be especially compelling (Diamond 1978). Because several species are generally absent among compared areas, however, it is especially critical that the potential for competition among specific sets of species be determined rather than simply inferred on the basis of the presence–absence patterns. This requires documentation that the species do indeed overlap in resource use in areas in which they co-occur.

Most studies of niche shifts or density compensation have satisfied this criterion, but few have established that the presence or absence of the putative competitor(s) is the only important difference between the com-

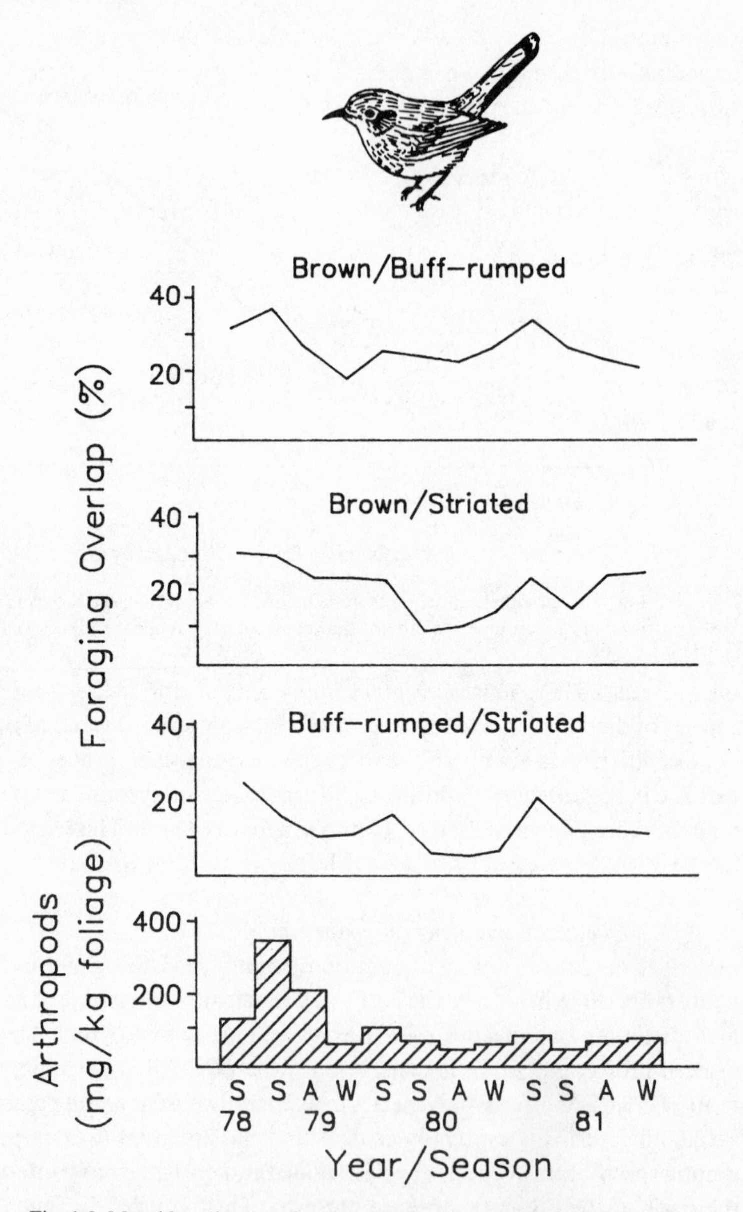


Fig. 1.5. Monthly estimates of arthropod abundance and foraging overlap between Brown, Buff-rumped and Striated thornbills (*Acanthiza pusilla, A. reguloides*, and *A. lineata*) in woodlands in eastern Australia. The histogram depicts arthropod abundance, while overlap (based on observations of birds not in mixed-species foraging flocks) is shown by the solid lines. Modified from Bell (1983) and Bell and Ford (MS).

pared areas. An acceptance of the *ceteris paribus* assumption is especially critical to such comparisons, but this assumption is unlikely to hold in most situations. Even if the presence or absence of a potential competitor is the major difference between several compared locations, a shift in niche parameters is not necessarily an indication of competitive effects. Niche-shift patterns may occur even if resources are not limiting. If species forage optimally, for example, any changes in the nature of the resource base may prompt a shift in behavior or prey choice, and the exploitation of the resources by one species may therefore produce a shift in the behavior of another whether or not resources are limiting (Thomson 1980). By the definition of competition used here, such shifts are not associated with competition. Alternative process hypotheses thus generate predictions of niche shifts matching those of competition, rendering the evidence rather weak in the absence of explicit tests of such alternatives. Patterns of density compensation are also predicted from a variety of alternative hypotheses (Chapter 3). Overall, these patterns provide only weak evidence of competition in the absence of additional enviromental measurements, and even then the evidence remains equivocal unless alternative explanations are considered and rejected.

***Aggressive interactions***

When members of different species aggressively interact with one another, the potential for interference competition is especially strong. One expression of interspecific aggression is simple dominance of one species by another. Hooded or Carrion Crows (*Corvus corone*), for example, are aggressive toward other corvid species (Bossema *et al.* 1976, Baeyens 1981, Waite 1984), and this has often been interpreted as interference competition related to food resources. In one study in agricultural areas of England during winter, displacement attacks by the crows on other species occurred at a higher rate in fields containing more of their favored earthworm prey (Waite 1984). The other corvid species, especially Rooks and Jackdaws, formed flocks, and crows aggressively chased such flocks more frequently than individuals or pairs (which occurred more frequently than larger flocks). The level of aggression, however, was unrelated to the degree of dietary overlap between the species. The rate of prey capture by crows was depressed in the presence of large flocks of other species (Fig. 1.6). Waite suggested that the flocks disturbed the earthworm prey, reducing their availability, and that the aggression of the crows toward such flocks was a response to the reduction in prey availability. In this situation, interspecific aggression is related to resource use, although the dominant aggressor appears to exhibit reduced prey intake in the situations in which aggression

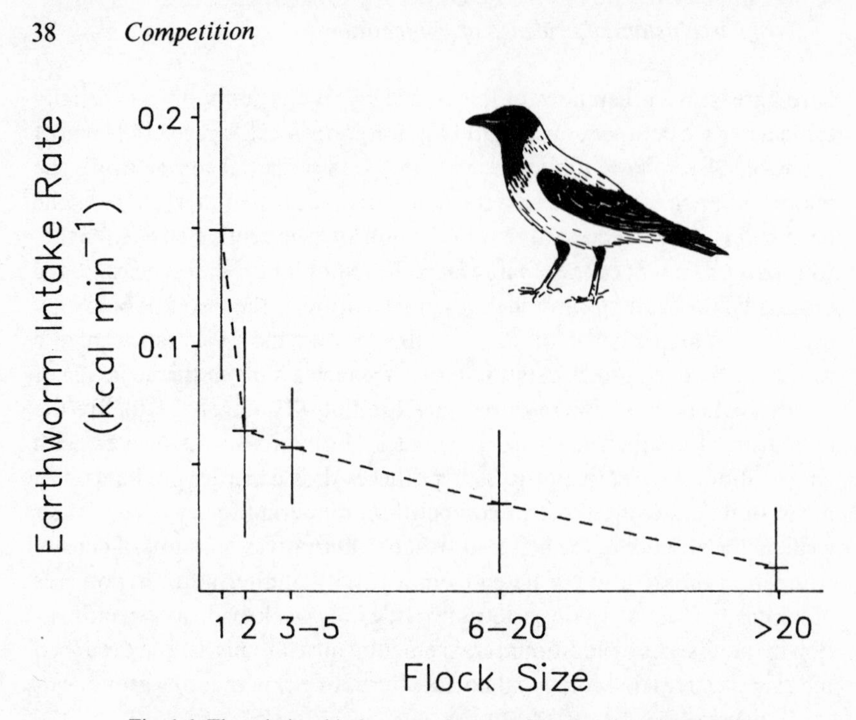


Fig. 1.6. The relationship between the rate of intake of earthworms by Eurasian Crows (*Corvus corone*) and the size of flocks of other species of corvids present on the crows' territories. After Waite (1984).

is greatest. Whether resource availability was actually limiting to any of the corvids at this time was not determined.

The relation between aggression and limiting resources is perhaps clearest with reference to nest sites for hole-nesting species. In areas in which both Prothonotary Warblers (*Protonotaria citrea*) and House Wrens (*Troglodytes aedon*) occur, wrens are both aggressive toward the warblers and predators on their eggs, often puncturing the eggs when the warblers are absent (Walkinshaw 1941). Marsh Wrens (*Cistothorus palustris*) exhibit similar aggressive behavior toward blackbirds sharing their marsh breeding habitat (Picman 1977, 1980, 1984). Starlings (*Sturnus vulgaris*) may aggressively evict Eastern Bluebirds (*Sialia sialis*) and other species from nest cavities (Zeleny 1978, Short 1979), and both temperate and tropical woodpeckers may often lose nest cavities that they have excavated to starling species (Short 1979, Collias and Collias 1984). In these situations, the resource associated with the aggression can be identified and the negative effects on the subordinate species determined, so the evidence of competition is reasonably strong. Where the resource is only inferred or the effects of the aggression on individuals are less clear, interspecific dominance by itself is suggestive of competition, but it is relatively weak evidence.

When interspecific aggression is associated with spatially defined territories, the observations constitute stronger evidence of competition. Interspecific territoriality potentially has a negative effect on individuals of the excluded species, by forcing them to locate territories in areas that may be less suitable. Because this pattern is superimposed on a system of intraspecific territorial defense, the occurrence of intraspecific competition is also likely. Whether or not either form of territoriality actually represents a competitive situation depends once again on the relations to resource availability. If the territorial aggression forces individuals into habitats in which their reproductive success or survival is reduced, the interactions are clearly competitive. Interspecific territoriality may often be associated with differences in habitat preferences of the species, however, and documentation of such negative effects on individual performance is therefore critical to interpreting the interaction as primarily competitive rather than an expression of differences in habitat selection in a patchy environment. If habitat preferences are based on subtle differences between areas, the latter explanation may be especially likely. Still, some instances of interspecific territoriality (e.g. blackbirds; Orians and Collier 1963, Miller 1968) provide convincing evidence of competition.

***Overview***

Considered separately, each sort of pattern that has been used as evidence of competition between species may be suggestive of that process but is not entirely convincing, as several of the criteria required to document competition (Table 1.4) are not met. Most categories of evidence are at least generally consistent with the patterns expected from competition theory, although in some instances the predictions of the theory are so broad that the patterns become ambiguous. Because comparisons are generally restricted to species exhibiting at least some degree of resource overlap (e.g. members of a guild), criterion 2 is usually met as well (Table 1.7). In most studies, however, the observations of patterns in morphology, distributional parapatry, niche shifts, and the like have been taken as firm evidence of competition in the absence of any consideration of the effects of the interaction on resource availability (criterion 4) or on individuals or populations of the other species (criterion 5). The occurrence of intraspecific competition (criterion 3) is usually inferred rather than assessed directly, and rarely have alternative process hypotheses (criterion 6) been considered, much less rigorously tested. The summary in Table 1.7, based on studies that have been the norm in avian community ecology, leads to the conclusion that much of the evidence of competition is weak but suggestive. When careful attention has been given to resources or to individual or

Table 1.7. *Summary of the degree to which various categories of evidence of interspecific competition meet the criteria necessary to document such interactions in most investigations of bird communities.*

Some studies provide stronger evidence for a given category either by documenting the patterns with particular care or by measuring other variables (e.g. resource levels, individual reproductive success); this tabulation indicates the success of 'average' studies.

| | Criteria[a] | | | | | |
|---|---|---|---|---|---|---|
| Category of evidence | 1 (Pattern consistency) | 2 (Species overlap) | 3 (Intraspecific competition) | 4 (Resource availability reduced) | 5 (Negative effects) | 6 (Alternatives considered) |
| *Morphology* | E | X | I | I | I | R |
| *Distributions* | | | | | | |
| Parapatry | X | X | I | I | I | S |
| Inverse density relationships | X | X | I | I | X | R |
| Habitat exclusion | X | E | I | I | I | R |
| *Niches* | | | | | | |
| Overlap | E | X | I | I | I | R |
| Overlap change | X | X | I | E | I | R |
| Shifts | X | X | I | I | I | S |
| *Aggression* | | | | | | |
| Dominance | X | X | I | I | X | R |
| Interspecific territoriality | X | X | X | I | X | R |

*Notes:*
[a] X = usually satisfied; E = predictions or evidence equivocal; I = usually inferred, not measured; R = rarely; S = sometimes.

population effects, the evidence has sometimes been consistent with our expectations from competition theory, sometimes counter to them, but in either case the arguments have been strengthened. Similarly, in situations in which several categories of evidence have been brought to bear on the issue (e.g. the finches studied by Grant and his colleagues, the *Parus* species investigated by Alatalo, Ulfstrand, and others, or the shrubsteppe birds we have studied), the conclusions drawn from them have been more convincing.

None of these categories of evidence can provide satisfying and convincing evidence of the operation of competition as a proximate process by itself. In the absence of measures of resource levels and an assessment of the supply–demand relationships between the populations and resources, statements about competition will always contain an undesirable amount of inference. Unless alternative hypotheses are given careful consideration, even evidence bolstered by measures of resource levels or of individual or population effects will remain equivocal. By combining careful documentations of the patterns themselves with appropriate measurements of environmental and population parameters for several categories of evidence, one may build a strong and convincing case for or against the operation of competition as a process in particular situations. In the end, the evidence remains circumstantial so long as it is based on comparative, correlative observations, although careful attention to the design of such studies may strengthen individual cases further (Diamond 1986).

**Experimental evidence of competition**

In some situations, circumstantial evidence of the presence or absence of competition as a proximate process may be strengthened by manipulating key variables in field settings. In well-designed experiments, the differences between treatments and controls can be attributed to the effects of the manipulated variables. Thus, if one removes a species from some plots in an area and a second species responds by increasing its niche breadth in comparison with control plots, the evidence of a competitive restriction of the second species is considerably stronger than that obtained from a comparison of areas in which the first species is naturally present or absent. In an experiment, the *ceteris paribus* assumption may be accepted with somewhat greater confidence, although only to the extent that the experiment is properly designed and conceived.

Good experiments are founded on good natural history. An experiment designed to explore the possibility of competitive interactions among species, therefore, should be conducted in situations in which previous observational studies have provided at least suggestive evidence of competition. To the extent that experiments are designed and conducted in this manner, the literature will contain an intrinsic bias favoring competition. This is because experiments are not likely to be conducted where there is little reason to expect it to occur, and experiments in which no evidence of competition is found may be regarded as 'failures' and not reported. Connell (1983) has discussed this bias in detail. A consequence of this bias is that attempts to discern the frequency of competition in nature from the frequency with which it is reported in experimental studies (e.g. Schoener 1983b, Connell 1983) are of doubtful value (see also Hairston 1985, Schoener 1985a, Underwood 1986b).

Experimental approaches to investigating competition among birds have been neither as frequent nor as successful as in some other groups of organisms. Schoener (1983b) listed nine experimental studies of competition involving birds (only seven of which addressed bird–bird interactions). Using more stringent criteria, Connell (1983) considered only three of these studies. In his listing of studies providing guidance to sound experimental design, Underwood (1986b) listed *no* examples dealing with birds, although investigations of other vertebrate groups were well represented. This would seem to confirm Grant's (1986b) view that experiments are of limited usefulness in studies of competition among birds. Nonetheless, experimental studies of bird communities have been conducted in which either resources or the densities of one or another bird species were manipulated, and it is instructive to consider some of these briefly.

***Resource manipulations***

Of the resources over which competition might occur, nest sites and food are perhaps most easily manipulated. Nest-site manipulations have focused on hole-nesters, for which the availability of nest sites can be increased by providing nest boxes of various designs. Whether the concern has been primarily with intraspecific (e.g. Krebs 1971, Krebs and Perrins 1978) or interspecific effects (e.g. Slagsvold 1978, Brawn *et al.* 1987), such manipulations have generally resulted in increases in breeding densities of at least one species of hole-nester, confirming the assessment that the natural availability of nest cavities was limiting. In one such experiment, Hogstad (1975) varied the number and availability of nest boxes in a 9-ha coniferous woodland in Norway over an 8-yr period and recorded the breeding densities of several hole-nesting and open-nesting species in relation to those in a nearby control area lacking nest boxes. Densities of the hole-nesters increased when nest boxes were available, largely because many Pied Flycatchers (*Ficedula hypoleuca*) immigrated into the woods. Densities of six species of open-nesters varied inversely with the combined density of hole-nesters (Fig. 1.7), although densities of only three of these species were significantly related to those of hole-nesters when temporal variations in the control area were considered. Thus, the experiment provides strong evidence of a negative interaction between hole-nesters (especially Pied Flycatchers) and *Phylloscopus trochilus, P. collybita*, and *Fringilla coelebs*. The other negative density relationships turn out to be nonsignificant, although, in the absence of information from the control area, they might well have been considered to be evidence of competition as well. This indicates once again the uncertainty of interpretations of negative density relationships in the absence of additional information.

In other experiments, food resource levels have been manipulated, almost always by providing additional food. These experiments have generally been interpreted in the context of food limitation and were discussed earlier in this chapter. In most studies, species have responded differently to the manipulation, and an interpretation of these results with respect to competition is difficult. If one species increases in density following food supplementation while another putative competitor does not (e.g. Great Tits versus Blue Tits in Källander's (1981) experiment), this might be because one species is food limited and the other not, because the increase in one species suppresses any increase in the other, or because the nature of the manipulation is more relevant to the ecology of one species than the other.

***Density manipulations***

Although manipulating the density of one or more species is usually more difficult than altering resource levels, it may have a more direct bearing on competition. Perhaps the most widely cited example of a density manipulation is Davis' (1973) investigation of the spatial relationships among wintering Golden-crowned Sparrows (*Zonotrichia atricapilla*) and Dark-eyed Juncos (*Junco hyemalis*). Davis found that the juncos were evenly distributed along the borders of an old field in California, whereas the sparrows were largely restricted to dense willow thickets in one part of the field. Davis removed juncos during one winter, but they were rapidly replaced by juncos from other nearby areas. The absence of a competitive effect of juncos on sparrows was inferred from the failure of the sparrows to expand their habitat occupancy during another winter when junco densities were naturally low. When Davis removed sparrows from the willow thicket, they were not replaced by conspecifics, however, and juncos expanded to use the willow thicket more; when the trapped sparrows were later released,

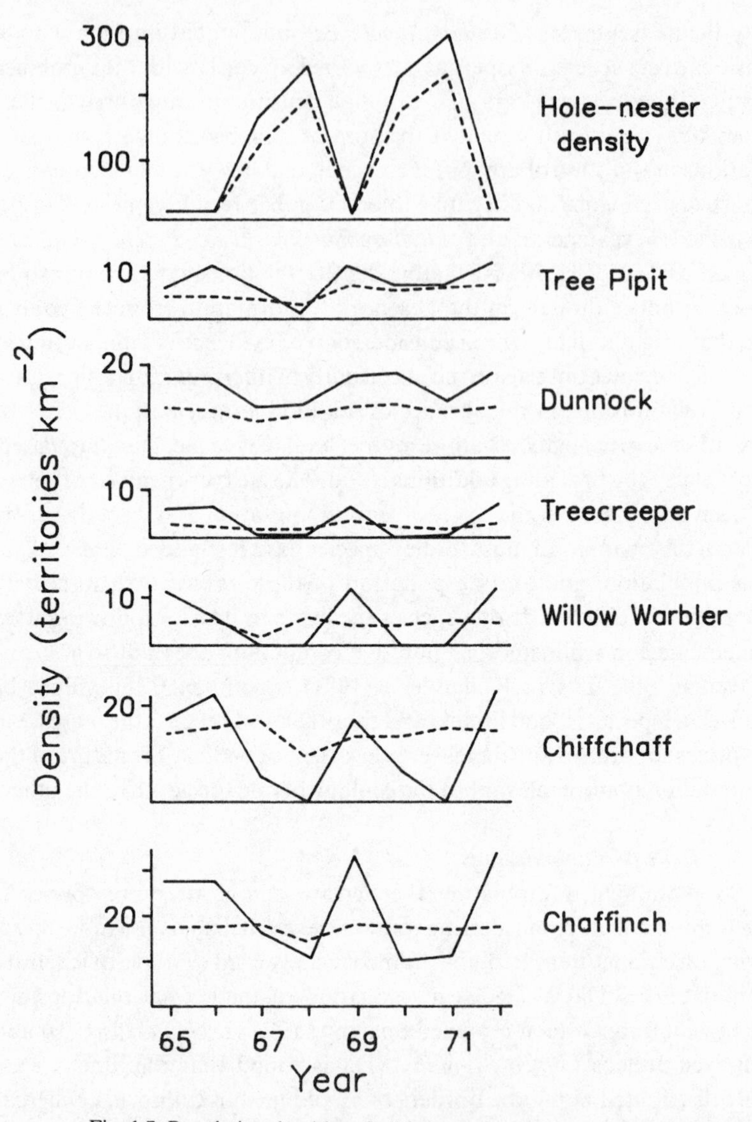


Fig. 1.7. Population densities of all hole-nesting species combined (top) and of six open-nesting species in areas supplemented by nest boxes (solid line) and in a control area (dashed line) in a Norwegian coniferous forest. During 1965 and 1966, no nest boxes were present; in 1967–68 the experimental area was supplied with boxes (35 and 15, respectively); in 1969 access to boxes was restricted by blocking entrance holes; in 1970, 70 boxes were available; in 1972, all boxes were again made inaccessible. The dashed line in the top panel illustrates the densities of Pied Flycatchers. Modified from Hogstad (1975).

use of the thicket by juncos diminished. The two species do not overlap in diet to any great extent during winter, and Davis attributed the apparent competition to the proximity of a water trough (an important water source for juncos) to the heavy cover preferred by the sparrows. Davis thought that the sparrows interfered with junco access to the water trough, although observations of interspecific aggressive encounters were few, even after the sparrows were released into the thicket at the conclusion of the exercise. The suggestion of an interspecific suppression of junco habitat occupancy by the sparrows is fairly clear. The experiment was conducted only over a 4-month period, however, and it lacked replication and included no control other than the state of the system prior to the manipulation.

The experiment conducted by Williams and Batzli (1979), in which Red-headed Woodpeckers (*Melanerpes erythrocephalus*) were removed from an upland forest during one winter, did include a control. Other members of the bark-foraging guild expended in habitat distribution and/or use of foraging sites in comparison with the control. There were no contemporaneous replicates of the treatment plot, although similar observations were obtained in another year in which woodpecker densities were naturally low in the upland forests. Together, these results provide reasonably convincing evidence of a competitive effect. Observations were restricted to a single month following the experimental removal, however, so the long-term consequences of the treatment could not be considered. The interaction between Red-headed woodpeckers and the other guild members is by direct interference, and the rapidity of the response of the other species to the removal of the woodpeckers is consistent with Miller's (1967) suggestions regarding interference competition.

Additions of potential competitors to a community, although more difficult to accomplish than removals, may provide additional insights into competition. Högstedt (1980a), for example, documented observationally that the diets of nestling Magpies and Jackdaws overlapped in southern Sweden and that, because many nestlings of the two species starve, food must be limiting and the overlap therefore of potential competitive significance. Jackdaws forage farther from nest sites, and, therefore, should be less strongly affected by the competition. Högstedt (1980b) tested the effects of these relationships by placing three nest boxes for Jackdaws in each of a number of Magpie territories. Other Magpie territories served as controls, and pairs of Magpies served as experimentals or controls in alternating years of the 2-yr experiment (thus controlling for individual differences in reproductive performance). Jackdaw foraging activity was substantially greater in the experimental than in the control Magpie territories, and,

Table 1.8. *The reproductive performance of Magpies in experimental and control groups.*

Jackdaw nest sites were placed within the territories of experimental pairs of Magpies, which produced increased interspecific aggression by the Magpies and reduced the amount of time they could devote to foraging. From Högstedt (1980b)

| | Experimental group | Control group | *P* (two-tailed) |
|---|---|---|---|
| Start of laying (median date) day 1 = 1 April | 20 | 21 | NS |
| Clutch size | 6.50 ± 0.30 | 6.48 ± 0.18 | NS |
| Mean weight of clutch | 64.7 ± 4.0 | 67.1 ± 2.3 | NS |
| Mean of mean nestling weight within brood, age 21–25 d | 182 ± 7 | 208 ± 4 | <0.02* |
| No. of started breedings (*N*) | 18 | 38 | |
| No. of hatched clutches (% of *N*) | 10 (56%) | 23 (61%) | NS |
| No. of broods with ≥ 1 fledged young (% of *N*) | 3 (17%) | 16 (42%) | <0.04† |
| No. of fledged young per successful brood | 2.0 ± 0.58 | 4.0 ± 0.26 | <0.05* |
| No. of fledged young per breeding attempt | 0.33 ± 0.20 | 1.68 ± 0.34 | <0.001 |

*Notes:*
Figures are given ± s.e. NS, not significant.
* Mann-Whitney U test.
† Fischer exact probability test.

although Jackdaws were not aggressive toward Magpies, Magpies did attack Jackdaws foraging in their territories quite frequently, especially during the nestling period. Magpies in experimental territories suffered significantly reduced reproductive success, producing fewer young with lower fledging weights compared with the controls (Table 1.8). Högstedt attributed this to both increased starvation, due to food depletion in Magpie territories by Jackdaws, and increased predation by Hooded Crows, perhaps a result of the increased time spent by adult Magpies foraging away from the nest and/or increased begging activity by the hungry young.

Other species manipulations have been directed toward understanding interspecific spacing patterns. In English woodlands, Blackcaps (*Sylvia atricapilla*) and Garden Warblers (*S. borin*) are interspecifically territorial

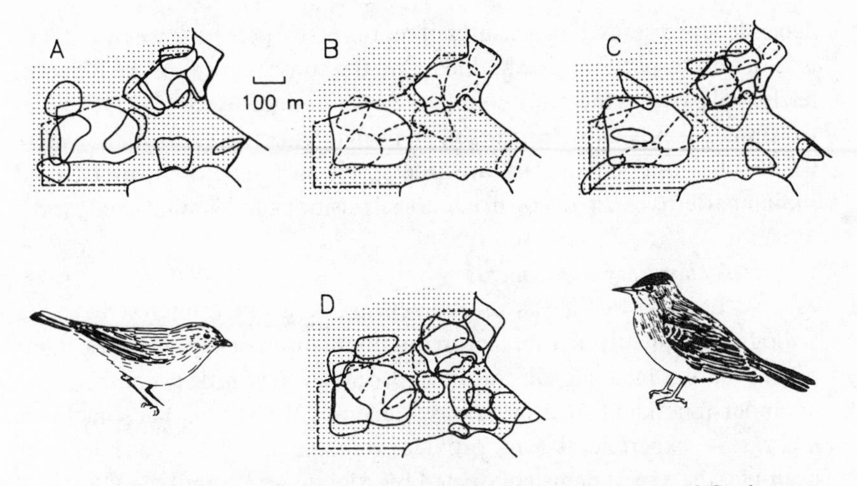


Fig. 1.8. The positions of territories of Blackcaps (solid lines) and Garden Warblers (dashed lines) in a woodland area in England. A. Blackcap territory positions during the preremoval period in 1979 before the arrival of Garden Warblers (15 April – 8 May); B. Positions of Garden Warbler territories during the period when Blackcaps were removed (9 May – 5 June); C. Positions of Blackcap and Garden Warbler territories during the period 6–15 June 1979, after the Blackcap removals had ended; D. Territory positions of the two species in 1980 (6 May – 6 June), when no Blackcaps were removed. The shaded area represents closed-canopy sycamore woodland, the clear area to the left open oak woodland, and the clear area in the upper right ash-elm woodland. After Garcia (1983).

during the breeding season. Blackcaps arrive earlier in the spring and establish territories before Garden Warblers arrive. To determine whether the spacing pattern is a response to subtle habitat differences or a result of interspecific exclusion of *borin* by *atricapilla*, Garcia (1983) removed *atricapilla* individuals from territories in one woodland before and during the period of territory establishment by *borin*. Although the species normally occupied nonoverlapping territories in somewhat different habitat situations at this site, *borin* individuals readily established territories in the removal area. After the removal period ended, some additional *atricapilla* individuals settled in the area, in some cases displacing established *borin* (Fig. 1.8). The results of this manipulation support a competitive explanation, although habitat differences also contribute to the pattern of *borin* territory placement (Fig. 1.8). Garcia included no real replication or control in his experimental design, observations in the years preceeding and following the experiment serving as 'controls' of a sort. Regional densities of *borin* were low in the experimental year, so it is doubtful that the expansion of the species in the experimental area was a consequence of intraspecific density-dependent pressures. No measures of reproductive performance by individuals of either species were obtained, and the consequences of the different territorial patterns therefore could not be determined. Reed (1982) conducted similar removal experiments in an insular situation off Scotland where Chaffinches and Great Tits are interspecifically territorial, and found similar patterns of expansion in territorial positions and habitat occupancy.

### *Laboratory experiments*

Because the effects of competition of greatest interest to avian ecologists frequently involve features of populations or communities, it has often been assumed that laboratory experiments have little to contribute to an understanding of such situations. Generally this is true, but sometimes laboratory experiments can provide valuable insights. Consider, for example, the experiments conducted by Alatalo and Lundberg (1983) on niche separation between Marsh (*Parus palustris*) and Willow tits. In northern Europe, Marsh Tits usually occupy rich deciduous woodlands, whereas Willow Tits are found in boreal forests dominated by conifers. By documenting the use of artificial spruce and oak 'habitats' by these species in the laboratory, Alatalo and Lundberg showed that Marsh Tits could forage efficiently in both substrate types, although they exhibited a preference for the deciduous growth form. Willow Tits, on the other hand, were more efficient than Marsh Tits at finding food in both substrate types. Thus, although there were differences in their foraging behavior in a controlled (albeit artificial) laboratory situation, the differences were seemingly not sufficient to account for the natural habitat distributions of the species on the basis of inherent species-specific foraging adaptations (such as have been shown in the laboratory for Coal and Blue tits; Partridge 1976a, b). Alatalo and Lundberg concluded that it is therefore likely that the natural habitat distributions of the species are restricted by interspecific competition, even though there were no differences in the dominance status of either species in the laboratory trials and the presence of one species did not affect the relative foraging efficiency of the other. Such experiments, although not resolving the role of competition in field situations, can shed light on some aspects of species' niches.

### *An overview of field experiments on birds*

These examples provide a sampling of the sorts of experiments that have been done with birds to test for the operation of competition. From a careful but incomplete scanning of the literature, I tallied some 35 experiments in which either resources (12 experiments) or species' densities (23) have been manipulated in a way that might reveal interspecific effects (Table 1.9). Of the resource-manipulation experiments, 6 produced fairly clear evidence of competition, 5 failed to indicate any competitive effects, and 1 was ambiguous. When densities were manipulated, 16 of the tests strongly implicated competition, 5 did not, and 2 were ambiguous. These experiments thus provide reasonably strong support for the operation of competition in situations in which previous observations suggested that it might occur, although this is not always the case. Manipulations of resources are apparently less successful at demonstrating competition than are density alterations. although the two techniques have not been applied to similar situations, so their comparability may be limited.

As with observational studies, however, the results of an experimental investigation of competition are only as sound as the methodology used to obtain them. Controls, for example, are essential; in their absence, there is nothing to compare the results of the manipulation with, and the 'experiment' becomes little more than a strictly observational study. Unfortunately, experiments with birds have not always included controls. Of the studies listed in Table 1.9, 12 (6 resource, 6 density) lacked proper controls. Well-designed experiments must also include replication of both control and experimental treatments. Only 5 of the studies I reviewed contained more than a single control plot or sample, and 19 of the studies (5 resource manipulations, 14 density alterations) included no replication of the treatments. Admittedly, it is often logistically difficult to replicate experiments in which birds are removed from a sufficiently large area to make the manipulation worthwhile, but the general lack of replication in most of these studies seriously compromises their value. Perhaps such experimental manipulations should not be done at all unless they can be done in a scientifically rigorous manner.

Because ecological situations involving interactions are subject to time lags of varying magnitudes (see Chapter 4), the duration of an experiment may also be important. Some responses may take a considerable length of time to be expressed, especially if they involve turnover in populations, and indirect effects or secondary interactions may become evident only some time after an experimental perturbation (Brown *et al.* 1986, Wiens, Rotenberry, and Van Horne 1986, Bender *et al.* 1984). Of the experiments tabulated here, the majority (8 resource, 14 density) lasted only one season or less, although several continued for more than 1 yr. Short-term experiments may be suitable for gauging immediate behavioral responses to density or resource manipulations in some situations, but even then a long-term perspective may provide additional insights.

Table 1.9. *Features of field experiments in which resources or densities*

| Reference | Species/ location | Variables manipulated | Response measure | Result |
|---|---|---|---|---|
| *Resource manipulations* | | | | |
| Thompson & Lawton 1983 | Wintering passerines/ England | Seed combinations | Species associations | No relationship to diet overlap; some − associations, more + |
| *Krebs 1971 | Great Tit-Blue Tit/England | Winter food | Densities | BT increased, no change in GT |
| *Källander 1981 | Great Tit-Blue Tit/Sweden | Winter food | Densities | GT increased (1 yr), no change in BT |
| *Jansson *et al.* 1981 | Crested Tit-Willow Tit/ Sweden | Winter food | Winter survival, breeding density, reproductive rates | Survival and breeding density increased, reduced reproductive rate in WT; less effect in CT |
| *Brömssen & Jansson 1980 | Crested Tit-Willow Tit/ Sweden | Spring and summer food | Reproductive parameters | Laying date advanced in both species, CT young heavier |
| Samson & Lewis 1979 | Chickadee-titmouse/ Pennsylvania | Winter food | Winter flock size, breeding density | Chickadee increased flock size, minor increase in breeding density; no response in titmouse |
| Slagsvold 1978 | Great Tit-Pied Flycatcher/ Norway | Nest box availability to PF | PF nest occupancy | PF not successful in displacing GT |
| Lyon *et al.* 1977 | Hummingbirds/ Arizona | Food dispersion | Territorial intrusions | Territorial exclusion decreased with increasing food dispersion |
| Pimm 1978 | Hummingbirds/ Arizona | Food predictability | Time spent in feeding and defense | Less predictable reduced defense by dominant, increased intrusion by subordinate |
| Pimm 1978 | Hummingbirds/ Arizona | Food predictability | Feeding times | Lower predictability reduced overlap |
| Pimm *et al.* 1985 | Hummingbirds/ Arizona | Food quality & density | Feeding activity | Dominant species caused density-dependent habitat shift in subordinate |
| Gill *et al.* 1982 | Hummingbirds-bees/Costa Rica | Food availability | Hummingbird feeding activity | Bees excluded hummingbirds from flowers |
| *Density Manipulations* | | | | |
| Davis 1973 | Juncos-sparrows/ California | Sparrow density | Junco winter distribution | Habitat shift by Juncos |
| Hogstad 1975 | Passerines/ Norway | Hole-nester density | Open-nester density | Inverse density relationship in 3 open-nesting species |
| Brawn *et al.* 1987 | Passerines/ Arizona | Hole-nester density | Open-nester density | No response (hole nesters increased) |
| Williams & Batzli 1979 | Wintering bark foragers/Illinois | Density of dominant species | Habitat distribution, foraging substrate | Other species expanded with removal of dominant |
| Dhondt & Eckyerman 1980 | Great Tit-Blue Tit/Belgium | GT excluded from winter roost sites | Breeding densities | GT density reduced, BT increased |
| Minot 1981 | Great Tit-Blue Tit/England | BT young removed | GT nestling weight | GT young heavier |
| Minot 1981 | Great Tit-Blue Tit/England | BT young added | GT nestling weight | GT marginally heavier |

*involving more than one species of bird have been manipulated*

| Control | Replication | Duration | Resource identified? | Exploitation or interference? | Conditions natural? | Alternatives considered? |
|---|---|---|---|---|---|---|
| 2 | 2 each treatment (7) | 1 week | Seeds | Both | Yes | Yes |
| 1 | None | 1 winter | Winter food | Exploitation? | No (nest boxes) | Yes (intraspecific effects) |
| 3 | 2 | 2 yr | Winter food | Not considered | ? | Yes (intraspecific effects) |
| 1 | None | 1 yr | Winter food | Not considered | ? | Yes (intraspecific effects) |
| 1 | None | 2 yr | Food | Not considered | No (nest boxes) | Yes (intraspecific effects) |
| 1 flock | None | 1 winter | Food | Exploitation | Yes | No |
| None | 2 (not considered as such) | 3 yr | Nest sites | Interference | No (nest boxes) | No |
| None | None | 1–2 day per treatment | Food/ territory | Interference | No (high food density) | No |
| None | 4–16 | 3 days/ experiment | Nectar | Interference | No (feeders) | Yes |
| None | 4–16 | 1 wk | Nectar | Interference | No (feeders) | No |
| None | Several | 1 wk/ experiment | Nectar | Interference | No (feeders) | Yes (intraspecific effects) |
| None | 2 | 3 days/ experiment | Flowers | Interference | No (artificial flowers) | No |
| None | None | 4 months | Water source | Interference | Yes | Yes (intraspecific effects) |
| 1 | None | 8 yr | No | Not known | No (nest boxes) | No |
| 1 | None | 3 yr | No | Not known | No (nest boxes) | No |
| 1 | None | 2 months | Habitat? | Interference | Yes | No |
| 1 | 2 (1 yr) | 2 yr | Winter roosts? | Both? | No (nest boxes) | No |
| 1 | None | 1 summer | No (nestling food?) | Exploitation | No (nest boxes) | No |
| 1 | None | 1 summer | No (nestling food?) | Exploitation | No (nest boxes) | No |

Table 1.9. (*cont.*)

| Reference | Species/ location | Variables manipulated | Response measure | Result |
|---|---|---|---|---|
| *Cederholm & Ekman 1976 | Crested Tit-Willow Tit/ Sweden | Each species removed in spring | Territory occupancy | Territories not occupied by either species |
| Alatalo *et al.* 1985 | Tits and Goldcrest/ Sweden | Willow and Crested tit densities reduced | Foraging sites of Coal Tits and Goldcrests | Coal Tits and Goldcrest increased foraging in inner tree canopy |
| Alatalo *et al.* 1987 | Tits and Goldcrest/ Sweden | Coal Tit and Goldcrest removed from flocks | Foraging sites of Crested and Willow Ttis | Both species shifted micro-habitat use in pine, CT in spruce |
| Samson & Lewis 1979 | Chickadee-titmouse/ Pennsylvania | Fall removal | Winter density | No replacement in chickadee, few in titmouse |
| Reed 1982 | Chaffinch-Great Tit/ Scotland | Chaffinch removal | GT territory occupancy | GT expanded |
| Garcia 1983 | Blackcap-Garden Warbler/ England | Blackcap removed | Territory, habitat occupancy | GW expanded |
| Payne & Groschupf 1984 | Indigobirds/ Cameroon | Territorial males removed | Subsequent territory occupancy | Vacancies occupied by conspecifics |
| Fonstad 1984 | Willow Warbler-Brambling/ Norway | Territorial males removed | Subsequent territory occupancy | Vacancies occupied by conspecifics |
| Högstedt 1980b | Jackdaw-Magpie/Sweden | Jackdaw density in Magpie territories | Magpie foraging and reproduction | Magpie attacked jackdaws, lower breeding success; jackdaws increased foraging time |
| Alerstam 1985 | Great Tit-Pied Flycatcher/ Sweden | Breeding densities | Nesting success | Both species' success reduced |
| Gustafsson 1987 | Great Tit-Blue Tit-Collared Flycatcher/ Sweden | Tit nesting densities | Flycatcher breeding performance | Flycatcher fledging number and mass increased |
| Sherry and Holmes 1988 | Least Flycatcher—Redstart/New Hampshire | Flycatchers removed | Redstart abundance and habitat use | Adult American Redstarts increased, yearling redstarts excluded by older redstarts |
| Montgomerie MS | Hummingbirds/ Mexico | Dominant species removed | Activity budget of subordinate | Subordinate switched to match dominant's resource-use pattern |
| Carpenter 1979 | Hummingbirds-Insects/ California | Hummingbird access to flowers | Foraging time of insects | Insects increased foraging time on flowers |
| Eriksson 1979 | Goldeneye-fish/ Sweden | Fish removed | Goldeneye use of ponds | Goldeneye used experimental lake more, but not treated section of another lake |
| Török 1987 | Great Tit-Blue Tit/Hungary | Tit breeding densities | Feeding and nesting success | BT fed nestlings larger prey when GT reduced; GT nestlings greater mass when BT reduced |

*Notes:*
* = Not intended as a competition experiment, but interspecific effects could be shown.

| Control | Replication | Duration | Resource identified? | Exploitation or interference? | Conditions natural? | Alternatives considered? |
|---|---|---|---|---|---|---|
| 1 | None | 1 summer | No | Interference | Yes | No |
| 3 | 3 | 1 winter | Foraging site (food) | Interference (and exploitation?) | Yes | No |
| 3 | 3 | 1 winter | Foraging site (food) | Exploitation | Yes | No |
| None | None | 1 winter | No | Both? | Yes | No |
| 1 | None | 1 summer | Territory? | Interference | Yes | No |
| 1 | None | 3 yr (experiment in only 1) | Territory? | Interference | Yes | Yes |
| None | 10 (territories) | 8 days | Social status | Interference | Yes | Yes (sexual selection) |
| None | 19 (territories) | Breeding season | Territory | Interference | Yes | Yes (habitat) |
| 38 | 18 | 2 yr | No (food?) | Both | No (high Jackdaw density) | No |
| None (before-after) | None | 2 yr | No | Exploitation (?) | No (nest boxes) | No |
| 1 (each of 2 density levels) | None | 1 summer | Food | Exploitation | No (nest boxes) | No |
| 1 | 2 (plus 2 plots in which flycatchers disappeared naturally) | 3 yr | Food | Interference | Yes | Yes |
| None (before-after) | 3 | 1 month | Nectar | Interference | Yes | No |
| > 1? | 5–7 | 2–3 days/ experiment | Flowers | Interference | Yes | No |
| 1 | None | 3 yr | Food | Exploitation | Yes | No |
| 1 | None | 3 yr | Food | Exploitation (?) | No (nest boxes) | No |

Finally, if one conducts an experiment as a means of inquiring into the possible operation of interspecific competition, the six criteria listed in Table 1.4 should be considered. Usually they are, although alternative explanations for the results are generally not considered. It is especially important to identify the resource over which competition is presumed to occur. In all of the experiments involving resource manipulations this has been done, although actual resource limitation has rarely been documented. Resources have been identified in fewer than half of the density-manipulation experiments; usually they are simply inferred from the results or from prior expectations.

Even if field experiments on competition are carefully designed and executed, however, several factors may complicate their interpretation (Mac Nally 1983, Underwood 1986b, Connell 1983):

1. If intraspecific competition is stronger than interspecific, it may override the effects of the interactions between species even if they are present. Thus, if one removes a species and the vacancies are filled by conspecifics (as happened when Davis removed juncos or in the experiments of Payne and Groschupf and of Fonstad; Table 1.9), the experiment has not really tested for interspecific effects, and one can draw no conclusions about the presence or absence of such competition.

2. If the experiment is conducted at the wrong time or place, no response may be evident, even though under somewhat different conditions competitive interactions might be clear. In some situations, individuals have been removed after the establishment of territories in the spring, and the vacancies have not been filled by either conspecifics or heterospecifics (e.g. Cederholm and Ekman 1976). Krebs and Perrins (1978) found that Great Tits were rapidly replaced by conspecifics if the removals were made before breeding, but if the removals were carried out after breeding began there were no replacements. Experiments conducted in environments that are temporally or spatially variable may produce inconsistent results, depending on their relationship to the fluctuating occurrence of resource limitation (Schoener 1983b). Few experimental studies of birds have been designed to reveal such variations, although Källander's (1981) food-supplementation tests produced quite different results in different years, depending on the availability of natural foods.

3. In some situations, the populations may lack the capacity to respond to a particular manipulation, especially if the features of behavior or ecology being examined are relatively fixed (e.g. the warblers studied by Emlen and Dejong (1981)). Rosenzweig (1979) and Cody (1979) suggested that where habitat selection by putative competitors is sharply differentiated (in rela-

tively stable environments, they claim), species-removal experiments will have little effect, as the other species lacks the capacity to expand its habitat occupancy. If past competition has produced optimally differentiated modes of foraging behavior among coexisting species, exploitation competition may also be difficult to document through perturbation experiments (Brew 1982, Rosenzweig 1985). Thus, the absence of patterns suggestive of competition in experiments only bears on the operation of competition as a *proximate* process. Such experiments cannot disprove its operation in the past, which may have contributed to inflexible differences between the species tested.

4. Experiments are usually conducted as if the few (usually two) species selected for close study were the only players on the ecological stage. Other species may also enter into interactions that, although less direct or of a different form, may nonetheless influence the outcome of a competition experiment. Suppose one studies a two-species system, removing species A to gauge its effects on species B. The removal might produce a decrease rather than the expected increase in B if A had been depressing the abundance of a third species (C), which, now released, competitively influences species B. Other sorts of counterintuitive consequences of indirect interactions have been considered theoretically by Levin (1974), Holt (1977), Colwell and Fuentes (1975), and Bender *et al.* (1984); Brown *et al.* (1986) have documented such effects in their long-term experimental manipulations of granivore assemblages. If indirect effects occur, a density gain by one species after the removal of its presumed competitor is not a necessary or sufficient demonstration of competition between the species, nor is a density loss necessarily indicative of the absence of competition (Tilman 1987). Unfortunately, most experiments with birds have been 'pulse' experiments, which are likely to yield information only on direct interactions (Bender *et al.* 1984).

Clearly, experiments may provide valuable insights into the operation and effects of competition between bird species in some situations. Avian ecologists have not always been attentive to the design of their experiments, however, and the degree to which the experimental evidence lends strong support to the occurrence or absence of competition among bird species is thus not so great as might first appear. Moreover, the results of experiments are situation-specific, and generalizations on the basis of limited experimental testing are risky (Underwood and Denley 1984, Underwood 1986b, Diamond 1986). Still, the few experiments conducted on birds do suggest that competition occurs and has major effects in some situations. They also indicate that it is absent or has negligible effects in other situations (Table 1.9).

### Competition among distantly-related taxa

Most observational and experimental studies of competition among birds focus on a small set of species that are ecologically or taxonomically rather similar. In adopting this approach, we presume that competition is likely to be most intense among such species. The possibility that important competitive effects might come from a broader array of species has been recognized in the idea of diffuse competition (MacArthur 1972), in which an observed pattern is related to interactions spread over some unspecified set of potential competitors. Diffuse competition has usually been offered as an explanation when evidence of direct pairwise competition is lacking or is inadequate to explain the observed patterns, but the idea has not been tested in a rigorous fashion. The notion of diffuse competition, although intuitively attractive, is supported only by weak and inferential evidence.

Diffuse competition is generally suggested to occur among species that are at least members of the same broad taxonomic group (order or class). As competition is defined in terms of the joint use of limiting resources, however, there is no reason why it might not occur between members of more distantly related taxa (Morse 1975, Reichman 1979). Because birds use a wide variety of resources, the possibilities for interactions with quite different sets of organisms are potentially great, although they have received relatively little direct attention.

Some of the clearest evidence of interactions between birds and distantly related taxa is provided by studies of nectarivorous birds and various insects. This work will be discussed in the following chapter, but one example can be given here. Although *Geospiza* finches on the Galápagos are primarily granivorous, some species may feed on nectar during the dry season. Schluter (1986) examined patterns of nectar use and body sizes of *G. fuliginosa* and *G. difficilis* on islands occupied or not occupied by a large carpenter bee, *Xylocopa darwini*, which is a major visitor to a wide range of plants. On the islands where *Xylocopa* was present, nectar comprised 4% of the diet of the birds, but where the bees were absent, nectar accounted for 20% of the diet. In addition, mean body sizes of the finches were smaller on islands lacking bees than where bees are present. Schluter suggested that the bees influence nectar use by the birds and that character displacement in finch body size occurs in response to the presence or absence of the bees. There were apparently no differences in the occurrence of flower types on the islands containing or lacking bees, but Schluter was unable to determine whether or not differences in nectar availability occurred. Acceptance of his interpretation of this pattern therefore requires acceptance of the *ceteris paribus* assumption.

In other situations, interactions between birds and mammals, fish, or lizards have been studied, with varying results. Aerial-feeding birds and insectivorous bats, for example, share a common form of foraging locomotion, and the potential for competitive interactions would seem to be great (Fenton and Fleming 1976). Shields and Bildstein (1979) examined this possibility by observing interactions between nocturnally foraging bats and Common Nighthawks (*Chordeiles minor*) feeding on insects attracted to the light of a large motel sign. When either taxon foraged in the absence of the other, it concentrated activity in the lower zone of the light cone, where insects were most abundant. When both bats and birds were present, bats frequently chased the nighthawks. Some of the nighthawks were forced into upper zones of the light cone, where insects were less abundant and foraging efficiency was presumably reduced. Nighthawks were also aggressive, although virtually all of their chases were directed at other nighthawks rather than at bats. The bats clearly interfered with nighthawk foraging activity, but there was no evidence that resource availability was limiting in this situation. In fact, it is likely that both bats and birds were attracted to the lighted area because insects were so abundant there. Nighthawks displaced to the upper zones of the light cone may have realized greater foraging efficiency there than in other, unlighted areas, and it is therefore not clear that this interaction really represents competition, in the strict sense.

Birds may also compete with insectivorous lizards in some situations. Wright (1979, 1981) suggested this possibility for birds and *Anolis* species in Panama and the Caribbean. He based his argument on his observations of dietary overlap between the species, of food limitation of at least one of the lizard species, of an apparent depression of arthropod levels by bird predation, and of negative correlations between bird densities and the densities, fecundity, and physiological state of the anoles. He suggested that on small islands where some bird species have become extinct there has been no ecological release among the remaining bird species; instead, lizards have increased in abundance, and, by consuming more prey, they inhibit recolonization of the islands by insectivorous birds. Wright's observations were obtained using rather coarse methods (e.g. mist-netting of birds, sweep-netting of arthropods, very general measures of diet composition), however, and were based on short visits to some islands at different times of year. As a result, there are many uncertainties in the patterns he reported. Waide and Reagan (1983) questioned the accuracy of Wright's diet overlap assessments and suggested that differences in predation pressures between islands were equally likely to account for the density patterns Wright observed (a possibility Wright considered but dismissed). The comparison is complicated by the possibility of predation by birds on the lizards as well; Schoener and Schoener (1978) found that anole survival was lower on a larger island containing more predatory birds than on a smaller island, although the number of lizard species and their relative abundances were similar on the two islands. Predation by rice rats (*Oryzomys palustris*) on Seaside Sparrows (*Ammospiza maritima*) likewise complicated an interpretation of aggressive interactions between these species in a different study of the use of shared salt-marsh habitats (Post 1981).

Aside from studies of nectarivores (Chapter 2), there have been few experimental tests of interactions between birds and other taxa. Eriksson (1979), however, used an experimental approach to investigate the effects of possible exploitation interactions between Goldeneyes (*Bucephala clangula*) and fish in Swedish lakes. General observations indicated that Goldeneyes (which feed on aquatic invertebrates) used lakes lacking fish more frequently than lakes with fish. Eriksson suggested that fish might depress prey densities, making lakes in which fish densities were high less suitable for the birds (see also Hunter *et al.* 1986). A net barrier was used to divide one lake into sections containing and lacking fish, and fish were removed from another lake by poisoning. In both situations, some prey taxa were more abundant in the treatment areas than in controls, although the effect was significant for rather few prey groups. Goldeneyes did not differ in their use of the halves of the divided pond but did use the treated lake significantly more than a nearby control in 2 of 3 yr. Eriksson's treatments were not replicated, he did not control for variations in habitat or in the abundance of the fish, and the diets of the fish and birds were not directly measured, so the conditions of resource overlap and exploitation competition between the taxa are not entirely clear. Eadie and Keast (1982) explored Eriksson's idea with a correlative study in southern Canada, finding that Goldeneye and perch densities were inversely related on nine study plots in 3 lakes where the species co-occurred. The relationship remained after the effects of habitat variations were removed through partial correlation procedures. Together with the documentation of high dietary overlap between the taxa, this lends strength to Eriksson's experimental documentation of an exploitation effect by fish on the birds. Bird-fish exploitation interactions have also been suggested in other studies (Springer *et al.* 1986, Safina and Burger 1985).

Not all investigators who have sought evidence of inter-taxa competition involving birds have found it, however. There is little indication of competitive effects between frugivorous birds and mammals, for example, although Pearson (1975, 1977) suggested that such interactions were responsible for the negative biomass relationships between birds and frugivorous primates in several neotropical locations. Tropical plants often specialize on either birds or mammals as dispersal agents and produce fruit specifically for the appropriate agent, and this pattern might therefore reflect differences between the sites in the relative proportions of bird- or mammal-dispersed fruits. Mares and Rosenweig (1978) found no evidence of ecological release in granivorous birds in South American deserts in which mammalian granivores were scarce in relation to North American Sonoran desert locations. They suggested that the birds were limited by physiological factors rather than food availability and were therefore incapable of responding to the absence of a suite of presumed competitors. In coniferous forests, cone crops may be exploited by a variety of taxa, but Smith and Balda (1979) proposed that the potential for intertaxa competition might be present only during poor years. At these times, however, the various groups of cone and seed feeders are likely to be limited by different factors, reducing the potential effects of competitive interactions.

The many differences between taxa as different as birds, bats, fish, or lizards make it difficult to evaluate the potential for competition among them and its possible effects. Not only are their population dynamics likely to be influenced by quite different factors, but the responses of individuals to variations in space and time may occur on quite different scales. Because of physiological differences, for example, an anole and a small insectivorous bird may scale time quite differently, and this influences considerations of energy demands, feeding rates, and the like. When one expands the scope of an investigation across broad taxonomic boundaries, trophic relationships may also become complex – possible competitors may also be involved in predator–prey relationships (Bradley 1983). We are thus confronted with a paradox: in order to consider the full range of potential competitors using a given resource, we may have to include organisms of quite different taxa. When we do so, however, the inherent differences between the taxa introduce major additional sources of variation into the comparison, and these render detection and interpretation of interactive effects difficult and equivocal.

## Where should we expect competition to occur?

Despite the weakness of a good deal of the observational (and some of the experimental) evidence, there are clear, strong indications that competition does occur as a proximate process among some groups of birds. The evidence seems strongest for groups that rely on a restricted resource (such as nectar) or are resident in restricted areas with variable resource conditions (such as the Galápagos Islands). Competition is certainly not a ubiquitous process, however, and it would be helpful if we could anticipate the situations in which it might be most likely to occur.

Diamond (1978) suggested that, because species packing is so much greater in the tropics than elsewhere and environments are relatively benign, we should expect competition to be most evident there. Connell (1975, 1983) emphasized the role of 'natural enemies' in reducing populations to levels at which they rarely pass a resource-limitation threshold and suggested that this predation effect might be most evident in relatively stable and benign environments. Competition, then, would be most likely to occur where the effects of natural enemies are reduced, as in variable, moderately harsh environments or on islands. Competition might also be likely to occur (or at least be most obvious) where episodes of resource limitation are interspersed with periods of superabundance, preventing the consumer-resource system from stabilizing. Such episodic limitation may be especially likely when resource variations are unpredictable, as in moderately harsh environments. If this is so, then the occurrence of competition bears no necessary relationship to the equilibrium status of a system. The more a system is closed to outside influences (e.g. islands), the more likely it is that environmental variations will lead to competitive interactions, as other forms of escape (e.g. immigration) are restricted.

Although competition has often been offered as an explanation for community-wide patterns, it seems unlikely that a single process would operate consistently over a broad array of species to produce such patterns. Moreover, indirect interactions become more probable as the number of species considered increases. A more appropriate focus for investigations may be on small, carefully defined guilds, in which the species follow a single major form of resource use whether or not they are taxonomically closely related. Interactions among such species may have important effects on guild structure (Landres and MacMahon 1983).

The strength of circumstantial evidence of competition also seems to be inversely related to the spatial scale of investigation. As the scale is expanded from a local to a biogeographic level, species are influenced by a greater variety of factors, and identifying the role of any single process in producing a pattern becomes more problematic. Observational comparisons conducted at a local scale are generally more tightly controlled, and well-designed experiments can be conducted only at that scale. In fact, the clearest evidence of competition has been obtained in studies focused on tightly defined guilds at a local scale. Spatial aspects of exclusion, such as interspecific territoriality or microhabitat partitioning, are also more likely to be expressed at a local scale (Connor and Bowers 1987). From an evolutionary perspective, Walter *et al.* (1984) have also argued for the importance of the local scale in competitive interactions.

In general, then, competition may be most likely to occur (a) in nonequilibrium systems in which consumers do not track resource variations closely, (b) in variable systems in which the consumers are capable of a rapid population response, or (c) in saturated communities with constant resource levels (P. Price 1984). Alternative (c) is the one assumed in most theory, but it is not likely to occur in many natural situations. Alternative (b) is unlikely to occur among many bird species, given their relatively low rates of population change and the time lags in responses produced by site tenacity (see Chapter 4). The critical requirement for competition is resource limitation, not equilibrium. I conclude that we should expect the strongest evidence of competition among species that are members of relatively restricted, resource-defined guilds and that are residents or occupy relatively closed, localized systems. Competition is most likely to be expressed as proximate exploitative or interference interactions in environments in which resources vary in availability and the frequency of resource variations is intermediate. These are not the sorts of situations in which competition is likely to be occurring whenever one chooses to look for it, so relatively long-term investigations are required.

## Conclusions

Documenting that competition occurs among species in a guild or community is not easy. It requires more than finding apparent agreement between the patterns one observes and predictions generated from formal theory or simple intuition. This is partly because the predictions are often derived from premises that are biologically unrealistic or oversimplified (Andrewartha and Birch 1984, Alley 1982), but it is also because it is not logically justified to infer the operation of a process only from observations of a pattern. A good deal of the evidence that has been offered in support of the view that competition plays a major role in producing community patterns does not go much beyond demonstrating an agreement between observed and expected patterns (Table 1.7). *If* these patterns can be produced by no process other than competition, such inferences are justified, but this is rarely the case.

The evidence that competition occurs among similar coexisting bird species, much less that it is responsible for producing the structural patterns of assemblages, is inconsistent and often weak, but it is not so wanting as some (e.g. Birch 1979) have claimed. In order to produce compelling evidence of competition, one must satisfy several criteria of increasing stringency (Table 1.4). Building a 'convincing case' for competition (Roughgarden 1983) involves more than gathering together bits and pieces of consistent evidence of marginal quality. The strength of a case rests in each situation on the strength of the evidence, and this is a function of the appropriateness of the methodology followed, the care with which a study is designed and implemented, and the degree to which the requisite criteria are evaluated. This is not to say that weak, circumstantial evidence is of no value – clearly, it is more suggestive of competition (or its absence) than no evidence at all. The danger, of course, is that conclusions based on weak or inferential evidence will be taken as general statements of how species interact or communities are organized, discouraging investigators from pursuing the more difficult but enormously more important detailed investigation of competition as a process.

In view of the difficulty of demonstrating competition according to rigorous criteria, one might be tempted to refute the competition hypothesis entirely, in the process perhaps making a new dogma of the rejection of competition (Martin 1986). It is important to remember, however, that all of the above statements apply with equal force to attempts to falsify the competition hypothesis, to document that competition does *not* occur in a given situation. Falsifications based on weak, fragmentary evidence or improperly designed tests are no more valid than are similar confirmatory studies. There is nonetheless a basic asymmetry between the weight of evidence required to falsify and that required to corroborate an hypothesis. After all, a careful documentation of a pattern that is clearly inconsistent with the predictions of competition theory may falsify the hypothesis that competition is currently operating in that situation, whereas the consistency of observations with predictions is not sufficient to corroborate the hypothesis in the absence of a consideration of the predictions of alternative hypotheses. Indeed, a failure to satisfy the requirements of any one of the criteria listed in Table 1.4 may cast doubt on the likelihood that competition will be found in a given situation, but strong corroboration requires that most of the criteria be satisfied.

Most evidence for or against competition is based on observations rather than experiments and is therefore weakened by the absence of any direct control over sources of variation in the comparisons and by a reliance on the *ceteris paribus* assumption. The limitations of uncontrolled observations are well-known, although they are sometimes forgotten when the observations are interpreted. One may strengthen observational evidence in several ways (Diamond 1986), perhaps most dramatically by conducting experimental manipulations. There is no doubt that experiments have often provided strong evidence of competition in situations in which it was suspected to occur. Many field experiments, however, suffer from inadequacies in experimental design, and the evidence must be interpreted in the context of these limitations. An experiment lacking replication is certainly not worthless, but the results must be viewed with considerable caution. An experiment lacking any sensible control *is* of very little value.

Documenting that competition occurs among species is difficult. Even if one does so, however, one cannot automatically conclude that competition is responsible for the community patterns observed. Competitive interactions might well occur between individuals at a certain time but have little effect on population dynamics or on long-term community attributes. To the extent that other processes may contribute to an observed pattern, its standing as evidence of competition is weakened. When assessing the existence and effects of competition in bird communities, one must evaluate alternative process hypotheses as well. Some of these alternatives are considered in Chapter 3.

# 2

## A case study: interactions among nectarivores

In the previous chapter, I described the difficulties of demonstrating interspecific competition and linking it with community patterns. Birds that feed chiefly upon floral nectar, however, are especially well-suited to investigations of competitive interactions, and it is therefore appropriate to consider their ecology in some detail.

In most parts of the world, a relatively small number of taxonomically related species comprise the chief members of the nectar-feeding guild: hummingbirds (Trochilidae) in the New World, honeyeaters (Meliphagidae) in Australasia, sunbirds (Nectariniidae) in Africa and parts of Asia, and honeycreepers (Drepanididae) in Hawaii. The species are often nectar specialists, and the resource of primary importance to them is thus clearly defined. Moreover, the abundance and availability of the nectar resource can be quantified with precision and related directly to the energetics of the birds. Many plants are adapted to pollination by nectarivorous birds (Stiles 1978, Feinsinger 1978, Feinsinger *et al.* 1982, Kodric-Brown and Brown 1979, Brown and Kodric-Brown 1979, Grant and Grant 1968), and in some cases the linkage between birds and plants suggests a tightly coevolved system. Because the flowers are adapted to make their nectar available to certain bird pollinators (Murray *et al.* 1987), measurements of nectar standing crop may accurately reflect resource availability as it is viewed by the foragers (Carpenter 1978). For these reasons, nectar-feeding birds and their resources have been popular subjects for investigations of time-energy budgeting (e.g. Wolf *et al.* 1975, Gill and Wolf 1977, Hixon *et al.* 1983), optimal foraging behavior (e.g. Pyke 1978), ecophysiology (e.g. Calder and Booser 1973), and species interactions.

Resource limitation can also be assessed more directly in nectarivore systems than is the case with most other systems, and the search for competitive interactions can thus begin from a firmer base. Small nectarivores such as hummingbirds have high metabolic demands that are often satisfied almost entirely from the nectar contained in the flowers on which they feed. Whereas the temporal variations in the availability of seeds or insects typically occur over seasons, nectar availability may change dramatically on diurnal as well as seasonal scales, due to exploitation by other species, changes in flowering phenology, or climatic events. These changes may substantially alter aspects of the foraging behavior, territorial defense, or occupancy of an area by individual foragers (Carpenter 1978, Gill 1978, Gill and Wolf 1979, Heinemann 1984). In parts of Australia, several species of honeyeaters maintain floral nectar at low levels, suggesting limitation (Ford 1979, Paton 1980, McFarland 1986a), and the total number of honeyeaters and the abundance of each species are positively related to seasonal variations in flowering intensity (Ford 1983, Collins and Newland 1986, McFarland 1986b). Short-term changes in nectar availability, either as a result of climatic events (Gass and Lertzman 1980, Heinemann 1984) or experimental manipulations (Hixon *et al.* 1983), may also produce rapid changes in the abundance or behavior of hummingbirds. Gass and Sutherland (1985), for example, documented that the amount of time that territorial Rufous Hummingbirds (*Selasphorus rufus*) spent foraging in patches of flowers was directly proportional to the number of flowers contained in the patches. When the nectar supplies of the flowers were supplemented experimentally without altering flower density in the patches, the birds remained foraging in the patches considerably longer. Because individual flowers vary in the amount of nectar they provide, foraging time by nectarivores may also increase nonlinearly with decreasing nectar per flower (Fig. 2.1). Thus, foraging time spent per flower may increase with reduced nectar availability when nectar supplies are limiting over an entire territory, but if certain patches within territories are 'hot spots' of relatively abundant nectar, birds also may forage longer in those patches. Although these patterns do not indicate direct effects of resource changes on individual fitness or population parameters (Carpenter 1978, Heinemann 1984), they do indicate the effects of varying resource levels on individual performance rather clearly.

## Interactions among hummingbirds

### *Ecomorphological patterns*

Among hummingbirds, features of morphology are closely linked to their ecology. Bill size and shape in particular are frequently associated with an ability to or a preference for feeding on flowers of a certain structure. Thus, long-billed species feed almost exclusively on flowers with long corolla tubes, and short-billed species feed almost exclusively on

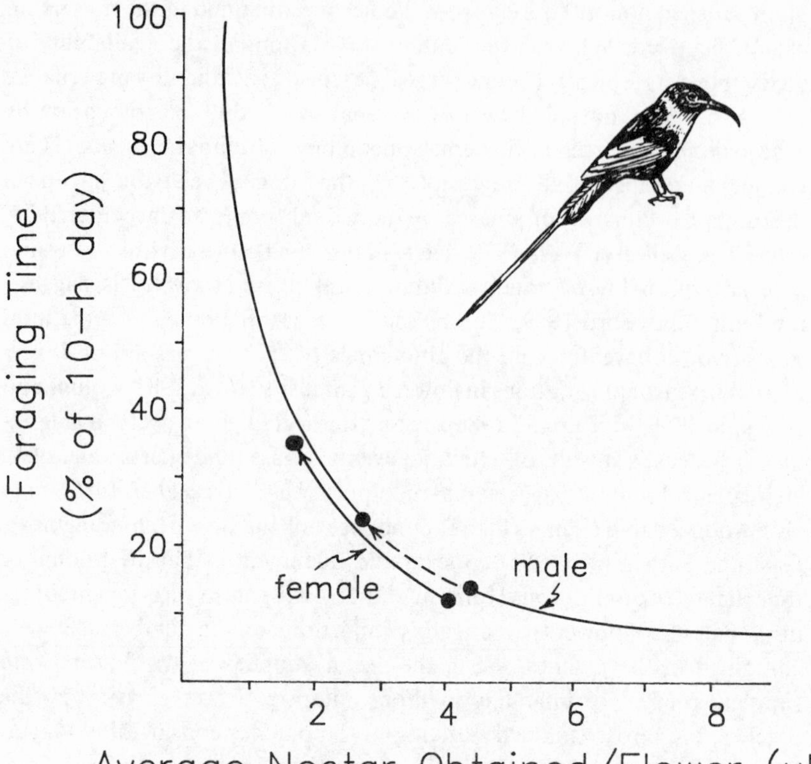


Fig. 2.1. Impacts of nectar losses to competitors on foraging time budgets of nonterritorial Golden-winged Sunbirds (*Nectarinia reichenowi*) feeding on *Leonotis* in Kenya. The foraging time budgets shown by the curved lines are the values required to maintain a balanced energy budget. Loss of nectar decreases the average nectar obtained per flower visit and increases the number of flowers that a sunbird must visit to obtain a specified amount of energy each day (including the costs of additional foraging). Because females are smaller, their costs are lower and they can forage less each day than males at a particular nectar volume. Females tend to lose more nectar to dominant individuals, however, and must forage longer each day to compensate for losses. From Gill and Wolf (1979).

flowers with shorter tubes (which often also secrete less copious nectar) (Wolf *et al.* 1976, Kodric-Brown *et al.* 1984). Some species with strongly decurved bills feed on flowers whose corolla structure precisely matches the birds' bill length and curvature (Grant and Grant 1968).

Given such a close coupling of morphology to resource use, one might expect ecomorphological patterns in nectarivore guilds to be more clearly expressed than in most other avian groups. Brown and Bowers (1985) examined this possibility by comparing the morphological patterns of regional assemblages of hummingbirds with those generated by several null models of community assembly. If one considers just the eight species whose ranges extend into temperate North America, the pattern appears to be one of convergence – the species are much more similar in bill length, weight, and wing length than any randomly drawn assemblage of eight species. They are also more similar to one another than each is to congeners found in subtropical or tropical locations. These species are all aggressive, tend to be spatially separated into different local habitats, and exploit flowers of quite different taxa that have generally converged in aspects of floral morphology.

In the Lesser and Greater Antilles, hummingbird size distributions are also nonrandom, but in a different way. On islands supporting two or more species, there is always one small and one large form (as Lack (1976) also found); where more than one large species is present, the species tend to be separated elevationally or by habitat (Kodric-Brown *et al.* 1984). Moreover, bill length increases more rapidly with increasing body weight among the Antillean species than in random samples of the world's hummingbirds, indicating that allometric patterns there are distinctive. Brown and Bowers attributed these morphological patterns to competition and mutualistic coevolution with plants, although their arguments are inferential, as are other dealing with such patterns.

***Community roles and resource defense***

Hummingbird species differ not only in morphology, but in their efficiency of exploitation of particular flowers and their level of aggressiveness (both of which are related to morphology) (Feinsinger 1976, Feinsinger and Colwell 1978, Colwell personal communication). Feinsinger and Colwell (1978) have defined several 'community roles' of hummingbirds based on these attributes. 'High-reward trapliners', for example, tend to be large, long-billed species that visit but do not defend dispersed nectar-rich flowers with long corollas. 'Territorialists', on the other hand, defend dense clumps of smaller flowers with lower per-flower nectar production and tend to be morphologically intermediate. Other community roles are similarly defined by variations in the floral resource exploited and the associated morphological features of the birds (Table 2.1, Fig. 2.2). The roles are not necessarily fixed species attributes, however. For example, *Amazilia tobaci*, which is normally territorial on Trinidad, may act as a generalist or even a low-reward trapliner when other species more suited to these roles are absent. Still, the definition of these roles and their relationship to floral

Table 2.1. *Community roles of hummingbird species in relation to their morphological characteristics.*

| Role | Size | Required power | Bill | Foot |
|---|---|---|---|---|
| Low-reward trapliner | Small | Low | Short/medium | Moderate |
| Filcher | Small | High | Short | Large |
| Generalist | Medium | Moderate | Short/medium | Moderate |
| Territorialist | Medium | Moderate/high | Short/medium | Moderate |
| High-reward trapliner | Large | Moderate/high | Long | Small |
| Marauder | Large | High | Medium | Small |

*Source:* From Feinsinger and Colwell (1978).

resource attributes raises the prospect that one might be able to generate 'assembly rules' for hummingbird communities, based on the spectrum of flowers available in a given situation. Feinsinger and Colwell (1978) suggested that simple communities on islands usually contain one low-reward trapliner or generalist and one territorial species and sometimes may support one high-reward trapliner as well. As communities become more diverse in mainland subtropical and tropical settings, additional species are added to this foundation, depending on the relative importance and constancy of various flower types.

Both foraging efficiency and the effectiveness of aggressive behavior in hummingbirds are closely tied to the power demands of hovering. This, in turn, increases with increasing wing-disc loading (the ratio of body mass to the area swept out by the wings in hovering flight) and decreasing air density at higher elevations (Feinsinger *et al.* 1979). Other things being equal, we might expect a species with a given wing-disc loading to become increasingly territorial at higher elevations, as the high cost of feeding by hovering at widely dispersed flowers (traplining) may become energetically infeasible with the greater power demands in the thinner air. If the species could dominate other species territorially, it would be able to obtain adequate nectar with reduced foraging time. In fact, *Colibri thalassinus* is rarely territorial where it occurs at low elevations in Costa Rica in the presence of other species with higher or lower wing-disc loading, foraging among dispersed flowers and emigrating during resource lows (Feinsinger 1976). At a higher-elevation site where no species with higher wing-disc loading and similar bill morphology occurs, *Colibri* defends territories. Feinsinger *et al.* (1979) suggested that, in general, one might expect increasing levels of interference competition at higher elevations. If elevational replacements of congeners occur, the higher-elevation species should have lower wing-disc

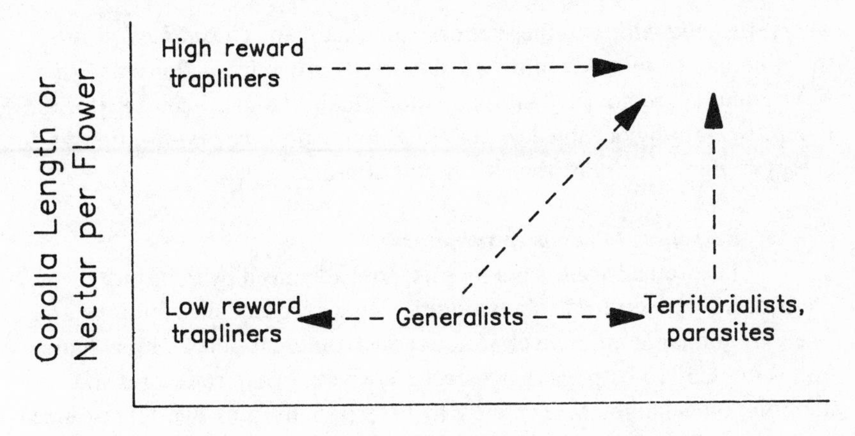


Fig. 2.2. Community roles of hummingbirds relative to flower dispersion, morphology, and nectar reward. From Feinsinger and Colwell (1978).

loading, and the hummingbird assemblages occurring at higher elevations should be characterized by lower mean wing-disc loading than should the lower-elevation communities. Their tests of these predictions were preliminary, and others (e.g. Kodric-Brown *et al.* 1984) have found contrary patterns, but the predictions provide a useful point of departure in analyses linking hummingbird energetics to community structure.

Because territories are so closely tied to resources, territorial species may play a major role in determining the overall composition and structure of hummingbird assemblages (Wolf *et al.* 1976, DesGranges 1979, Carpenter 1978, Feinsinger 1976). In successional tropical habitats in Costa Rica, for example, the territorial species, *Amazilia saucerottei*, aggressively excluded other species from rich food sources. Another species, *Chlorostibon canivetti*, which was excluded from the concentrated nectar supplies by *Amazilia*, foraged by traplining dispersed flowers, where it aggressively dominated in interactions with other hummingbirds. The remaining 12 species that occurred in the assemblage specialized on floral resources not used by the dominants or they opportunistically exploited the heterogeneous habitat mosaic of the area. Resource levels varied in time, and only *Amazilia* had the potential to respond to these changes while maintaining territories. Several of the subdominant species emigrated during periods of low nectar availability. Here, then, the behavior of the territorial species dictated much of the organization of the assemblage as a whole. Montgomerie (unpublished) studied a similar situation in western Mexico, where *Amazilia rutila* was the territorial dominant and *Cynanthus latirostris* the itinerant trapliner feeding on dispersed, low-reward flowers. When Montgomerie removed *Amazilia* individuals from their territories, *Cynanthus* rapidly occupied the areas, adopting an *Amazilia*-like time and energy budget to exploit the rich nectar sources.

***Responses to resource variations***

Theoretical arguments such as those advanced by Feinsinger *et al.* are necessarily based on the assumption that resource distributions and levels are similar at different elevations and at different times. This assumption is unlikely to be precisely correct anywhere, but nectarivores such as hummingbirds might be expected to be especially sensitive to resource variations. The theory in its general form is therefore probably more heuristically than empirically useful. Some indication of the effects of resource variations on hummingbird guilds of different sizes emerges from studies on Trinidad and Tobago (Feinsinger and Swarm 1982, Feinsinger *et al.* 1985). The area investigated on the North Range of Trinidad contained 11 hummingbird species and a bananaquit (*Coereba flaveola*), whereas Tobago supported only 5 hummingbird species and *Coereba*. The basic patterns of nectar use by the guilds were similar on the two islands despite the differences in guild size. There were also few differences in the apparent intensity of competition: one species, *Amazilia tobaci*, won most interspecific aggressive encounters on both islands. The differences in guild size apparently influenced the plants, however. Feinsinger *et al.* (1982) observed that plants on Tobago received fewer pollinator visits and that pollinators carried mixed rather than single-species pollen loads. In at least some plant species, flowers on Tobago produced more nectar than on Trinidad.

During the 13-month period of study, nectar supplies varied more than 100-fold on each island, being superabundant from December through August and quite low in September–November, when all nectar produced was completely used. Despite reductions in guild size through emigration at this time, the ratio of guild demand for nectar to nectar supply was considerably greater during this lean period than at other times of the year (Fig. 2.3). On Tobago, only three nectarivores were present at this time, and these species represented quite different morphological types. On Trinidad, on the other hand, seven nectarivore species persisted, and although the diet breadth of the guild as a whole did not change much, *Amazilia*, the competitive dominant, restricted its foraging to a small set of flowers that matched its bill morphology. *Amazilia* retained a rather broad foraging

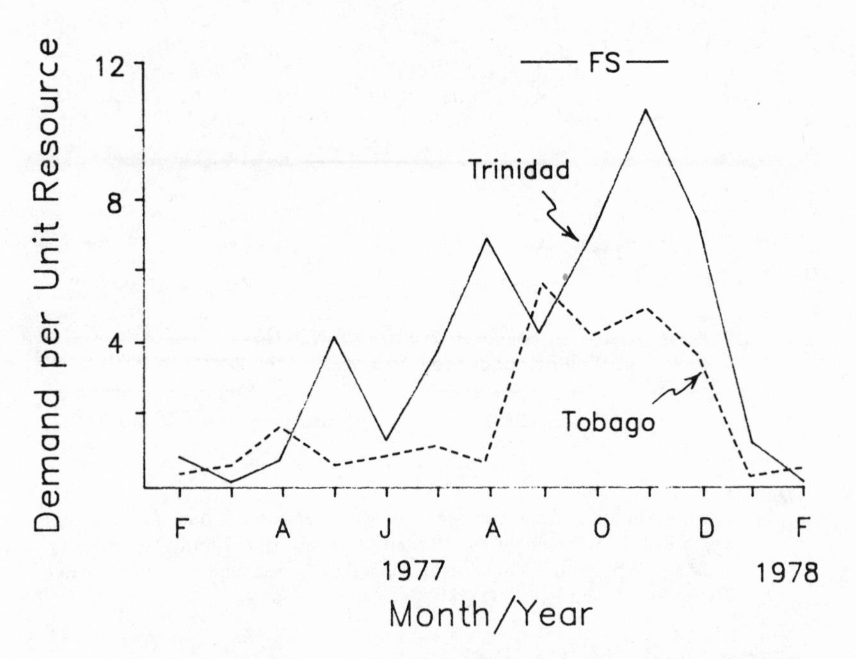


Fig. 2.3. Ratio of the nectarivore guild demand for nectar (measured as the estimated number of flowers visited per day, weighted by nectar value) to supply of nectar (flower abundance, weighted by nectar value) at locations in Trinidad and in Tobago. FS = period of food shortage. After Feinsinger *et al.* (1985).

niche on Tobago, suggesting a pattern of ecological release. This pattern, however, was only evidenced during the 3-month period of resource scarcity.

Hummingbirds are bellicose creatures, and aggression is a primary form of interaction both within and among species. It is frequently expressed in simple displacements at localized nectar sources (e.g. Colwell *et al.* 1974), but it may be formalized as active territorial defense as well. Both the defense and size of territories are generally closely related to resource levels. Rufous Hummingbirds migrate through western temperate North America in late summer, establishing territories in montane areas where they accumulate fat and energy reserves for the next migratory flight. In this species, territory size is related to flower density within the territory: if the number of flowers falls below the minimum number required to support the individual, the territory is abandoned, whereas above a certain size territories are no longer maintained, regardless of flower densities, because it requires too much time and/or energy to repel intruders from the larger area (Fig. 2.4, Kordric-Brown and Brown 1978, Carpenter and MacMillen 1976).

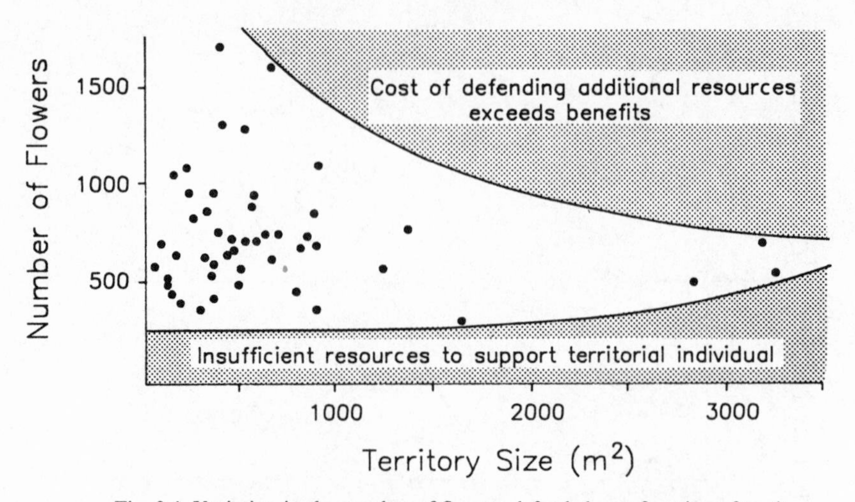


Fig. 2.4. Variation in the number of flowers defended as a function of territory size in Rufous Hummingbirds in eastern Arizona. All territories lie between thresholds representing the limits of economically defendable resources. After Kodric-Brown and Brown (1978).

There have been several experimental tests of the relationships between territorial defense and resource levels. Lyon *et al.* (1977) clustered feeders together in an area in southeastern Arizona, inducing a Blue-throated Hummingbird (*Lampornis clemenciae*) to establish a territory around them. Conspecific males and individuals of other hummingbird species were aggressively driven from the food sources. Lyon and his colleagues then arrayed the feeders in three progressively more dispersed patterns, each treatment lasting 1–2 days. As the food supply became more dispersed, the Blue-throated first permitted Black-chinned Hummingbirds (*Archilochus alexandri*), then Rivoli's Hummingbirds (*Eugenes fulgens*), and finally other male Blue-throated Hummingbirds to forage within the territory. The bird finally abandoned the territory. Pimm (1978) conducted similar experiments in the same area, manipulating the predictability and quality of resource supplies and quality in feeders. Once again, *Lampornis* was dominant at the feeders, but its defense faltered with increasing resource unpredictability in time and space, and *Archilochus* invaded. By placing feeders of differing food quality in patches, Pimm *et al.* (1985) tested the effects of resource dominance in a patchy environment (c.f. Rosenweig 1979, 1981. 1985, Pimm and Rosenzweig 1981). Feeding activity by *Lampornis* in the richer food patch decreased with an increase in the overall activity levels of conspecifics, whereas feeding activity of *Archilochus* and *Eugenes* in the poorer patch type increased with increasing *Lampornis* activity (Fig. 2.5).

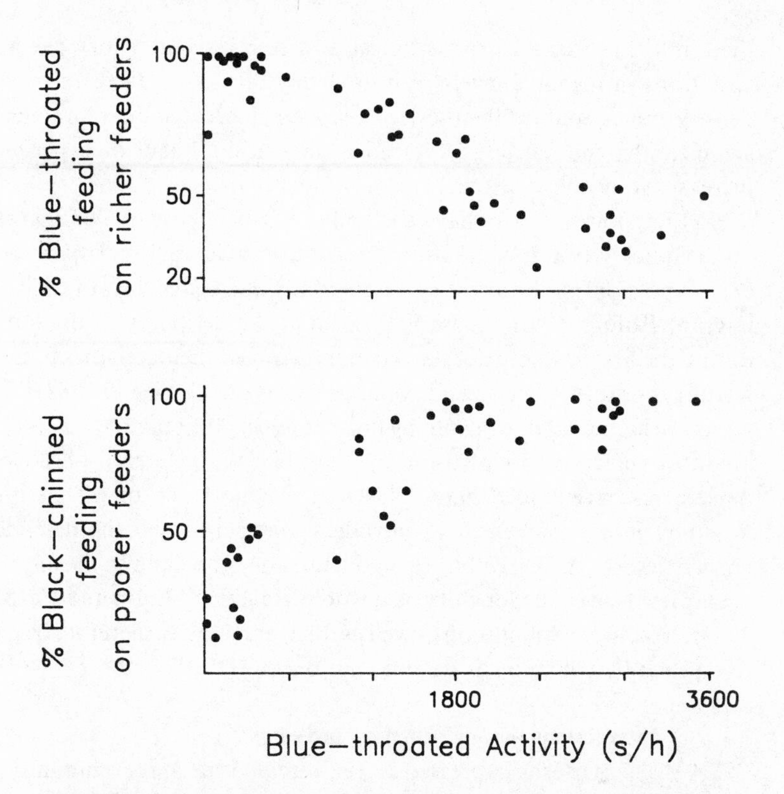


Fig. 2.5. Above: the percentage of feeding time spent by Blue-throated Hummingbirds on the richer of two food-patch types (feeders supplied with sucrose solutions) as a function of overall activity level of Blue-throated Hummingbirds (an indirect measure of their density); below: the percentage of Black-chinned Hummingbird feeding time spent on the poorer food patches in relation to Blue-throated density. Blue-throated feeding distributions are influenced by intraspecific competitive pressures, whereas the distribution of Black-chinned feeding is governed by interspecific effects. After Pimm *et al.* (1985).

There was thus no evidence of an interspecific effect on Blue-throateds, but the other hummingbird species were strongly affected by the dominance of the Blue-throateds, shifting to complete use of lower quality 'habitats'.

Variations in the types or quality of nectar resources and in the energy demands of individual hummingbirds may produce variation in territory sizes between the limits set by overall economics (Gass *et al.* 1976, Kodric-Brown and Brown 1978; Fig 2.4). Carpenter *et al.* (1983) suggested that individuals adjust their territory size over time in order to maximize their rate of weight gain. Hixon *et al.* (1983) conducted experimental removals and additions of flowers to territories of migrant Rufous Hummingbirds in the California Sierra Nevadas to test how responsive territory size was to variations in resource levels. Within 1 day of a 50% decrease in flower density, males doubled their territory size and increased their foraging time significantly; the patterns reversed again within 1 day of experimental increases in flower densities.

Hummingbirds clearly have the ability to alter territory configurations rather quickly to track variations in resource availability. They may also exploit portions of the territory differentially during the day. In California, migrant Rufous Hummingbirds foraged at the periphery of the territory during the first few hours of the day but restricted their activities to just the territory core in afternoon and evening (Paton and Carpenter 1984). Nectar levels are highest at dawn, and, by first exploiting the territory edges (which are most vulnerable to intrusion by other birds), the territory-holder may deplete resource levels there. This reduces the value of the periphery, lessening its attractiveness to intruders and permitting the defender to concentrate its aggressive efforts in the territory core later in the day, when nectar levels over the territory as a whole are lower (Paton and Carpenter 1984). Resource exploitation as well as interference may therefore be part of the means by which birds defend territories.

**Interactions among other nectarivores**

The patterns expressed in the interactions and community relationships of other nectarivorous species – honeyeaters, sunbirds, Hawaiian honeycreepers – are often remarkably similar to those found in guilds of hummingbirds. Because some of these species (especially some honeyeaters and sunbirds) rely on insects and carbohydrates other than nectar to a greater extent than do hummingbirds, however, their linkages with nectar resource dynamics are often not so dramatic. By virtue of their dietary flexibility, these species may be able to cope with nectar scarcity in ways that are not available to most hummingbirds. As a result, nectar resources may less often be limiting to these nectarivores than to hummingbirds (Carpenter 1978).

This does not mean that nectar limitation and resource-related aggression do not occur among such species, however. Carpenter's (1978) energy-budget calculations for Hawaiian honeycreepers, for example, indicated that nectar was limiting in only one of three summers. When resources were abundant, the three honeycreeper species exploiting nectar overlapped considerably in their foraging, although they differed in foraging positions in the trees and in the degree to which individuals were sedentary. Little interspecific aggression was evident. In the lean year, all three species were sedentary and attempted to feed from the same flowers high in the canopy. Aggressive encounters between the species were frequent and intense. The largest species, Iiwi (*Vestiaria coccinea*), dominated these encounters and established territories about rich nectar sources from which the other species were excluded. The intermediate-sized Apapane (*Himatione sanguinea*) was displaced to patches of intermediate flower density. There, it aggressively displaced individuals of the smaller Common Amakihi (*Hemignathus virens*) but did not occupy stable territories. Amakihis were relegated to areas with few or no flowers. Pimm and Pimm (1982) independently documented similar patterns of resource use and aggressive exclusion among these species in the same area. On Maui, Carothers (1986) found that Crested Honeycreepers (*Palmeria dolei*) dominated Iiwis but both species defended canopy territories in ohia trees (*Metrosideros collina*) from which they excluded Apapanes and Amakihis. The Apapanes often fed in flocks that moved about nomadically and overwhelmed the aggression of the dominant species to gain access to nectar resources in their territories. Amakihis, on the other hand, persisted by foraging secretively as individuals. When nectar is limiting, then, the system is reminiscent of that Feinsinger (1976) documented among tropical hummingbirds.

The relationship of Australian honeyeaters to rescource levels is often rather similar to that of hummingbirds and honeycreepers. In many habitats, nectar is obtained from a wide variety of plants. Species of *Banksia* are the most important in coastal areas of southern Australia, but *Eucalyptus, Grevillea, Dryandra*, mistletoes, and epacrids may be more important elsewhere (H. Ford, personal communication). Carpenter (1978) found little evidence of nectar limitation and no overt nectar-related aggression among the honeyeater species she studied in an eastern Australian heath, but elsewhere nectar may be scarce, at least at some times of the year (Collins and Briffa 1982, McFarland 1986a, Collins personal communication). Paton (1985) used mass-dependent metabolic equations to estimate the energy requirements of the honeyeaters occurring in a Victoria eucalypt woodland and compared these estimates with measures of nectar production and availability (production minus consumption by insects). Nectar levels varied dramatically over the $2\frac{1}{2}$-yr study and fell below the projected limitation threshold of the birds during both autumns (Fig. 2.6). McFarland (1986a) used similar procedures to document energetic limitation of honeyeaters by low nectar supplies on 16 of 65 sampling dates in woodlands in the New England tablelands. Here, days of low nectar production were associated with adverse weather, especially low overnight temperatures (McFarland 1985a).

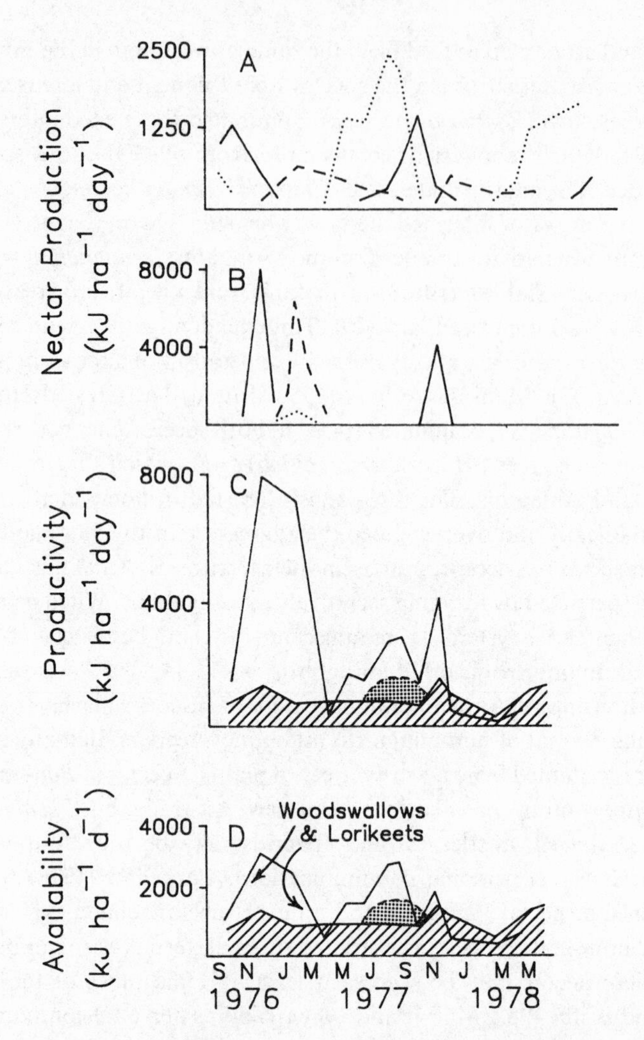


Fig. 2.6. A. Production of nectar within a 20-ha study area of eucalypt woodland in Australia by *Banksia marginata* (dashed line), *Astroloma conostephioides* (dotted line), and *Grevillea aquifolia* (solid line). B. Nectar production by *Callistemon macropunctatus* (solid line), *Eucalyptus obliqua* (dashed line), and *Eucalyptus melliodora* (dotted line) in the same area. C. Comparison of the estimated energy requirements of the assemblage of honeyeaters of this area (hatched zone; the stippled area represents the additional energy required during breeding) with total nectar production. D. Relationship between honeyeater energy demands (as in C) and nectar availability (production minus consumption by insects). During the 1976–77 summer, woodswallows (*Artamus* spp.) and lorikeets (*Glossopsitta* spp.) were abundant in the area and also fed on nectar; these species probably consumed most of the 'surplus' nectar available during this period. Modified from Paton (1985).

The flowering patterns of the nectar-producing plants in Australia may change dramatically in time and space (e.g. Fig. 2.6). As a consequence, the abundance and availability of nectar in a locality tend to be quite variable. In response to local variations in flowering phenologies, honeyeaters may shift between different resource patches (Collins *et al.* 1984, Collins 1985, Pyke and Recher 1986) or may move over large areas (Ford and Paton 1985, McFarland 1985b, Stewart and Craig 1985, Pyke 1983, personal communication). Honeyeater densities in local areas are generally positively correlated with variations in nectar production rates or inflorescence density rather than in arthropod abundances (Pyke 1985, Collins and Newland 1986, McFarland 1986b), although these correlations are weakened by the large number of transient individuals whose movements are often unrelated to local nectar levels.

As in many other nectarivores, access to nectar resources by co-occurring honeyeater species is often determined by interspecific aggression in which larger species dominate the smaller species (Collins 1985, Collins and Newland 1986). On Kangaroo Island, Ford and Paton (1982) compared three sites that differed greatly in nectar-production rates. The largest of the meliphagids present occurred only at the richest site, where it defended flowers. A medium-sized species also occurred there and at the intermediate-production site, where it defended territories. The smaller honeyeater species also fed at the latter site but were often chased, and the smallest species was most common at the poorest site, where few of the other species occurred. McFarland (1985b, 1986a) observed a similar size-dependent dominance hierarchy among several honeyeater species in a New England forest. Nectar levels varied considerably on a day-by-day basis in this area, and the smaller species were able to persist during days of low nectar production because of their greater foraging efficiency and their secretive behavior. The large species, on the other hand, relied on fat reserves accumulated during periods of greater nectar abundance, but if low nectar supplies continued for very long the larger species left the area. An intermediate-sized species, the New Holland Honeyeater (*Phylidonyris novaehollandiae*), occupied territories whose size varied as a function of inflorescence density (McFarland 1986c). The birds defended these territories by a combination of intra- and interspecific aggression and nectar depletion, which McFarland argued made the territory less attractive to intruders. In this system, as in other meliphagid assemblages (Craig and Douglas 1986), aggression was most evident at intermediate resource densities (c.f. Fig. 2.4).

An indication of the dynamic relationship between nectar production and aggressive interactions between species is provided by Paton's studies in a small area of *Banksia marginata* heathland near Melbourne (Paton 1979, Ford and Paton 1985). There, New Holland Honeyeaters established territories as inflorescences began to appear in early June (Fig. 2.7). As nectar production increased, a larger species, the Little Wattlebird (*Anthochaera chrysoptera*), established territories and excluded many of the New Holland Honeyeaters. As nectar production passed its peak, the wattlebird territories were expanded so that few New Hollands remained. By early September, the wattlebirds left the area, and the honeyeaters returned to re-establish territories (Fig. 2.7), In this situation, the capacity of the honeyeaters to track variations in nectar production was severely restricted by territorial aggression from the wattlebirds (Ford and Paton 1985).

Interspecific aggression by honeyeaters is not confined to interactions with other nectarivorous species. Honeyeaters often feed on lerps [a carbohydrate exudate of psyllid (hemipteran) larvae that feed on eucalypt foliage] and may be aggressive toward other lerp-feeding birds. Pardalotes (*Pardalotus*) in particular feed extensively on lerps. Woinarski (1984) estimated that as much as 5% of the time and 9% of the energy expended by pardalotes might be spent avoiding aggressive honeyeaters, and he speculated that the hole-nesting habits of pardalotes might represent an evolutionary adjustment to escape such aggression.

Among the meliphagids, Bell Miners (*Manorina melanophrys*) are notable in the extent to which they feed on psyllid larvae and lerps. Miners are communal breeders and occupy large group territories from which they aggressively exclude a wide variety of other insectivorous as well as nectarivorous species. In some areas of southeastern Australia, Eucalyptus forests occupied by the miners are heavily infested with psyllids, perhaps because insectivorous birds are excluded by the miners. Loyn *et al.* (1983, Loyn 1985) examined this possibility by experimentally removing miners from a psyllid-infested area of forest. There was a rapid invasion of the area by other bird species and a concomitant decline in psyllid abundance on the foliage (Table 2.2), and these conditions persisted for the duration of the observations. The ability of the insectivorous species to control the psyllids in the absence of the miners indicates the value to the miners of territorially excluding these species.

Territoriality is also a major form of resource defence in sunbirds, especially where nectar sources are patchily distributed. In Kenya, Gill and Wolf (1977, 1978, 1979) documented that territorial Golden-winged Sunbirds lost 8% of the average nectar supply per flower to intruding

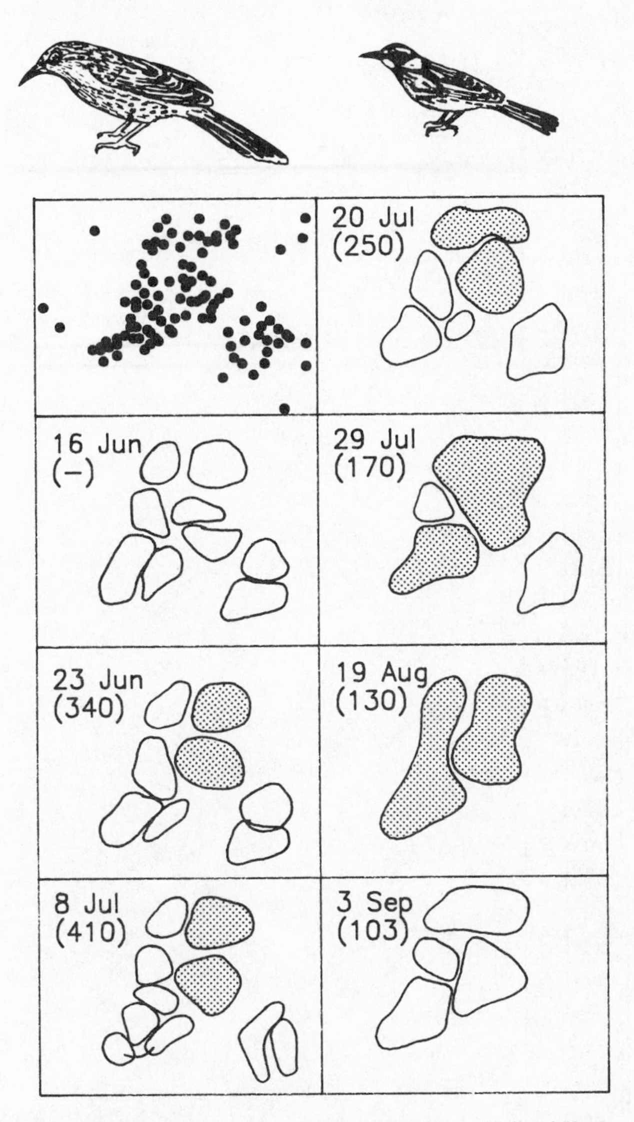


Fig. 2.7. Pattern of territory distributions of New Holland Honeyeaters (clear areas) and Little Wattlebirds (stippled areas) in a *Banksia marginata* woodland in Victoria, Australia, over a 3-month period. The number of productive inflorescences in the site at each time period is given in parentheses. The top left panel indicates the distribution of *Banksia* plants. From Ford and Paton (1985).

Table 2.2. *Numbers of birds and psyllids observed on a eucalypt woodland study area in southeastern Australia before and after the experimental removal of Bell Miners.*

Numbers of birds are expressed as mean number seen per 10 20-min searches.

| Group[a] | Date | | | |
|---|---|---|---|---|
| | April–July 1981 | Aug–Sep 1981 | Oct–Dec 1981 | Jan 1982–Mar 1983 |
| | (Bell miners removed July/August) | | | |
| *(A)* | | | | |
| Bell Miners | 188 | 7 | 0 | 0 |
| *(B)* | | | | |
| Crimson Rosella (*Platycercus elegans*) | 8 | 26 | 9 | 6 |
| Eastern Rosella (*P. eximius*) | 0 | 9 | 10 | 0 |
| Striated Thornbill | 0 | 78 | 24 | 14 |
| White-naped Honeyeater (*Melithreptus lunatus*) | 5 | 90 | 6 | 5 |
| Spotted Pardalote (*Pardalotus punctatus*) | 3 | 17 | 4 | 5 |
| Striated Pardalote (*P. striatus*) | 0 | 13 | 33 | 1 |
| *(C)* | | | | |
| White-throated Treecreeper (*Climacteris leucophaea*) | 0 | 10 | 3 | 5 |

| | | | | |
|---|---|---|---|---|
| Red-browed Treecreeper (*C. erythrops*) | 0 | 4 | 4 | 0 |
| Varied Sittella (*Daphoenositta chrysoptera*) | 0 | 24 | 2 | 0 |
| Psyllids on foliage | abundant | declining rapidly | very few | hardly any |
| Psyllid lerps collected in 24 trays (mean per month) | 1920 | 211 | 6 | 3 |
| Epicormic crown cover (%) | 32.5 | | 33.3 | 37.3 (Feb 1982) |

*Notes:*
[a] Group *A* (Bell Miners) were feeding mainly on psyllids and their associated lerps and honeydew, often taking the psyllid nymph but not its protective lerp. Group *B* fed mainly on psyllids and lerps while they were in the study area. Group *C* were bark-gleaners. Birds that fed in the understorey or on other foods in the canopy, and uncommon species are not shown. Data from control plots and other years showed seasonal variation but no peak in bird numbers in Aug/Sept.
*Source:* From Loyn (1985).

Table 2.3. *Nectar uptake and losses to competitors by Golden-winged Sunbirds feeding at* Leonotis nepetifolia *in Kenya.*

Losses are expressed as a percentage of the total potential nectar uptake ±one standard error. Standard errors of per cent loss data were calculated using arcsin transformations.

| | Territorial | Nonterritorial | |
|---|---|---|---|
| Nectar | Males | Males | Females |
| *Potential uptake (μl/fl)* | | | |
| Mean ± SE | 3.89 ± 0.34 | 4.35 ± 0.24 | 4.10 ± 0.32 |
| N, Range | 15, 2.2–6.8 | 30, 2.6–7.7 | 20, 2.0–7.9 |
| *Actual uptake (μl/fl)* | | | |
| Mean ± SE | 3.54 ± 0.37 | 2.67 ± 0.13 | 2.26 ± 0.24 |
| N, Range | 15, 1.2–6.4 | 30, 1.1–4.9 | 20, 1.1–4.4 |
| *Losses (% potential) to* | | | |
| *N. reichenowi* ♂♂ | 2.2 ± 0.48 | min. 9.7 ± 1.35<br>max. 20.2 ± 2.01 | 22.8 ± 1.93 |
| *N. reichenowi* ♀♀ | 4.1 ± 0.97 | 8.7 ± 0.92 | min. 0.0<br>max. 8.4 ± 1.21 |
| *N. famosa* | 0.9 ± 0.27 | 9.1 ± 1.17 | 11.0 ± 1.30 |
| *N. venusta* | 0.7 ± 0.31 | 7.6 ± 1.03 | 10.4 ± 1.61 |
| Other species | 0.4 ± 0.25 | 0.6 ± 0.22 | 0.4 ± 0.15 |
| *Total* | 8.3 ± 1.23 | min. 35.8 ± 2.37<br>max. 46.4 ± 1.96 | min. 44.0 ± 2.35<br>max. 52.4 ± 2.85 |

*Source:* From Gill & Wolf (1979).

conspecifics and individuals of other sunbird species (chiefly *N. famosa* and *N. venusta*). Nonterritorial male *reichenowi*, on the other hand, lost an estimated 36–46% of their potential nectar uptake to competitors, and losses by females were even greater (Table 2.3). Gill and Wolf estimated that territorial and nonterritorial males would have to forage 17% and 72% more each day to maintain an energy balance because of these losses. Exploitation of nectar therefore has a nonlinear effect on foraging time (Fig. 2.1), and territoriality is clearly advantageous, at least in appropriate resource situations.

There are therefore substantial similarities in the behavior and ecology of nectar-feeding birds in different parts of the world. There are also some differences, perhaps most conspicuously in body sizes. Brown *et al.* (1978) proposed that there should be an upper size limit of nectar-feeding birds of around 80 g. They based their argument on the supposition that there are increasing costs and diminishing returns to plants in producing the quantities of nectar necessary to attract large pollinators; 80 g approximates a threshold in the balance between benefits to the birds and costs and benefits to the plants. Brown and his colleagues developed their thesis on the basis of their experiences with North and Central American hummingbirds and the flowers they feed from and pollinate. Elsewhere in the world, however, the situation is quite different. There are many species of honeyeaters that weigh considerably more than 80 g, and the maximal size of nectarivores in Hawaii is intermediate between the Australian honeyeaters and the American hummingbirds (Craig and MacMillen 1985). This gradient in body sizes appears to be associated with differences in the magnitudes of nectar production by the plants that attract nectar-feeding birds (greatest in Australia) and in the frequency of food limitation in the communities (greatest in America) (Carpenter and MacMillen 1978, Pyke 1980). Using estimates of the nectar production in small patches of typical habitat in North America and Australia and of the energy demands of nectarivores of various sizes, Craig and MacMillen (1985) determined that the flower density and quality in an Australian heathland habitat would readily support a bird as large 120 g on a plot as small as 20-m radius, whereas a plot of the same size in a North American montane habitat would be capable of supporting only a single 5-g hummingbird (Fig. 2.8).

**Interactions with distantly related taxa**

Nectarivorous birds share a distinctive food resource that is often limiting with other nectar-feeding organisms, expecially insects. The potential for competitive interactions among these taxa is therefore great. In some instances, the birds have negative effects on the availability of nectar to the insects. In the Sierra Nevada Mountains of California, migrant territorial Rufous Hummingbird aggression toward hawkmoths is associated with a decreased rate of visitation to shared flowers by the moths (Carpenter 1979). When Carpenter enclosed some plants in screening to remove hummingbird effects but permit access by the moths and by bees, the rate of foraging on these plants by the insects increased. Others (Lyon and Chadek 1971, Primack and Howe 1975, Boyden 1978) have noted similar aggressive interactions between hummingbirds and large insects.

More often, the interaction is mediated by nectar exploitation by the insects. Although their capacity for direct aggressive interference with birds is limited (but see Gill *et al.* 1982), insects have the advantage of being small. Their individual energy requirements are therefore low and, by foraging economically, insects can reduce nectar availability to levels that will not support larger animals (Kodric-Brown and Brown 1979). Insect nectarivores can also swamp the resource-defense systems of birds by sheer

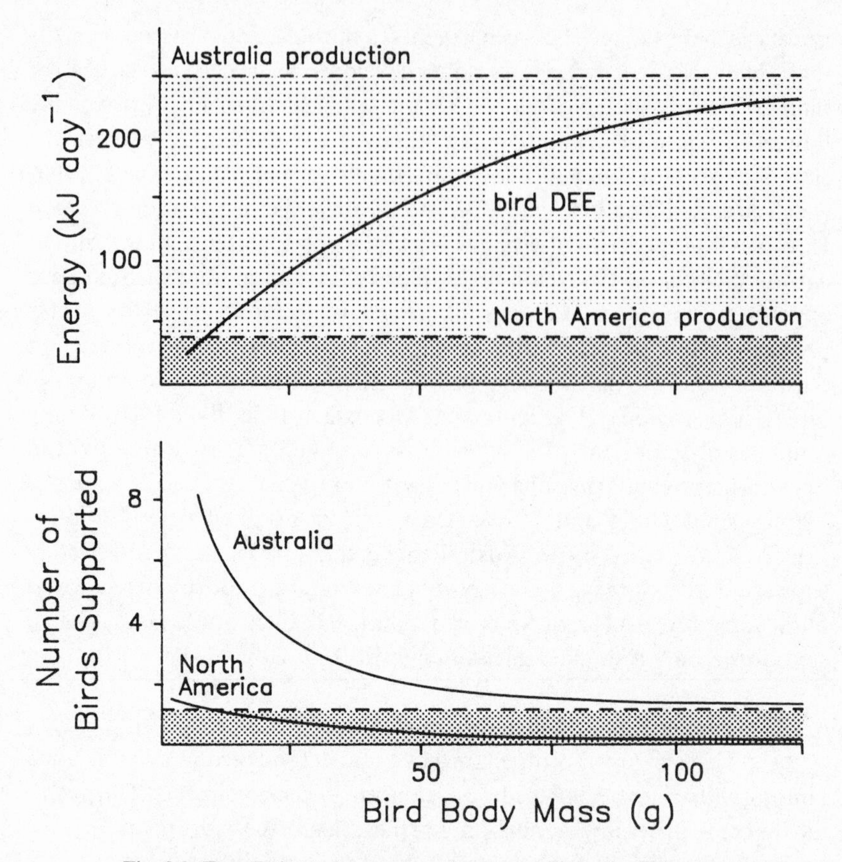


Fig. 2.8. Top: Estimated energy requirements of hypothetical nectarivorous birds ranging in size from 5 to 120 g in relation to the nectar production in typical honeyeater and hummingbird habitats (represented by circular plots of 20-m radius). Bottom: Projected carrying capacities of these nectar productivities for birds of different body sizes. The stippled area indicates the zone in which the nectar production is insufficient to support one individual in the 20-m radius area. From data in Craig and MacMillen (1985).

force of numbers. In Western Australia, for example, the levels of nectar of *Calothamnus quadrifidus* are reduced within a few hours after dawn and remain low throughout the day. Both honeyeaters and bees contribute to this resource depletion, but bees have a greater overall impact. Their exploitation of this resource therefore reduces its availability to the birds (Collins *et al.* 1984). Ford (1985) suggested that there are few nectarivorous birds in Europe relative to numbers elsewhere in the world because the pollinator role in Europe may have been pre-empted evolutionarily by social bees.

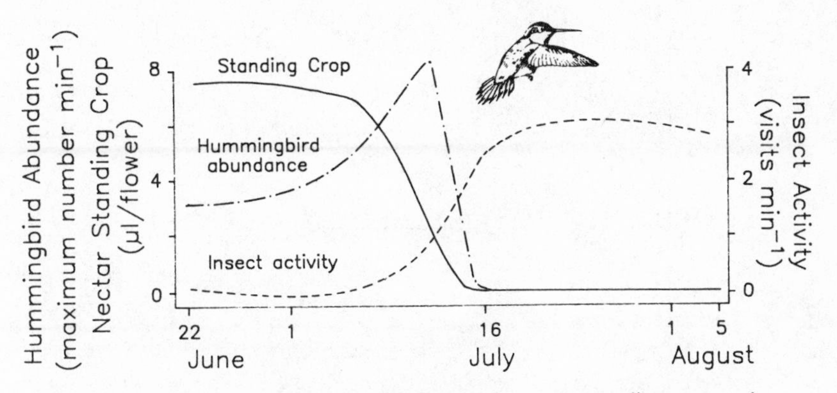


Fig. 2.9. Temporal patterns of insect activity, nectar standing crop, and hummingbird abundance in a large patch of *Scrophularia* in a montane area of New Mexico during 1982. From Heinemann (1984).

On of the most detailed studies of bird-insect interactions was conducted by Heinemann (1984, unpublished) in the mountains of central New Mexico. There, breeding Broad-tailed Hummingbirds (*Selasphorus platycercus*) establish territories about clusters of *Scrophularia montana*. In early July, migrant Rufous Hummingbirds arrive and rapidly displace the Broad-tails, establishing territories in *Scrophularia* patches that are tenaciously defended. Several weeks later, however, all birds abrupty leave these patches, although Rufous Hummingbirds continue to be seen at other flowers in the area well into September. Heinemann measured the standing crop of nectar in *Scrophularia* inflorescenses and the nectar production rates of flowers bagged to prevent access by nectarivores during two breeding seasons. He also monitored levels of insect (primarily hymenopteran) activity and the time budgets and territorial behavior of the hummingbirds.

Hummingbird abundance increased from early June to a peak just before their abandonment of the resource in mid-July in 1982. Nectar availability began to decrease precipitously a few days before hummingbird abundance peaked (Fig. 2.9). Coincident with the sudden decrease in nectar levels there was an increase in insect activity, which remained at high levels for the remainder of the summer even though nectar levels continued to be low. The timing of this pattern was different in the following year, but once again the abandonment of the *Scrophularia* patches by the hummingbirds coincided with a decrease in nectar standing crop and an increase in insect visitation rates to the flowers. In both years, territorial hummingbirds occasionally attempted to chase hymenopterans from flowers, but the number of insects was too great for this aggression to be effective.

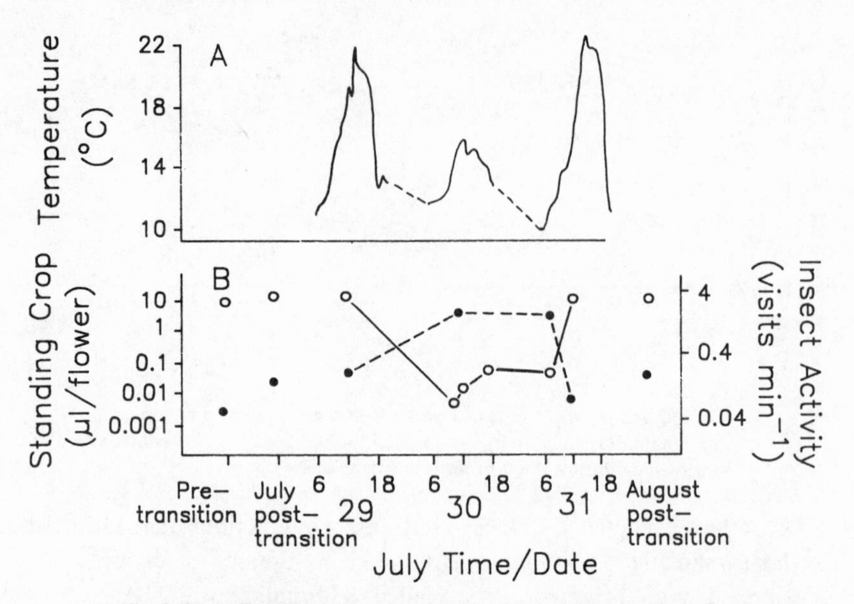


Fig. 2.10. A. Temperature regime for 29–31 July 1982 in a montane meadow location in New Mexico where hummingbirds and insects interacted over nectar resources. B. Insect activity (open circles) and nectar standing crop in *Scrophularia* flowers (solid circles) during the same period. Insect activity levels and nectar supplies before (pre-transition) and after (post-transition) the departure of hummingbirds from the resource patch are also indicated. Hummingbirds established territories in the location early on 30 July, but abandoned the resource by 1300 on 31 July. After Heinemann (1984).

That the relationship between insect activity levels, nectar standing crop, and hummingbird territoriality is more than chance coincidence is shown by a 'natural experiment' that occurred during late July 1982, after hummingbirds had abandoned the *Scrophularia* patches. Weather conditions on 30 July were unusually cold and wet, and insect activity did not reach levels typical of this time of year. Nectar standing crop increased to levels typical of those found before abandonment, and by mid-day hummingbirds began establishing territories in the area. By early evening, territory density was high (Fig. 2.10). This situation persisted into the following morning, but temperatures that day were normal and insect activity levels increased rapidly. Nectar levels fell, and by early afternoon all territorial hummingbirds had once again abandoned the resource.

Heinemann postulated that this sort of scenario is driven by the exploitation of the nectar by the hymenopterans, which depresses nectar standing crop below levels at which it is energetically feasible for hummingbirds to occupy the area and defend territories. To test this proposition in another way, he constructed a model of hummingbird energetics from which he could estimate the nectar standing crop required to meet a bird's demands for existence and a minimum fat-deposition rate. This model indicated that birds varying in size from 3 g (lean) to 5 g (ready to migrate) require a mean standing crop of 0.47–0.80 $\mu$L/flower. When standing crops measured in the field are related to this energy-demand threshold (Fig. 2.11), it is apparent that abandonment was closely related to a failure of the resource levels to meet the birds' minimal energy requirements. Abandonment in fact appeared to occur 1 day after the threshold had been passed, suggesting that the birds may persist in defending territories even after resource availability has dropped to unacceptable levels, at least for a short time (Heinemann unpublished).

Such close linkages between the exploitation of nectar by insects and the use of resources by hummingbirds may not be unusual among obligate nectarivores. Brown *et al.* (1981) and Laverty and Plowright (1985) described similar temperature-dependent effects in insect activity levels and exploitative depletion of nectar used by hummingbirds.

## Conclusions

The evidence that competition occurs among nectarivore species in local assemblages is convincing. Interactions do not always occur, but variations in the presence and intensity of interactions can often be related directly to measures of nectar abundance and availability, and these measures can in turn be compared directly with estimations of the energy demands of the birds. Such calculations (as well as the more obvious behavior of individuals) demonstrate that nectar is limiting to the birds at least at certain times. Most of the criteria listed in Table 1.4 are met in many of the studies described in this chapter. The species clearly overlap strongly in their use of nectar resources, intraspecific interactions often occur, the patterns are consistent with the expectations of competition theory, and use of nectar by individuals of one species leads to depletion and a reduction in resource availability to other individuals. It is less frequently documented that one or more of the species suffers negative effects from the interactions. This seems likely, however, when individuals must expend time and energy responding to aggression or are displaced from access to a known resource supply and must search for nectar elsewhere (especially if the nectar sources are patchily distributed). The evidence supporting an interpretation of the patterns of interactions or of community structure based on competition often seems so compelling that alternative explanations are not considered, however. Some of these alternatives may shed additional light on broader,

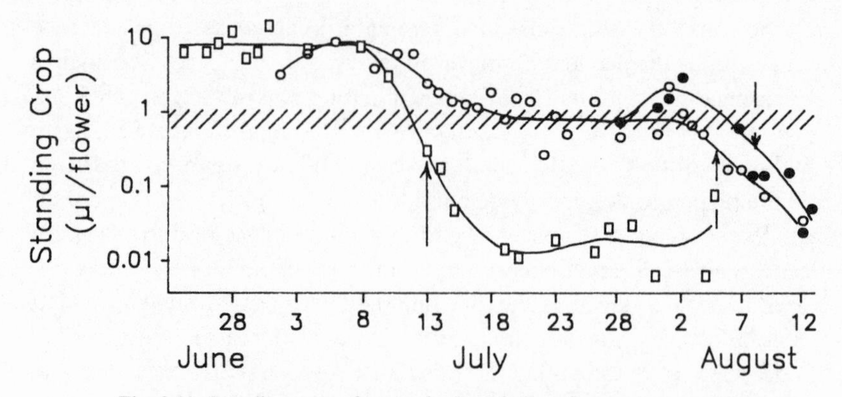


Fig. 2.11. Standing crop of nectar in *Scrophularia* flowers in a montane location in New Mexico in 1982 (squares) and in two plots in 1983 (open and solid circles). The hatched zone indicates the range for the predicted minimum standing crop required to meet the energy demands of a hummingbird. The vertical arrows indicate the times at which hummingbirds abandoned territories in the *Scrophularia* patch in the two years. After Heinemann (1984).

biogeographic patterns of nectarivore community structure, but they do not contribute very much to a better understanding of the more proximate behavioral interactions among nectarivores.

Is competition really more prevalent among nectarivores than among most other birds, or is it just more easily documented? This question is difficult to answer, for studies of non-nectarivore systems have generally inferred rather than measured resource levels relative to demands. Some of the information necessary to evaluate the possibility that competition is occurring is therefore usually missing from such studies. It also seems likely, however, that the distinctive features of nectar as a resource and the high degree of specialization of many nectarivores predispose these systems to be competitive. Because individuals can selectively deplete the resource, exploitation may be an effective means of defending space, and because nectar supplies in blooming flowers are rapidly replenished, both exploitative and aggressive defense of the resources are adaptively feasible. Competition is therefore evidenced especially strongly in overt aggression, which makes its occurrence and its consequences much more obvious than may be the case in other systems founded on resources that have quite different temporal and spatial dynamics.

Nectarivores provide the confirmatory evidence of the importance of competition that advocates of competition theory seek, but the characteristics of these systems are distinctive enough that it would be a mistake to generalize from them to other sorts of bird communities.

3

# Beyond competition: other factors influencing community structure

Patterns in avian communities have usually been explained as the results of interspecific competition, and other processes or factors that might contribute to the patterns have been given only superficial consideration or ignored. Finding that individuals of an island population of a species differ in behavior from their mainland conspecifics, for example, ecologists often have not asked what factors might account for the differences but instead have explained the patterns solely in terms of the absence of presumed competitor species on the island. In the previous chapters, I have argued that this approach is overly simplistic and sometimes logically invalid. Even if competition is shown to occur, this does not mean that other processes or factors are absent or unimportant. Only if the patterns predicted on the basis of these other factors differ from those expected from competition can one justly conclude that competition is the primary cause of the patterns and that the contributions of other processes are unimportant. Testing such alternatives is a pivotal feature of any logical approach to scientific explanation (see Volume 1).

According to Schoener (1982), the challenges to the competition paradigm that developed during the late 1970s followed several avenues: (1) mathematical modification of the theory, (2) statistical re-evaluation of the patterns through the use of null models, (3) consideration of the influences of environmental variability, (4) emphasis of the role of predation, and (5) determination of the meaning of resource overlap. Of these areas of concern, only the fourth really emphasizes an alternative process; the first and fifth involve adjustments of competition theory, the second deals with the patterns, and the third explores the constancy or intermittency of the competition process. Although it is certainly appropriate to reconsider the reality of patterns, the details of competition theory, and the ubiquity of competition as a process, it is equally important to determine the possible effects of other factors on community patterns.

This emphasis is by no means new. In 1954 Andrewartha and Birch proposed that competition occurs infrequently and that weather, predation, and spatial heterogeneity of habitats are the primary influences on populations. In their more recent development of a 'theory of environment' (1984), mates, predators, resources, and 'malentities' are the primary factors that act *directly* on the distribution and abundance of animals; other factors influence populations only indirectly, by affecting one or more of these primary factors. Competition is thus considered either as a directly acting malentity (i.e. aggressive interference) or, more commonly, as an indirectly acting component influencing resources (i.e. exploitation).

Whether one follows Andrewartha and Birch's classification of environmental components or some other arrangement of processes and influences, there are clearly several factors in addition to competition that may influence community characteristics. Some of these were recognized by individuals whose work is closely associated with the development of the competition paradigm, although they did not receive much emphasis. MacArthur (1972), for example, considered reduced predation and limitations on dispersal as possible contributors to enhanced densities of species on islands, and he proposed that the northern limits of species' ranges (in the Northern Hemisphere) were most likely determined by climate rather than biotic interactions. Cody included a chapter on 'Alternatives to competitive displacement patterns' in his book on bird communities (1974) and, although most of his emphasis was on behavioral manifestations of competitive exclusion, he did deal briefly with the consequences of nonlimiting resources and predation. Lack (1971) noted that, beyond competition, factors such as habitat or climate might influence the ranges of species, but he then suggested that the habitat differences reflected competition in the past and he discounted climatic limitation as having much real effect on endotherms such as birds. More recently, alternative processes or factors have been included with greater frequency in explanations of patterns. The studies of Grant and his colleagues in the Galápagos and of Järvinen and his associates in northern Europe are especially notable in this regard, but other examples are scattered through Volume 1.

In order to advance beyond the traditional views of communities, we must consider the effects of factors such as predation, parasitism, individualistic responses to environments, climate, resource patchiness, and disturbance. Because these effects occur over time, history and chance may also leave an imprint on the patterns. In this chapter I will appraise these processes and factors and consider the ways in which they may affect community patterns. Variations in these influences in time or space may have especially profound effects on such patterns; these are mentioned briefly here and are then considered in greater detail in Part II.

**Individualistic responses of species**

The foundation of a competitionist view of communities is that species' responses to environmental situations are strongly governed by competitive interactions with other species. Species may instead respond to environmental circumstances in a species-specific fashion, in accordance with their characteristic adaptations and constraints. Of course, these adaptations and constraints may have evolved in circumstances in which interspecific competition was a potent selective force, even though it may no longer be occurring. Once present in a species, these traits may dictate its *proximate* responses to environmental conditions in a species-specific, individualistic manner (e.g. Sherry 1984). The hypothesis that community patterns are produced by the summation of independent, species-specific responses by members of the assemblage is an alternative to the hypothesis that the patterns are a result of proximate competitive interactions, but it does not exclude the possibility that the traits that determine the species' responses were influenced by competition in the evolutionary past.

An emphasis on such individualistic, species-specific responses is the cornerstone of Gleason's (1926) view of communities and is the foundation of the approach taken by Andrewartha and Birch (1954, 1984) as well. Many 'random' models of community assembly (e.g. Connor and Simberloff 1978, Strong *et al.* 1979, Simberloff 1978, 1983b) are also held to be essentially individualistic models. If species are allocated to model communities on the basis only of their specific habitat affinities, physiological tolerances, dispersal capabilities, and the like (e.g. Graves and Gottelli 1983) this may be so, but when species are simply drawn from a pool without regard to these traits (e.g. Connor and Simberloff 1978, Strong *et al.* 1979) the individualistic attributes of species are randomized as well. Roughgarden's (1983) complaint that the null models are 'empirically empty' because they contain no biological processes then has merit.

Strong *et al.* (1979) claimed that the null hypothesis of no species interactions (the individualistic hypothesis) has 'logical primacy' over other models of community assembly and composition. Although Simberloff (1983a) later conceded to Roughgarden's (1983) challenge of this view, Strong (1983) did not. In the sense of formal logic, the individualistic hypothesis may have no special priority over other process explanations, but it nonetheless makes sense to consider species-specific responses to environmental conditions as the infrastructure on which other, interactive processes operate. The ability of species to occupy a particular place is ultimately restricted by those species-specific features of physiology, ecology, morphology, and behavior; whether or not a species actually occurs there may be further constrained by biological interactions of various sorts. This is precisely the relationship that Hutchinson (1957) envisioned between individualism and competition in his conceptualization of fundamental and realized niches. Hutchinson's model was incomplete, in that he considered only the effects of interspecific competition in reducing the fundamental niche to the realized niche, but it clearly specified that the individualistic-response hypothesis and the competition hypothesis are a hierarchically nested set, not mutually exclusive alternatives. The individualistic responses of species to environments determine the basic patterns of their distributions and abundances, which then may be modified by competition or other processes to produce a different pattern (Fig. 3.1).

In some situations, species-specific responses to climatic factors may have clear effects on population levels, distributions, and community patterns (see Volume 1, Chapter 8). Not surprisingly, some of the variation in species diversity in communities is related to the effects of weather (Kricher 1975, Rotenberry 1978, Rotenberry *et al.* 1979). Climatic/physiological limitations take on greater importance when they are considered at a finer scale, to explain why some species are absent or occur in low densities in local communities in seemingly suitable habitat. In northern Denmark, for example, long-term population fluctuations of Coal Tits are strongly influenced by winter climate, those of Great Tits less so, and those of Blue Tits scarcely at all (Bejer and Rudemo 1985). In northern Finland, the reproductive success and population fluctuations of hole-nesting species may be linked to the rigorous and variable climate there, but in different ways (Järvinen 1983). Adults of the resident *Parus cinctus* are relatively insensitive to cold, but in some years with late cold spells the nestlings suffer increased mortality. Adults of *P. major* breed more successfully than *P. cinctus* but are less able to cope with the arctic winter. Redstarts (*Phoenicurus phoenicurus*) and Pied Flycatchers are both migrants, but the Redstart is a northern native and breeds successfully under most conditions. The flycatchers, on the other hand, have recently expanded into the region from the south and suffer severe nesting losses due to the harsh climate, especially in cold summers (Järvinen and Väisänen 1984). Population fluctuations of the redstarts are apparently caused by events occurring on the wintering grounds, whereas flycatcher recruitment is so low that the

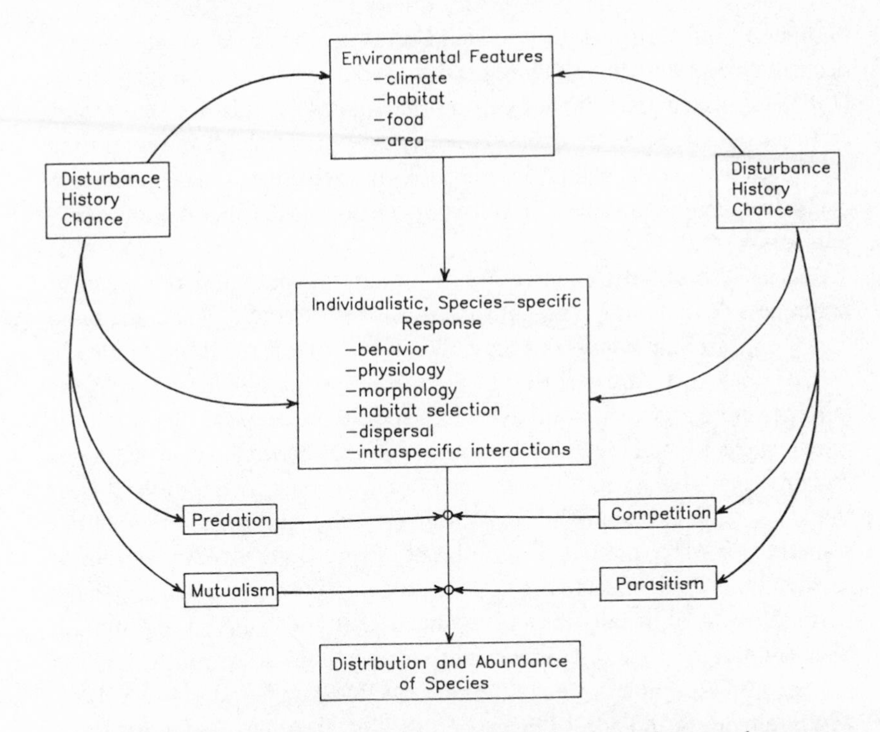


Fig. 3.1. Generalized scheme of the relationships among factors and processes influencing the distribution and abundance of species, and thus community attributes. The primary determinants are individualistic, species-specific responses to environmental factors, which may be modified by processes such as predation, competition, parasitism, or mutualism. All of these components and their relationships may be altered by disturbances; through time, this creates an historical aspect to distribution-abundance patterns, in which chance effects may play an important role.

population is maintained only by continued immigration from the south. In the exceptionally cold and rainy summer of 1981, flycatcher breeding success was normal in southern Finland but extremely poor in the north. Some other species (e.g. *Fringilla montifringilla*) also produced very few young in the north, but others (notably redstarts and *Luscinia svecica*) bred normally (Hildén *et al.* 1982).

The consequences of climatic extremes and variations on populations thus may be severe, but they differ among species. The occurrence or relative abundances of species in the communities may change between locations or between years at a given location as a result of the climatic effects. Changes in other community patterns, such as niche relationships or morphological features, may then follow. Hildén *et al.* (1982) calculated that periods of exceptionally bad weather during the breeding season have occurred in Finland once or twice per decade during recent times, and Grant (1985) estimated that finches on the Galápagos Islands are reasonably certain to encounter at least one extremely dry year and one wet year during their lifetime. Extreme climatic conditions in such environments may thus occur frequently enough to have major, persistent effects on community patterns.

Species also differ in their responses to habitat characteristics, area, or patchiness (Lynch and Whigham 1984, Simberloff and Abele 1982, Sabo and Holmes 1983, Wiens and Rotenberry 1981a). Changes in habitats in time or space therefore may lead to changes in species' abundances or niche patterns quite apart from any effects of species interactions. These effects are most clear when the habitats compared differ markedly in structure or resource availability, but they may be present in comparisons among areas that are held to be similar in habitat as well. Although attempts are sometimes made to control for habitat differences in these comparisons or to segregate their effects through statistical procedures such as partial correlation, it is usually simply assumed that the habitats are similar. Species differ in subtle ways in their responses to habitat variation, however (Wiens and Rotenberry 1981a, Wiens 1985b, Sherry and Holmes 1985, Mountainspring and Scott 1985, Abbott 1980), and coarse categorizations of habitat features are likely to miss such details. This may lead to a premature rejection of the hypothesis that the observed patterns are associated with habitat differences to which species respond individualistically. In most comparative studies, this hypothesis represents a valid alternative to that of competition (or any other process) and should be tested carefully.

Species-specific characteristics may influence community patterns in other ways, of course. If the density of a species in an area increases, increased intraspecific competition may force an expansion of the population into additional habitats, a broadening of individual foraging behaviors, an increase in dietary diversity, or the like (Svärdson 1949, Fretwell and Lucas 1969). Should such density changes happen by chance to coincide with the reduction in density of another species (perhaps because environmental changes favor one species but harm the other), the patterns of niche change would parallel those expected from interspecific competition, although the process responsible would in fact be individualistic. Species may also differ in the ways in which spacing patterns of individuals are related to resource availability or dispersion (Waser and Wiley 1979). As a consequence, the social behavior of species and the forms of resource utilization associated with them may change in different ways with changes in resource conditions, producing different patterns of niche shifts in the different species.

Despite the potential effects of species-specific responses on the patterns of local communities, individualistic hypotheses have generally not been popular explanations of community patterns (but see Rice *et al.* 1983a, Robinson and Holmes 1984, Rotenberry and Wiens 1980a, Wiens and Rotenberry 1981a, James *et al.* 1984, Mountainspring and Scott 1985, Wiens 1986). It is easy to see why. If one accepts the notion that individualistic responses are of primary importance in determining community patterns, then as environmental conditions change in space or time the different members of a community are likely to respond to the changes in different ways. These biological idiosyncrasies produce excessive 'noise' in the patterns and thereby thwart attempts to develop general theory that applies to the details of local assemblages. Generality may be achieved only by expanding to a broader scale, over which the local idiosyncrasies of species-specific responses are statistically averaged (e.g. Brown 1984).

**Predation**

Birds are subject to predation from a variety of sources, and many of their traits (e.g. flocking, nest-site locations, injury-feigning displays, coloniality, drab or cryptic coloration) have been interpreted as adaptations to reduce predation. If the effect of predators on members of a community is nonrandom, community patterns may be altered from those expected in the absence of predation. For example, where house cats (*Felis catus*) have been introduced to islands previously lacking carnivores, they have devastated many bird populations, altering the species composition and relative-abundance distributions of the island communities (e.g. Merton 1975, Fitzgerald and Veitch 1985). On Herekopare Island off New Zealand, cats were introduced in *c.* 1925; by 1970, six landbird species had become extinct and a vast breeding population of diving petrels and thousands of prions were probably exterminated as well. Other species, however, persisted. On Guam, predation by the introduced brown tree snake (*Bioga irregularis*) apparently has led to a very rapid reduction in range or extinction of 10 native species of forest birds, although bird populations in savannah habitats, which the snakes avoid, have remained relatively stable (Savidge 1987).

Traditionally, predation has been considered to influence community structure by depressing populations of competitors beneath resource-limitation levels, modifying the outcome of competitive interactions or eliminating them entirely (Connell 1975, Holt 1984, Andrewartha and Birch

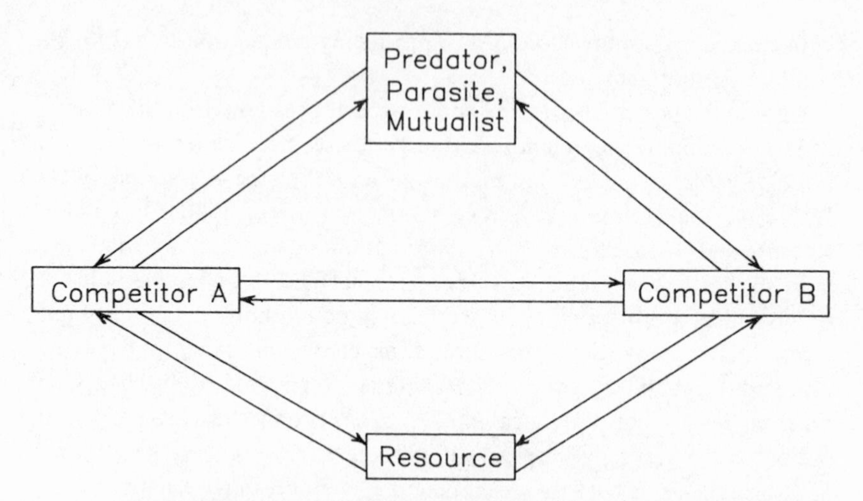


Fig. 3.2. Pathways of potential interactions between two competing species (A and B), the resources over which they compete, and the mediating interactions with a predator, parasite, or mutualist. Modified from Price *et al.* (1986).

1984, Tilman 1986). Thus, two competitors (A and B) might interact by direct interference or through exploitation of a shared resource (Fig. 3.2). If a predator reduces the population size of species A, either or both avenues of interaction with B may be affected and the competitive effects on B reduced. By reducing competition, predation facilitates coexistence and diversity increases, especially if predation effects are greatest upon the competitively dominant species (Paine 1969, Roughgarden and Feldman 1975, Glasser 1979). MacArthur (1972: 191) noted such effects, observing that 'if abundant predators prevent any species from becoming common, the entire picture changes. Resources are no longer of any concern.... More correctly, resources are still a concern, but their manner of subdivision is irrelevant.' Judged by the amount of space he devoted to it, MacArthur did not consider these effects of predation to be commonplace. Connell (1975), on the other hand, viewed predation as being of primary importance in structuring communities, with other factors entering in only where the effects of predation are reduced. Holt (1984) also advocated the pre-eminence of predation.

Rates of predation depend on the functional, numerical, and developmental responses of predators, and these in turn vary with differences in habitat, climate, the availability of refuges for the prey, the presence and diversity of alternative prey, and the impact on the predators of higher-order predators, parasites, competitors, and the like (Taylor 1984, Sousa 1984). Predator densities in an area may be especially important in deter-

Table 3.1. *Slopes of species-area relationships (*S *vs log* A*) for guilds of birds in central Arizona montane forests defined on the basis of nesting substrate.*

| Guild | Slope |
|---|---|
| Cavity nesters | 2.92 |
| Foliage nesters | 4.53 |
| High foliage nesters | 1.82 |
| Ground and low foliage nesters | 2.63 |

*Source:* Modified from Martin (1988b).

mining overall community densities (Fretwell 1972, Tomiałojć and Profus 1977). The actual importance of predation in any given situation thus depends on a wide variety of factors.

The effects of predation on birds are most evident during the breeding period. Because activities are centered on the nest, the disappearance of eggs, nestlings, or attending adults is usually obvious and readily attributable to predation (Collias and Collias 1984). The magnitude of nest losses to predation varies considerably (Ricklefs 1969), but it may be extremely high, especially in the tropics (Skutch 1985). In a given habitat, predation upon open-nesting species generally is greater than on cavity-nesters, and ground-nesters tend to suffer more than species nesting higher in the canopy. This produces considerable differences in predation vulnerability among members of a community or guild that have different nesting habits. As a consequence, community patterns may vary. In a study in Arizona, for example, Martin (1988b) found that the slope of the species-area curve differed among guilds that were defined by nesting substrates (Table 3.1). Martin argued that, if one assumes that predation risk is greater in small forest blocks than in larger tracts because of the increased proportion of edge habitat, then the guilds that are most vulnerable to predators (species nesting on the ground or in low foliage) should exhibit the lowest slopes, as predators are more effective in preventing some species from occupying small areas. Alternatively, one might expect predation to enhance diversity, in which case the more vulnerable guilds should have greater slopes. Martin did not consider this possibility, but his observations (Table 3.1) are generally in accord with the first hypothesis.

Even within a group of species having similar nesting habits, predation probabilities may differ among species. Over a 9-yr period in southern Sweden, for example, Nilsson (1984) found that predation (chiefly by woodpeckers) was the major cause of nest failure in several cavity-nesting species. Species that nested at greater heights, however, suffered lower losses than those occupying lower cavities. Marsh Tits were apparently relegated to low holes by competitive pressures from other species, such as Starlings, Blue Tits, and Nuthatches (*Sitta europaea*), and they incurred greater losses to predators than did the other species. Nilsson also compared nest losses from natural cavities with those from nest boxes: among the *Parus* species, losses were nearly three times greater from the boxes. This finding indicates once again that populations breeding predominantly in nest boxes may not be subject to the same sort of mortality pressures that characterize natural populations (see Chapter 1) and that predation susceptibility may vary within a species in accordance with nest-site or habitat selection. Nilsson reported that in several species individuals that nested higher were preyed upon less often than those occupying lower nest sites. Elsewhere, Dickcissels (*Spiza americana*) breeding in native prairies in Kansas suffer lower nest predation than individuals breeding in nearby old-field habitats, primarily because the latter habitat supports more predators (snakes) (Zimmerman 1984). Tomiałojć (1979), Gibo *et al.* (1976), Angelstam (1986), and Møller (1987) also reported large differences in predation losses by populations of a species occupying different habitats, perhaps due to differences in the complexes of predators present in the areas.

Such differences in predation patterns between habitats types may be an important determinant of community patterns. Indeed, Martin (1987) has suggested that, if predators respond to nests in a density-dependent fashion and can specialize on nest types (both of which are known to occur), then predation pressures may be a major selective force promoting partitioning of nesting heights or microhabitats by species with similar nest types. By segregating nesting areas, such species can reduce the functional response of nest predators or diminish the possibilities for predator foraging by local enhancement. Martin's preliminary analyses indicate that species with similar nesting habits are dispersed more evenly in the vertical vegetation profile than would be expected on the basis of their foraging behavior, lending greater support to the predation hypothesis than to a competition hypothesis. Martin's hypothesis merits broader tests.

Both the abundance and diversity of predators and the abundances and types of species on which they prey usually differ between situations, and comparisons of predation effects are therefore difficult. One way to standardize such comparisons is to use artificial nests with the same kinds of eggs placed in the locations to be compared. Such experiments have been conducted to explore differences in predation levels between habitat types

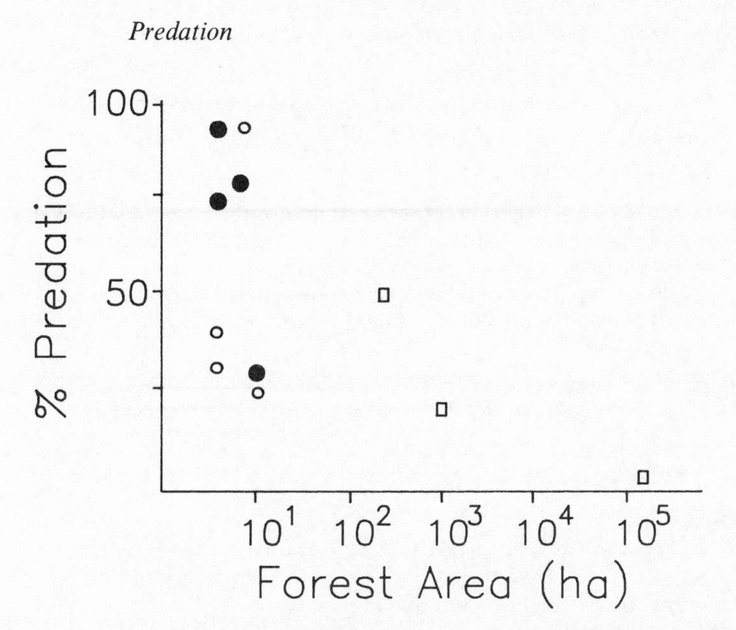


Fig. 3.3. Predation rates on experimental nests in forest tracts of different sizes in eastern USA. Solid circles = forests located in urban areas; open circles = forests in rural areas; squares = large forest tracts. After Wilcove (1985).

or between island and mainland locations. Wilcove (1985), for example, used such procedures to test the proposition that nest predation has been a major factor causing the decline of breeding populations of migratory songbirds (especially neotropical migrants) in small woodlots in eastern North America (Robbins 1980, Ambuel and Temple 1983). He found that predation rates on experimental nests were indeed considerably greater in small forest fragments than in large tracts (Fig. 3.3), primarily because predator densities were greater in the smaller areas. Small woodlots located close to urban areas had higher predation rates than areas of similar size in more rural settings (Fig. 3.3). Ground nests were also considerably more vulnerable than those placed 1–2 m up, and open-cup nest types suffered greater predation than simulated cavity sites. In another study, artificial ground nests were placed at varying distances from the edges of plots in several habitat types. The predation rate on these nests did not vary systematically with distance from the plot edge, although nests in clear-cut plots were disturbed less than those in mature woods (Yahner and Wright 1985). On Mt. St. Helens, Washington, on the other hand, experimental nests in an area totally devastated by a volcanic eruption suffered nearly an order of magnitude greater predation than nests in a less spectacularly disturbed timber blowdown area (Anderson and MacMahon 1986).

Table 3.2. *Results of experiments in which artificial nests were placed in habitats on the Baja California mainland and on an island (Coronados Island) in the Sea of Cortez.*

In Experiment 1, nests were placed in locations similar to those used by Black-throated Sparrows for nesting; each artificial nest contained two House Sparrow (*Passer domesticus*) eggs and was checked every 3 days. In Experiment 2, nests were visited every 6 days and naphthelene was spread around half of the nests to deter mammalian predators. All nest failures were attributable to predation.

| | Experiment 1 | | Experiment 2 | |
|---|---|---|---|---|
| | Successful[a] | Failed[b] | Successful | Failed |
| Island | 16 | 0 | 9 | 1 |
| Mainland | 4 | 12 | 1 | 9 |

*Notes:*
[a] Number of nests that were not preyed upon.
[b] Number of nests in which at least one egg was removed by predators.
*Source:* From George (1987).

Differences in predation rates between island and mainland situations have also been tested using experimental nests. On Barro Colorado Island in the Panama Canal, populations of small mammals and snakes have flourished since the island was formed during construction of the canal, and they occur there at greater densities than on the adjacent mainland. This has been attributed to reduced human hunting pressure on the island (which is a reserve) and to the absence of large predators (Karr 1982a). Predation on small birds might therefore be expected to be greater on the island than the mainland. Loiselle and Hoppes (1983) tested this hypothesis by placing open and closed artificial nests in rainforest habitat in both areas. Over the 2-day test period, 35% of the island nests suffered predation, compared with 4% of the mainland nests. Losses from open nests were generally greater than from closed nests, and elevated nests suffered less predation than nests placed on the ground. George (1987) conducted similar tests on an island and mainland location in the Sea of Cortez. There, densities of several songbirds are greater on the island (see Volume 1, Chapter 11), and one possible contributor to this pattern might be reduced predation on the island. Predation was responsible for most nesting failures in both areas, and overall nesting success of Black-throated Sparrows (*Amphispiza bilineata*) was lower on the mainland than on the island. George's artificial nest experiments clearly indicated a substantially greater level of predation on the mainland (Table 3.2). Nilsson *et al.* (1985) found that predation on experimental nests on small islands in a Swedish lake did not differ from that in an adjacent mainland area, although small mammals were the major nest predator on the mainland versus Carrion Crows on the islands.

Overall, these artificial-nest experiments provide some interesting insights on apparent differences in predation rates among areas or species. Because the experimental nests often do not match natural nests in either construction or placement, however, they may not provide an accurate measure of true predation rates. Despite their seeming greater conspicuousness, predation on artificial nests may actually be lower than that on natural nests in some cases (Martin 1987, Willebrand and Marcström unpublished), perhaps because the artifical nests are more attractive to avian predators whereas natural nests generally suffer greater mammalian predation (Andrén *et al.* 1985, Angelstam 1986, Sugden and Beyersbergen 1986, Storaas 1988).

Although it is usually less obvious than nest predation, predation during winter may also have important effects on bird populations and community patterns. Ekman (1986) determined the predation patterns of Pygmy Owls (*Glaucidium passerinum*) on birds in a boreal forest in Sweden over five winters by analyzing the contents of regurgitated pellets. The owls fed more often on small species that were energetically less profitable (e.g. Goldcrests (*Regulus regulus*), Coal Tits) than would be expected on the basis of the abundances of these species and took fewer larger, more profitable prey (e.g. Great Tits) than expected. Ekman related these differences to differences in the vulnerability of the prey, the species occurring in the outer branches of the trees being more exposed to owl predation than those frequenting the inner branches or the trunks. Some of the differences in the foraging niches of these species (e.g. Alatalo 1982) may therefore be related to differences in predation vulnerability and the importance of seeking cover.

In these and many other examples, the loss of individuals to predators is clear. What is less clear is whether or not such losses have significant consequences on population dynamics. The difficulty of linking nest predation to population dynamics is illustrated by events in Wytham Woods in England, where Lack, Perrins, and their colleagues have been studying tit population dynamics for decades. During most of this time, the numbers of Sparrowhawks (*Accipiter nisus*) in the woods were quite low due to the use of pesticides in the area. With the cessation of pesticide use, the hawks again became common, and 6–8 pairs have nested in the woods in recent years. The hawks prey upon adult tits during the nesting period and capture many young birds during the weeks following fledging; Perrins and Geer (1980) estimated that 18–34% of the young are taken by the time the young of the hawks have become independent and the predators shift to other prey. Despite such a clear change in the level of predation intensity on the tits, Perrins and Geer were unable to demonstrate that increased hawk numbers were associated with reduced population densities of tits, although there is evidence that the dispersal patterns and age structure of the tit populations changed.

Variations in predation rates on bird populations may depend on the population levels of alternative prey species as well as the densities of predators and of primary prey. In the English woods, for example, weasels (*Mustela nivalis*) may take up to $\frac{2}{3}$ of the tit young hatched from nest boxes in some areas in some years, but losses are slight in other years. Over several years, the predation rate on tits by weasels was inversely related to small rodent densities at the same time (Dunn 1977, Perrins 1979). Mustelids also prey on hole-nesting passerines in northern Finland, but there such predation occurs only in years of quite low small rodent densities (Fig. 3.4; Järvinen 1985). Järvinen suggested that the mustelids increase numerically with increasing rodent densities and then turn to small passerines as prey when the rodent populations crash. No predation on hole-nesters was recorded following the crash in 1976, however (Fig. 3.4). Hole-nesting species were especially scarce then, and the predators may not have formed a search image for prey in nest boxes.

In boreal Fennoscandia, synchronous population fluctuations of several bird and mammal species are associated with varying predation pressures prompted by changes in availability of cyclic voles, the predators' primary prey (Angelstam *et al.* 1984). In more southerly Scandinavian locations, voles are less cyclic and a greater diversity of alternative prey is available to generalist predators. As a consequence, numbers of predators fluctuate less and pressures on the alternative prey species are greater and more even over time (Erlinge *et al.* 1983, 1984). There is no synchrony in population changes of the alternative prey (Angelstam *et al.* 1985).

Andrén *et al.* (1985) tested the proposition that the disappearance of synchronous cycles in prey populations in the south was due to the increased predation pressure by placing artificial grouse nests in three locations on the gradient from noncyclic to cyclic regions. After 8 days of exposure, 6% of the nests in the north had suffered losses to predators, versus 56% in the south; the differences in another year were similar but less dramatic. The patterns of predation were correlated with the abundance of corvids, which are major nest predators. Corvid abundance, in turn, was

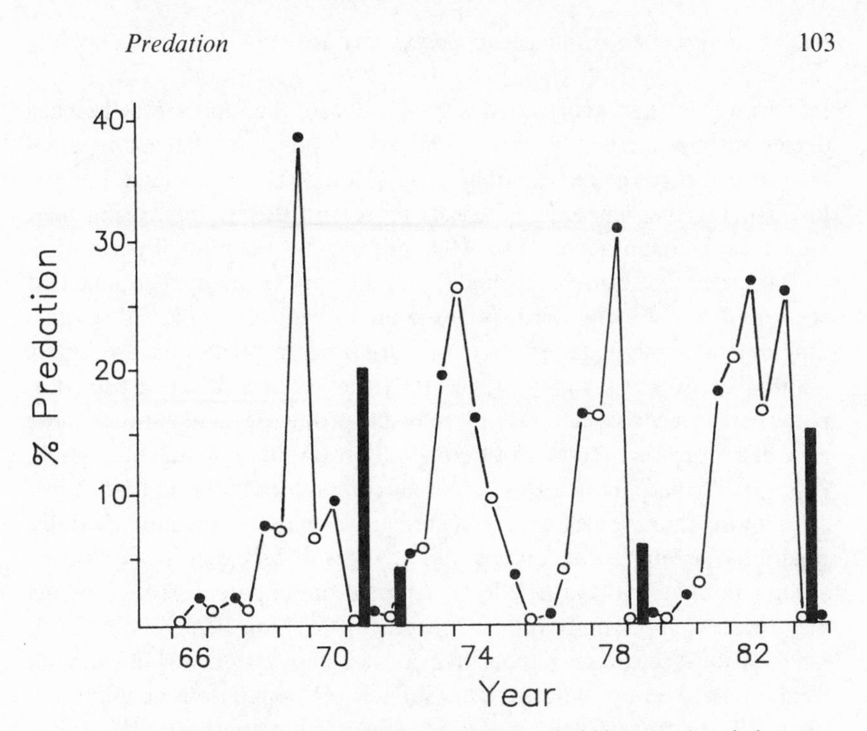


Fig. 3.4. Predation rate (robbed nests in %, vertical histograms) on hole-nesting passerines by mustelids in Finland. The relative densities (individuals/100 trap nights) of small rodents in mid-June (open circles) and in early September (solid circles) in a nearby area in 1966–84 are also shown. After Järvinen (1985).

positively related to agricultural development, the degree of forest fragmentation, and human density.

The pattern of increased predation on birds such as grouse during low phases of the vole cycle in northern areas is consistent with the hypothesis that large populations of vole predators turn their attention to birds as alternative prey at such times. Widén *et al.* (1987), however, suggested an alternative hypothesis: at population peaks the voles may deplete the vegetation, leading to a reduction of food available to herbivorous birds following the vole crash. Breeding hen grouse might then be in poorer condition and would be forced to leave their nests more often to forage, increasing their risk of predation. Widén *et al.* tested this hypothesis by analyzing patterns of Goshawk (*Accipiter gentilis*) predation on female grouse in Swedish boreal forests. Because Goshawks do not feed on voles, their predation on grouse should not be driven by changes in vole availability as food. Indeed, the proportion of female grouse in the diets of the raptors was inversely correlated with vole density, supporting the herbage depletion hypothesis. One could also argue, however, that mammalian predators turned their attention to grouse nests when vole abundance was low, forcing the hens off their nests more and thereby increasing their vulnerability to Goshawks. The two hypotheses are not mutually exclusive.

These examples show that predation may vary in intensity in time and space and that different bird species in an area may be subjected to quite different rates of predation. If these variations in predation rates affect population sizes of the prey species or their microhabitat use or behavior, patterns of species relationships may be produced that parallel those expected from the operation of competition (Holt 1984, Schluter 1984). Holt (1977, 1984) termed this 'apparent competition' (a term that obfuscates more than it clarifies). Depending on whether predators exhibit functional or numerical responses to changes in prey densities or switch among alternative prey, the effects of predation on potentially competing prey species may be quite varied (Murdoch 1969, Holt 1977, Gilpin 1979).

Predation is seemingly important in at least some instances, but it has not been found to be a significant factor influencing populations or communities in all situations where it has been considered. Abbott *et al.* (1977), for example, found no clear patterns of variation in finch abundance or diversity in relation to the number of predator species present on islands in the Galápagos, and Nilsson *et al.* (1985) documented no differences in overall predation rates on islands in a Swedish lake and on the nearby mainland. Landres and MacMahon (1983) concluded that the effects of predation on the community patterns they documented in oak woodlands were 'uncertain'. The possible effects of variations in predation rates on community patterns have not been considered very often, however, so it is difficult to gauge the overall importance of this process in bird communities. To conclude that 'predation cannot be regarded as being of primary importance either in directly determining species composition . . . or in preventing competitive exclusion' and that it 'might be a proximate controlling factor of species richness in some instances, but competition must still be operating as the ultimate structuring mechanism' (Giller 1984: 78) is premature until the effects of predation have been considered more explicitly in a larger number of studies.

### Parasitism

Birds are subjected to two different sorts of parasitism, the conventional array of internal and external parasites that infect most vertebrates, and brood parasites, other species of birds that lay eggs in the nests of host species. To the extent that either of these forms of parasitism influences the behavior of individuals or the dynamics of populations, it has the potential to alter niche patterns and the distribution and abundance of species in communities. There is little direct evidence of such effects, but these effects are difficult to demonstrate and parasitism has not received much attention as a process influencing community features (but see Price *et al.* 1986, Holmes and Price 1986). Several examples show that, under certain circumstances, brood or internal parasites may be of considerable importance.

***Brood parasitism***

Many brood parasites remove host eggs when they deposit their own eggs in host nests, and in most parasitic species the nestlings either aggressively evict host chicks from the nest or grow more rapidly, diminishing growth rates and survival of host young. Brood parasitism thus usually reduces the reproductive output of parasitized host individuals, an effect that can vary between habitats for a single species. Parasitism of Dickcissels by Brown-headed Cowbirds (*Molothrus ater*) is greater in prairie habitats in Kansas than in nearby old-field habitats, for example. These effects are offset by the greater predation pressure in the old fields, however, and birds in the two habitats actually realize equivalent per-nest production (Zimmerman 1984).

An indication of the importance of brood parasitism on the population dynamics of the hosts is provided by the various anti-parasite behaviors that have evolved in many host species (Friedman 1929, Welty 1982). Potential host species differ in their susceptibility to parasitism, however. In North America, Yellow Warblers (*Dendroica petechia*) frequently build replacement nests on top of nests that are parasitized by Brown-headed Cowbirds, reducing the effects of the parasitism. Kirtland's Warblers (*Dendroica kirtlandii*) lack any anti-parasite behaviors and are heavily parasitized, to the extent that cowbird parasitism has been identified as a possible cause of the population decreases of this species in recent decades (Mayfield 1978). Decreases in the abundance of Golden-cheeked Warblers (*D. chrysoparia*) in Texas have also been attributed to cowbird parasitism (Oberholser 1974, Pulich 1976), as has the extirpation of Bell's Vireo (*Vireo belli*) from central California (Gaines 1974, Goldwasser *et al.* 1980).

The effects of brood parasites on host populations are especially clear where the parasite has recently expanded its range to bring it into contact with new host species, as is the case in central California and the Sierra Nevadas (Gaines 1974, Rothstein *et al.* 1980, Verner and Ritter 1983). In the Sierras, relative abundances of Warbling Vireos (*Vireo gilvus*) are negatively correlated with cowbird abundance, suggesting an inhibitory effect of the brood parasite on population levels (Verner and Ritter 1983). An especially clear example of such relationships comes from the Caribbean, where Shiny Cowbirds (*Molothrus bonariensis*) have recently expanded their range from South America into the West Indies. Some of the bird species in recently invaded areas such as Puerto Rico are aggressive toward the cowbirds and suffer little parasitism, but many of the species lack such defenses and are quite susceptible. Yellow-shouldered Blackbirds (*Agelaius xanthomus*), for example, were subjected to parasitism at 94% of the nests that Cruz *et al.* (1985) examined in Puerto Rico, whereas the closely related Yellow-hooded Blackbird (*A. icterocephalus*) in Trinidad (where the cowbird has long been present) sustained only 45% parasitism. Cruz and his colleagues reported that 61% of the resident passerine species on Puerto Rico were parasitized by the cowbird, in comparison with parasitism rates of 20–40% in South America. Many of the host species suffer reduced nesting success as a consequence of the parasitism. In Puerto Rico, for example, parasitized nests on average hatched 12% fewer eggs and fledged 67% fewer of their own chicks than nonparasitized nests (Wiley 1985). Post and Wiley (1976, 1977) suggested that the abundance of the Yellow-shouldered Blackbird may be limited by the effects of cowbird nest parasitism.

It is clear that brood parasitism affects the species in these island communities unequally and that changes in the relative abundances of the species may follow as a consequence. Such changes may affect competitive relationships among species and produce changes in niche patterns as well (Fig. 3.2). The population effects of brood parasitism may depend on a variety of factors (May and Robinson 1985), but if parasitism leads to the extinction of local populations, it will certainly affect other community patterns, such as species-area relationships, species-abundance distributions, or distributional patterns.

***Internal parasitism***

The possible effects of internal parasites on population and community patterns in birds are less clear, largely because few studies of parasites of birds have been conducted in an ecological context. Bethel and Holmes (1973, 1977) demonstrated that the behavior of amphipod prey of surface-feeding and diving ducks is altered when the amphipods are parasitized by acanthocephalan worms. The effect of the behavioral changes is to make those amphipods that serve as intermediate hosts for acanthocephalans that have diving ducks as definitive hosts more susceptible to predation by these ducks. The amphipods that carry worms that have surface-feeding ducks as definitive hosts behave in a way that enhances their likelihood of being eaten by surface-feeding birds. Moore's (1984) studies showed similar effects of behavioral changes in parasitized isopods; here Starlings were the definitive host. These studies demonstrate that the diet selection of a species may be shifted if some of its prey are parasitized, and diet niche relationships among species or dietary changes between areas may be sensitive to such effects. They also raise the prospect that other behavioral differences between different bird species or between individuals of the same species in different areas might be caused by parasites (see Milinski 1984, 1985), although at present there is no direct evidence of such influences.

Perhaps the clearest demonstration of the effects of an internal parasite on bird population and community patterns is provided by studies of avian malaria on the Hawaiian Islands (Warner 1968, Laird and van Riper 1981, van Riper *et al.* 1986). The mosquito *Culex quinquefasciatus* was accidentally introduced to the islands in 1826 and spread rapidly to lowland areas of all of the high islands. It is the vector for *Plasmodium relictum capistranoae*, which causes malaria in both native and introduced bird species and which also is not native to the Hawaiian Islands. Warner (1968) examined the distributional patterns of the parasite and vector in relation to the distributions of native and introduced birds and concluded that the extant native birds are restricted to elevations above 600 m because they are highly susceptible to malaria. Warner's surveys suggested that mosquitoes are largely restricted to lower elevations, and therefore the incidence of malaria is greater there. Introduced bird species are largely resistant to malaria and are chiefly distributed below 600 m elevation. Warner attributed the extinction of several native honeycreepers to malaria. The parasite appears to have changed the composition of the avifauna of the islands and caused an altitudinal restriction of the endemic bird species that seems to be largely complementary to that of the exotic species. This is an especially neat and clear example of community patterns that one might attribute to competition but that instead are apparently produced by an entirely different process.

The situation is actually more complex than Warner indicated, however, van Riper *et al.* (1986) conducted intensive studies over 3–4 yr on elevational transects in dry and wet forest habitats on Mauna Loa (Hawaii) and found that larvae and pupae of the mosquito vector were present from sea level to 1500 m elevation throughout the year and occasionally were found at even higher elevations. Adult mosquitoes were present at greatest

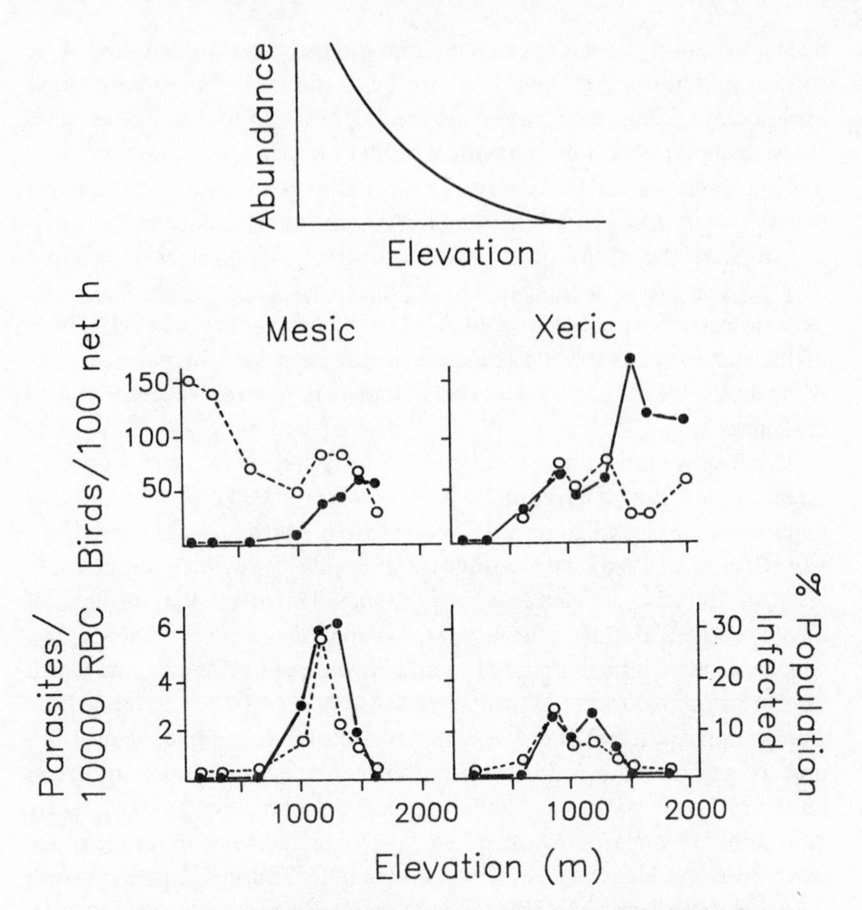


Fig. 3.5. Distribution of mosquito vectors of avian malaria on elevational transects on Mauna Loa, Hawaii (top) in relation to the distribution of native (solid circles) and introduced (open circles) bird species (middle) and the incidence of the malarial parasite in birds (as measured by number per 10 000 red blood cells (solid circles) and the percentage of the bird population infected (open circles)) in mesic and xeric forests. Modified from van Riper *et al.* (1986).

densities at the lower elevations and were sparse at higher elevations, but the distribution clearly extended well above 600 m (Fig. 13.5). The distributional limits of native and introduced bird species likewise were not as sharp as Warner suggested. Native species were restricted to elevations above 1000 m in the mesic forest but ranged lower in the xeric habitat; in both habitats, densities increased with greater elevation (Fig. 3.5). In the mesic habitats, densities of introduced species were greatest in the lowlands and gradually decreased with increasing elevation; they were more or less evenly distributed throughout the elevation range in the xeric woodlands (Fig.

3.5). Contrary to the pattern reported by Warner, breeding populations of the mosquito vector were present throughout much of the extant native bird habitat, and infected birds were found at all elevations. The incidence of the malarial parasite, however, was much greater in the 900–1500-m range, where distributions of the vector and the native birds overlapped most strongly. Malarial incidence was also considerably greater in the mesic forests (Fig. 3.5).

These patterns make sense if the native species are susceptible to the parasite and the exotic species are refractory, as Warner suggested. Van Riper *et al.* conducted laboratory tests to determine the levels of immunity in a number of native and introduced species. In general, the native species had lower levels of immunity than did the introduced species, some of which were totally refractory to infection. The duration of the primary attack period of the parasite was also three times longer in the native species. In both groups, however, there was between-species variation in infection rates, and among the native species the degree of susceptibility was closely related to the present-day abundance of the species. The Iiwi, for example, was the most susceptible of the native species from the main islands, and it has also undergone the greatest population declines since the turn of the century; the most resistant of the native birds, the Common Amakihi, is still abundant throughout the islands. There are differences in resistance between populations of amakihi, however. Birds from Mauna Loa mesic forests are highly resistant to the parasite, whereas those from xeric habitats on Mauna Kea are extremely susceptible. Populations in both areas are quite sedentary, and malaria is absent from the dry forests on Mauna Kea. Apparently, the birds on Mauna Loa, which are more or less continuously exposed to malaria, have developed immunogenetic capabilities absent from the Mauna Kea population, and there is little genetic interchange between the two populations to erode these local differences.

Mesic forests occupy a larger area on Hawaii and have greater insect standing crops and nectar production rates than do xeric forests (Mueller-Dombois *et al.* 1981). One would therefore predict that bird densities should be greater in mesic forests at any given elevation. For the native species this is not the case (Fig. 3.5); both Iiwi and Apapane, for example, are 'wet-forest birds' (Berger 1981), but on the transects surveyed by van Riper *et al.* (1986) as well as in the more extensive surveys of Scott *et al.* (1986) these species were most abundant in the xeric habitats.

Many of the native species are nectarivores, and nectar production during the autumn is greatest at lower elevations. To obtain maximum quantities of nectar, birds should therefore move to these areas at this time.

In so doing, they encounter an expanding vector population and the potential for malarial transmission is enhanced. Not unexpectedly, the incidence of malaria among the native species is greatest at this time. Some of the species, however, have apparently developed behavioral mechanisms for reducing contact with the vectors. Both Apapane and Iiwi undertake daily elevational movements, gradually moving downslope during the day and then, just before dusk, assembling from the lower elevations (*ca.* 1200 m) to migrate upslope to common roosting areas at *ca.* 1700 m elevation (MacMillen and Carpenter 1980). The timing of these movements brings the birds into the lower elevations when the daily activity levels of the vector are lowest, and the birds leave these elevations to return upslope before the vectors again become active.

Although the evidence is largely correlative, it appears that both the elevational distributions of the native bird species and their relative abundances in mesic and xeric habitats are influenced to a major extent by the malarial parasite. Because the species differ in their susceptibility to the parasite, however, the extent of this influence varies among species. The introduced species, being largely resistant to the parasite, are distributed more or less independently of it and reach greatest densities in the lowland mesic forests where food resource levels are generally greatest. The partial elevational 'exclusion' between the groups appears to be more strongly mediated by the effects of parasitism than by interactions among the bird species.

This is not to say, however, that levels of parasitism are the only factor influencing distributional patterns of the native Hawaiian species. Habitat availability is clearly an important factor, and historical changes in habitat distributions (especially of native flora) have led to changes in the distributions of several bird species (Scott *et al.* 1986). In addition, several of the species are physiologically adapted to particular climatic circumstances, and this may limit distributions. The native Palila (*Psittirostra bailleui*), for example, occupies cool montane habitats at high elevations and has an upper critical temperature that is 7°C lower than that of the congeneric Laysan Finch (*P. cantans*), which is confined to a low, treeless atoll (Weathers and van Riper 1982). The Palila may thus be physiologically restricted to the cool higher elevations. This was strikingly demonstrated when the captive individuals that van Riper had used in the physiological tests were transferred to the Honolulu Zoo at the conclusion of the work. Following 3 days of temperatures in excess of 32°C, all of the birds died of heat stress (van Riper personal communication). Weathers and van Riper (1982) had predicted 31°C as the upper temperature threshold for the species.

**Commensal and mutualistic interactions**

Birds may also engage in commensal or mutualistic interactions with other species, although these are generally even less obvious than predator–prey or parasite–host interactions and thus are even less studied. The formal distinction between commensalism and mutualism breaks down in practice, and I combine them here to include interactions in which one of the species benefits; the other may also, but in either case it suffers no negative effects.

Some cases of mutualism are fairly clear – relationships between hummingbirds and flowers pollinated by the birds are one example (Chapter 2; Grant and Grant 1968). Some instances of bird–insect mutualisms are also fairly clear. Smith (1968) described a complex set of interrelationships between Chestnut-headed Oropendolas (*Psarocolius wagleri*), Giant Cowbirds (*Scaphidura oryzivora*), botflies, and wasps in Panama. The cowbirds are brood parasites of the oropendolas, yet individuals in some colonies of oropendolas do not resist cowbird intrusions and nest parasitism whereas individuals in other colonies are highly aggressive to the cowbirds and suffer little parasitism. Smith found that oropendola chicks are frequently infested with botfly larvae. Nests containing cowbird young fledge more host chicks than do unparasitized nests in the same colony, because the cowbird nestlings groom the host nestmates, removing the botflies that otherwise would cause high chick mortality. The relationship between the brood parasite and the host in these colonies is clearly mutualistic. Why, then, do some host colonies actively discourage an association with the cowbirds? Smith discovered that all of these colonies were located in trees containing wasp nests. The wasps reduce botfly activity about the nests, removing a major source of nestling mortality. Under these conditions, the behavior of the oropendolas toward the cowbirds is radically different, and the behavior of the cowbirds changes as well. A similar, though less complex, relationship between nesting Bananaquits (*Coereba flaveola*) and wasps was described by Wunderle and Pollock (1985). In this situation, the birds associated with wasps suffered significantly lower nest predation than did birds without such associates. Similar bird–insect nesting associations have been described for species in a diverse array of families (Wunderle and Pollock 1985).

There are fewer documentations of commensal or mutualistic relationship between bird species. In areas where Yellow-bellied Sapsuckers (*Sphyrapicus varius*) occur in North America, however, a variety of mammals and birds are attracted to their drilled holes, which exude large amounts of tree sap (Foster and Tate 1966). Hummingbirds are especially frequent at these drillings. Tree sap is similar to floral nectar in its chemical composition; by exploiting sapsucker drillings, hummingbirds might therefore be able to occupy areas in which nectar is scarce or does not appear until well after the birds have arrived in the spring. Miller and Nero (1983) suggested that this commensal relationship with sapsuckers has permitted Ruby-throated (*Archilochus colubris*) and probably Rufous hummingbirds to occur in high-latitude areas of North America. In Norway, species nesting in association with colonies of Fieldfare (*Turdus pilaris*) may be less susceptible to predation by Hooded Crows than are birds nesting apart from such colonies (Slagsvold 1980). Similar benefits from nesting in association with species that mob predators have been found among species in larid colonies.

In most cases, determining that mutualistic interactions have population consequences is reasonably straightforward – that, after all, is a necessary part of determining that the relationships *are* mutualistic. Extending these consequences to the community level is more difficult. If a species benefits from nesting in association with wasps, the presence or absence of wasp colonies becomes an important factor in that species' habitat selection and may therefore influence its habitat relationships with other bird species. The dynamics of hummingbird communities may be closely tied to the distribution and phenology of the flowers pollinated by different species (Chapter 2). In the situation studied by Slagsvold, passerine community diversity was increased in areas lacking Fieldfares and thus subjected to greater predation pressure. In other situations, indirect (coevolutionary) mutualism may result when a predator exploits the avoidance behavior evolved by prey against another predator to increase its own feeding efficiency (Sherry 1984). Although they may be relatively infrequent, such interactions may influence community patterns where they do occur.

### Disturbance and chance

The effects of these processes and environmental factors on the distribution and abundance of species vary in time and space, largely as a result of disturbances (Fig. 3.1). Broadly viewed, a disturbance is 'any relatively discrete event in time that disrupts ecosystem, community, or population structure and changes resources, substrate availability, or the physical environment' (White and Pickett 1985: 7). Disturbances are responsible for a change in the state of a system, and systems that are not in equilibrium may therefore be disturbed just as readily as equilibrial systems (Sousa 1984).

Table 3.3. *Characteristics of disturbances in ecological systems.*

| Descriptor | Definition |
|---|---|
| Distribution | Spatial distribution, including relationship to geographic, environmental, and community gradients. |
| Frequency | Mean number of events per time period. Frequency is often used for probability of disturbance when expressed as a decimal fraction of events per year |
| Return interval, cycle, or turnover time | The inverse of frequency; mean time between disturbances |
| Rotation period | Mean time needed to disturb an area equivalent to the study area (the study area is arbitrarily defined; some sites may be disturbed several times in this period and others not at all – thus, 'study area' must be explicitly defined) |
| Predictability | A scaled inverse function of variance in the return interval |
| Area or size | Area disturbed. This can be expressed as area per event, area per time period, area per event per time period, or total area per disturbance type per time period. Frequently given as a percentage of total available area |
| Magnitude | |
| Intensity | Physical force of the event per area per time (e.g. heat released per area per time period for fire and windspeed for hurricanes) |
| Severity | Impact on the organism, community, or ecosystem (e.g. basal area removed) |
| Synergism | Effects on the occurrence of other disturbances (e.g. drought increases fire intensity and insect damage increases susceptibility to windstorm) |

*Source:* From White and Pickett (1985).

Major, catastrophic disturbances of systems such as fires or floods are easily recognized, but many disturbances are more subtle. If the density of a predator population is reduced by the introduction of a pathogen, the pressures that it places on populations of primary and alternate prey may be changed, resulting in changes in their population sizes and interrelationships (Fig. 3.2). The disturbance (the sudden appearance of the pathogen) has a direct effect on the predator but, in addition, it has indirect effects on other populations.

Disturbances vary not only in magnitude, but in their frequency, predictability, spatial distribution, and duration (Table 3.3; Sousa 1984, White and Pickett 1985). Populations of different species also vary in the time they require to respond to a disturbance and to recover from it.

Because virtually all environments and ecological systems are subject to frequent disturbance of at least some of their components, they exhibit a shifting, dynamic mosaic of patches in time and space. The occurrence and relative abundances of species and the operation of various processes may thus differ considerably between patches (Wiens 1976b, 1985b, den Boer 1981, Pickett and White 1985). Because disturbance may be expressed on a variety of scales, it may produce extremely complex effects. It is little wonder that most theory ignores disturbance or considers it a random process.

Birds may be less sensitive than some other organisms (especially sessile ones) to small-scale disturbances. Most birds are highly mobile and can move to avoid localized disturbances (Cody 1981, Karr and Freemark 1983). During a bush fire in open *Eucalyptus* woodland in Western Australia, for example, resident family groups of Splendid Blue Wrens (*Malurus splendens*) were seen flying from their territories into nearby unburned areas just ahead of the advancing flames. Some birds returned to their territories in the still smouldering bush within hours, and all of the groups had reoccupied former territories within several days (Rowley personal communication). In other situations, however, birds may not respond to disturbances immediately. In our studies of breeding birds in North American shrubsteppe habitats, we have observed that densities of territorial individuals may remain unchanged or change only slowly in response to major changes in habitat structure produced by rangefire (Rotenberry and Wiens 1978) or physical removal of shrubs (Wiens 1985b, Wiens and Rotenberry 1985, Wiens *et al.* 1986). There may be similar time lags in responses to flooding in riparian habitats (Knopf and Sedgwick 1987).

The combination of time lags in responses to disturbances with mobility may confer on bird populations and communities a degree of resilience to disturbance. Wright *et al.* (1985) argued that this may lead to greater predictability in the composition and structure of bird communities. They compared islands of the Pearl Archipelago off Panama to derive community patterns differentiating small from large islands and found that the patterns were persistent despite disturbances such as fire, windstorms, and drought that occurred between 1971 and 1978. They concluded from this that the communities were predictable and highly structured, that they were governed by deterministic processes, and that disturbance was of minimal importance. The conclusions, however, were based upon the persistence of general correlative patterns across island types and upon censuses conducted with mist nets at different times of year on different islands in different years by different observers over periods of a few days to 2 weeks. Only three sites were actually censused twice, and the effects of disturbance were not actually measured directly. The conclusions drawn by Wright *et al.* are premature.

In other situations in which the effects of disturbances have been considered less inferentially, clear influences have been found. The fragmentation of large forest tracts into smaller woodland patches, for example, is often accompanied by the loss of forest-interior species (especially neotropical migrants) and by increases in the density or occurrence of edge species (Whitcomb *et al.* 1981; see Chapter 5). The isolation of Barro Colorado Island from the mainland by the construction of the Panama Canal had major effects on guild composition and diversity (Willis 1974, Karr 1982a). In prairie habitats, grazing alters habitat structure and is accompanied by changes in the composition of breeding bird communities (Risser *et al.* 1981). Logging of forests has profound effects on both the distribution of species and their behavior, changing foraging niche relationships among species (Szaro and Balda 1979). On a smaller scale, the formation of gaps in forest canopies by treefall creates local patches that are often occupied by different species than occur in an unbroken forest (Schemske and Brokaw 1981, Wiens 1985a, Wunderle *et al.* 1987). The list could easily be extended.

Climatic variations may also act as disturbances. The work of Grant and his colleagues on the Galápagos Islands has documented the effects of droughts on diets, niche relationships, population densities, survivorship, and morphology of the finches (Boag and Grant 1981, 1984, Price *et al.* 1984, Grant 1985, Gibbs and Grant 1987a, b, c). Droughts have been a recurrent phenomenon in the history of the Galápagos, and Grant (1985) suggested that these climatic disturbances have had unequal effects on the distribution and abundance of different finch species, perhaps leading to the extinction of some species from some islands. Cavé (1983) and den Held (1981) attributed variations in the survival rates of Purple Herons (*Ardea purpurea*) in The Netherlands to drought conditions in the wintering areas in the Sahel.

Intense storms or unusually wet conditions may also alter population levels or community composition. A hurricane in southern Florida in 1935, for example, completely destroyed a population of Cape Sable Seaside Sparrows (*Ammodramus maritimus miribilis*) (Bent 1968), and a severe tropical storm probably eliminated the Puerto Rican Bullfinch (*Loxigilla portoricensis grandis*) from St. Kitts (Raffaele 1977). Laysan Island originally supported an endemic subspecies of the Apapane (*Himatione sanguinea freethii*), but the population declined following the introduction of rabbits (which destroyed the vegetation) and the few remaining birds 'perished during a three-day gale that enveloped everything in a cloud of swirling sand' (Wetmore 1925).

On the Galápagos, wet years have occurred 1 of every 4 yr on average during this century; droughts are less frequent (1 in 8 yr; Grant 1985). The wettest years have been associated with the El Niño Phenomenon, in which ocean-circulation patterns off the coast of Peru change, bringing unusually warm water of low salinity into the area (see Barber and Chavez 1983, 1986, Cane 1983, Rasmusson and Wallace 1983). In wet years, seed production is high and recruitment and survival are correspondingly good; during El Niño events, breeding activity may be extended over most of the year (Grant 1986a, Gibbs and Grant 1987c, Grant and Grant 1987). It is likely that during such years the finch populations may not be food-limited at any time of year. Interestingly, Lack's (1945) original description of the ecology of the finches was based largely on observations he made during 1939, an El Niño year. El Niño events may also have dramatic but quite different effects on breeding seabirds. During the 1982–83 El Niño, there was a nearly complete disappearance of breeding seabirds from Christmas Island and a total reproductive failure of those populations (Schreiber and Schreiber 1984). By 1985, some of the species had recovered to former population levels and were breeding successfully, but other species had yet to reappear on the island, despite the apparent recovery of the food resource base (Schreiber and Schreiber 1987). This El Niño event was also associated with severe reductions in breeding populations of several seabirds on the Galápagos Islands (Valle 1985, Gibbs *et al.* 1987) and in their abundances in feeding areas at sea (Duffy and Merlen 1986).

Over a longer period, disturbances may follow a directional trend, leading to systematic changes in populations and communities. During the period 1850–1970, for example, northern Europe added an average of 2.8 species and lost 0.6 species per decade per country as a consequence of increasing forest fragmentation, increased hunting pressure, habitat changes associated with agricultural practices, and habitat changes elsewhere in the species' ranges (Järvinen and Ulfstrand 1980). Different groups of species were affected differently by these changes; turnover was greatest, for example, among species derived from steppe regions. Haila *et al.* (1979; Järvinen and Haila 1984) documented similar long-term changes in the composition and relative abundances of birds in the Åland archipelago. Such temporal dynamics of communities are considered in more detail in Chapter 4. The obvious consequence of the differing responses of species to an assortment of disturbances occurring over time is that the web of potential interactions in the community is quite dynamic.

Disturbances increase the heterogeneity of environments and of the communities that occupy them. These temporal and spatial variations are expressed over a spectrum of scales, and they have important theoretical and operational effects on how we consider communities (see Part II). One consequence of this increased heterogeneity is an increase in both alpha and beta diversity. The mosaic structure of environments in time and space provides a variety of niche opportunities for different species and fosters local or regional coexistence. The relationship between disturbance and diversity is not necessarily linear, however. If a homogeneous system is disturbed at a moderate rate or intensity, diversity may be increased. If disturbance is frequent and intense, however, its effects on some populations may be so severe that they are unable to persist in the area, and diversity is reduced (Denslow 1985, Wiens 1985a, Wiens *et al.* 1986). Although these consequences of disturbances on community diversity are conceptually straightforward and are supported by studies of plants and sessile animals (Denslow 1985, Sousa 1979, 1984), there are no quantitative treatments of this relationship for bird communities (but see MacArthur 1975). Deriving such relationships requires that 'disturbance' be quantified, and, because the different aspects of disturbance (Table 3.2) interact and complement one another, deriving biologically meaningful measures of disturbance is difficult.

Although the overall occurrence of disturbances such as fires, windfall, or climatic extremes may in some cases be predicted for a region over some span of time, there is considerable uncertainty about exactly when or where disturbances will occur at a local scale. Disturbance thus introduces chance effects into the determination of community patterns. Different species may respond to disturbances in a characteristic, species-specific manner, so even if the occurrence of a disturbance is to some extent stochastic, the community response to a disturbance when it does occur may be largely deterministic. If different species respond in similar ways, however, changes in community patterns may be more probabilistic. For example, if a local disturbance eliminates several species from a community and the community is then recolonized, exactly which species are added may be determined not only by how well the ecological requirements of the species are met by the environmental situation but also by which species arrive first. There is thus a stochastic element to community assembly or readjustment following disturbance, as has been envisioned in the 'lottery hypothesis' of the assembly of coral reef fish communities (Sale 1978, Sale and Dybdahl 1975, Chesson and Warner 1981, Chesson 1986, Roughgarden 1986) or of trees colonizing treefall gaps in forests (Brokaw 1985, Denslow 1985).

That chance should play a part in both the local occurrence of disturbances and in population and community responses to the disturbances increases the likelihood that the patterns observed in a community at any point in space and time contain at least some stochastic effects as a result of past disturbances. Environments and the communities they contain differ in the importance of such stochastic effects or the degree to which they may be overridden by stronger or more persistent deterministic effects. The stochastic effects, however, are difficult to identify. Simply attributing unexplained variance in a statistical analysis to undefined stochastic effects is unsuitable. This variance might be associated in a strictly deterministic manner with other variables that have not been included in the analysis. Moreover, because the response of individuals or populations to changes in variables is not instantaneous, variance of this sort may be produced by time lags in responses to environmental changes. 'Unexplained' variance is therefore likely to include both deterministic and stochastic components.

**The role of history**

Whether they are deterministic or stochastic, events that occur at one time may affect the structure and composition of the community at some later time, perhaps long after the processes or factors have ceased having any direct, proximate influence. Communities have a history, and their attributes at any time bear the imprint of that history. Whether or not a species is present in a local community may be as much a consequence of features of its dispersal or sensitivity to disturbances in the past as of proximate ecological conditions (Endler 1982a). Such historical effects may range from events occurring in a previous year (e.g. the lag in recovery of populations from disturbances such as drought; Wiens 1977, 1986, Grant 1985, 1986b, Smith 1982) through changes occurring decades ago to events in some previous geological period (Davis 1986, Van Devender 1986). Järvinen and his associates, for example, have attributed some features of the present composition of bird communities in northern Europe to past environmental changes or disturbances coupled with dispersal features of the species (Järvinen and Ulfstrand 1980, Järvinen and Haila 1984, Haila *et al.* 1983). The distributions of several species in Britain have also been related to long-term climatic changes (Williamson 1975); other examples are noted in Chapter 4.

On a longer time scale, human activities occurring on islands before their colonization by Europeans may have affected the composition of present-day communities in fundamental ways. On Henderson Island in the South Pacific, for example, examination of archeological material has revealed that at least one-third of the landbird species present when the Polynesians first arrived became extinct shortly thereafter (Steadman and Olson 1985). On the Hawaiian Islands, perhaps half of the native species have suffered extinction since the arrival of Polynesians (Olson and James 1982). As a consequence of these massive eradications, such community patterns as species-area relationships may be quite different from those of even a few hundred years ago. Distributional patterns derived from modern records may also be influenced by such historical changes. Steadman and Olson found evidence that two pigeon species, one small (either *Ducula aurorae* or *D. pacifica*) and one large (*D. galeata*), were eaten to local extinction on Henderson Island by the Polynesians. *Ducula galeata* now occurs on only one island in the Marquesas, where it was thought to have evolved as an endemic. The modern distribution of *D. pacifica* is restricted to Tahiti and Makatea, whereas *D. aurorae* is widely distributed through the southwestern Pacific region. The archeological findings from Henderson Island and elsewhere indicate that both the small and the large species were formerly much more widely distributed than they are now, apparently coexisting on some islands. This example indicates that one should be cautious in interpreting distributional patterns such as checkerboards (e.g. Diamond 1975a) as equilibrial or as indicative of evolved patterns in the absence of historical information.

Long-term environmental changes may also produce historical vestiges in contemporary communities. Pregill and Olson (1981) described how the environment of the West Indies was considerably more xeric 10 000–12 000 years ago during the Pleistocene than it is now. Savannah-scrub woodlands were widespread then, and many of the bird species were affiliated with this habitat. Savannah-scrub habitat is now quite restricted, but many of the bird species in contemporary communities are nonetheless scrub-adapted species, and within the northern Lesser Antilles the same species are present on most islands. Terborgh *et al.* (1978) found such patterns 'paradoxical', but they become much clearer when the effects of Pleistocene history are considered. Among islands in the West Indies, patterns of replacements of species of congeners are frequent (Lack 1976), and there is also an historical component to these patterns. With changing conditions during Pleistocene glacial and interglacial periods, bird populations were repeatedly isolated, and both speciation and extinction were frequent. This created a mosaic of relict distributions superimposed on one another, which has been modified by extinctions and distributional shifts associated with the contraction of xeric habitats and increased island isolation since the Pleistocene (Pregill and Olson 1981). Proximate species interactions may have something to do with these distributional patterns, as Lack suggested, but the historical contributions are too great to be ignored.

The effects of Pleistocene events have also been applied in explanations of contemporary community patterns in the Andes (Vuilleumier and Simberloff 1980), in North American hardwood forests (Sabo and Holmes 1983) and oak woodlands (Landres and MacMahon 1983), and in South African woodlands (Cody 1983). The distributional patterns of birds and other organisms in the Amazon basin of South America, however, have generated special interest. There, several 'areas of endemism' or high species diversity exist within apparently continuous lowland forest (Haffer 1969, Prance 1982). Haffer (1969, 1982) and others have proposed that these distributional patches represent areas that remained as forest refugia during Pleistocene climatic variations that fragmented forest cover and caused the replacement of forest by savanna, caatinga, and cerrado vegetation in much of the region. Some populations within these refugia differentiated, leading to endemism and an increase in the regional diversity. There is a certain circularity in the hypothesis, in that former refugia are generally defined on the basis of contemporary areas of high endemism (Endler 1982b, Beven *et al.* 1984, Salo 1987). Also, some of the areas of endemism are considerably older than the Pleistocene and may instead be related to the physiographic evolution of the South American continent (Cracraft 1985b) or geomorphological differentiation among areas (Salo 1987). Haffer (1985) and Mayr and O'Hara (1986) have vigorously defended the concept against such criticisms, but the idea remains controversial. In any event, it highlights the potential importance of events on a geological time scale in influencing contemporary patterns.

Because historical events influence the patterns of phylogenetic differentiation in lineages, there is potentially a close coupling of evolutionary patterns to past geomorphological and/or ecological-climatic barriers and contemporary distributions of species. Brooks (1985) explored these relationships in a general way, and Cracraft (1986) developed them more specifically with reference to avian distributional patterns. Using cladistic analysis in conjunction with information on the contemporary distributions of species or well-differentiated subspecies in several genera of Australian birds, he demonstrated not only that the phylogenetic patterns provide insights into the genesis of the distributions, but that congruences in the phylogenetic-distributional patterns of several genera indicate areas of endemism that can be associated with known or postulated vicariant events in the past. In the genus *Psophodes*, the wedgebills and whipbirds, there are six differentiable taxonomic units (Fig. 3.6). One closely related pair, *olivaceus* and *lateralis*, is distributed disjunctly in eastern coastal woodlands. This pair is the sister-group of the *nigrogularis-leucogaster* species pair, which is discontinuously distributed in southern Australia. The third

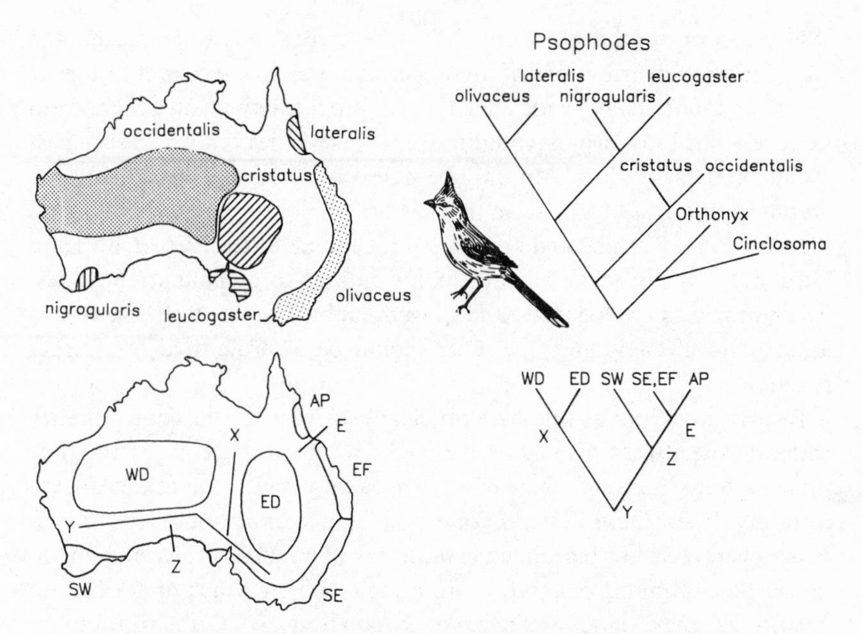


Fig. 3.6. Top: Distributions and phylogenetic relationships of differentiated taxonomic units of *Psophodes* in Australia. The phylogenetic tree is rooted by the closely related taxa *Orthonyx* and *Cinclosoma*. Bottom: Hypothesized barriers separating areas of endemism related to the distribution of *Psophodes* (left) and the postulated sequential arrangement of those barriers (right); WD = western desert, ED = eastern desert, SW = southwestern corner, SE = southeastern corner, EF = eastern rainforest, AP = Atherton Plateau. Modified from Cracraft (1986).

species pair, *cristatus* and *occidentalis*, is distributed parapatrically in the arid interior of the continent (Fig. 3.6). Other genera, such as the emu-wrens (*Stipiturus*) and the quail-thrushes (*Cinclosoma*), exhibit largely congruent phylogenetic-distributional patterns. Cracraft postulated that these patterns reflect a series of isolating events, in which the southern and eastern elements were first separated from the interior biota by climatic and environmental gradients that date from the mid-Miocene (barrier Y, Fig. 3.6). Separation of the eastern and western desert forms was a consequence of environmental and geomorphological gradients and barriers (barrier X) that were established from the Oligocene through the Pleistocene, and the separation of the southern forms (barrier Z) may have been produced by climatic and geological changes occurring until the late Pliocene. Finally, the uplifted Atherton Plateau of the northeastern coast was separated from the mesic forests of southeastern Australia (barrier E) by a wide area of savannah vegetation that probably developed during the Pleistocene.

Whether or not this historical scenario is correct, it is apparent that the contemporary distributions of these species are consequences of historical events and that phylogenetic analysis may provide important insights into these relationships. In most situations, of course, little is known of such past events, and their effects can only be inferred at a rather broad, biogeographic scale. It is at this scale that historical biogeography (Nelson and Platnick 1981, Platnick and Nelson 1978, Cracraft 1983, 1986, Brown and Gibson 1983) can make valuable contributions to community analyses. Extending both phylogenetic and biogeographic insights to the interpretation of local community patterns should be a fertile area for further research.

Events occurring at another time leave an imprint on contemporary community patterns. Events occurring elsewhere in space may also affect these patterns, although these effects are equally difficult to establish with certainty. In northern Finland, for example, breeding populations of Pied Flycatchers have low reproductive success, and population levels are maintained by continuing emigration from the south (Järvinen and Väisänen 1984). On the Åland archipelago, populations of Ortolan Buntings (*Emberiza hortulana*) and Wheatears (*Oenanthe oenanthe*) have decreased considerably over the period from the 1920s to the 1970s. Haila *et al.* (1979) attributed these changes to habitat changes in the tropical wintering areas of these species. Interestingly, populations in Sweden also declined during this period but those in Finland did not, suggesting that the two populations may have been subjected to different pressures on the wintering grounds. In North America, populations of forest insectivores that are neotropical migrants have declined markedly in some areas since 1950, and some people have attributed these changes to the widespread devastation of wintering habitats in the neotropics (Whitcomb *et al.* 1981, Terborgh 1980). Many of these species occupy forest interior habitats, and the fragmentation of breeding habitats may also have contributed to the population changes (Forman *et al.* 1976, Whitcomb *et al.* 1981), as may predation (Wilcove 1985). Such spatial influences on communities are considered more fully in Chapter 5.

### Multiple pathways to community patterns

Although ecologists have shown a fondness for single-factor explanations of community patterns, the processes and factors that affect communities rarely act in isolation or in a manner that produces patterns unique to a particular process. Rather, they overlap both in their influences and in their outcomes. This overlap has two consequences.

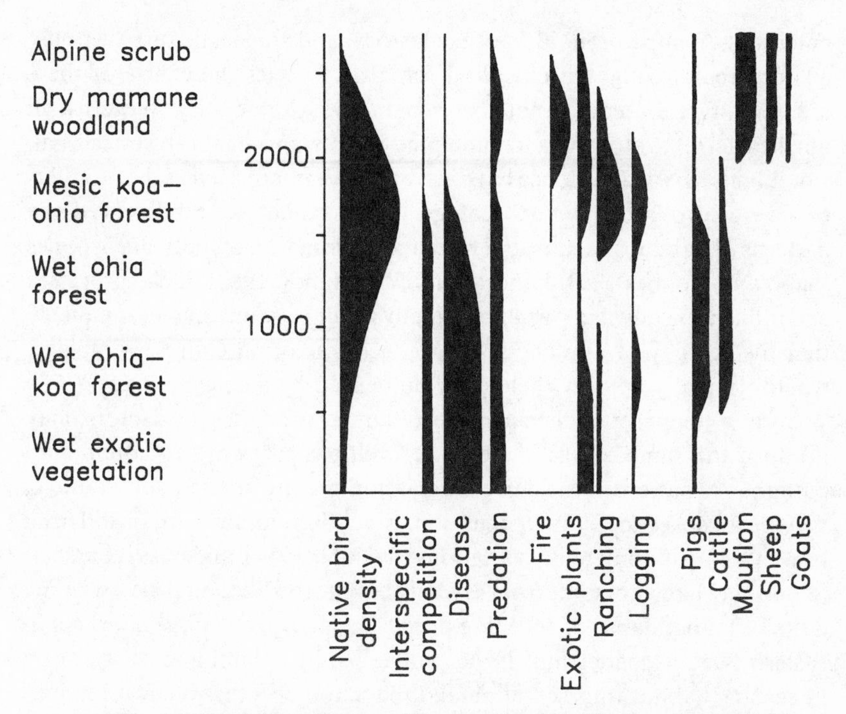


Fig. 3.7. Generalized diagram of the elevational distribution of major limiting factors influencing native bird population densities and distributions in the Hawaiian Islands. The breadth of the black area indicates the relative importance of a particular factor. Elevational habitat zones are listed on the left. After Scott *et al.* (1986).

First, a given community pattern may be the result of the combined effects of several forces. On the Hawaiian Islands, the elevational distributions and the abundances of native species are influenced by a wide array of factors, all of which vary in the intensity of their effects with elevation (Fig. 3.7). Species differ in their sensitivity to these factors, but all species are influenced by more than a single factor (Scott *et al.* 1986). Mountainspring (1986) constructed a model to analyze the contributions of each of these factors to recent changes in the distribution or abundance of the native species. Averaged over all species, habitat loss resulted in a mean reduction of the original ranges of 74%, parasitism and disease in an eradication from 46% of the available habitat, activities of feral ungulates in a 22% reduction in density, and competition from introduced birds in a 9% decrease in abundance. In considering the causes of distributional patchiness in puna and páramo birds in South America (see Volume 1, Chapter 8), Vuilleumier and Simberloff (1980) noted that incomplete sampling, habitat patchiness, competition, Pleistocene history, and stochastic turnover may all contribute to the patterns, and for many species the effects of these different factors could not be separated. Community patterns in northeastern hardwoods and subalpine forests and in southwestern oak woodlands in North America have likewise been interpreted as the products of several evolutionary and ecological events that have influenced the patterns of habitat selection and resource exploitation of individual species (Sabo and Holmes 1983, Landres and MacMahon 1983). If the processes and influences shown in Fig. 4.1 normally occur in nature, it seems unlikely that the ecology of any one species, much less of an entire assemblage, would normally be overwhelmingly influenced by a single process.

Even if they act independently, different processes or factors may produce the same effects – there are multiple pathways to community patterns. A pattern of negative association among species, for example, may be a consequence of competition, predation, mutualism, or differing responses to environmental variations (Schluter 1984), and positive associations may likewise result from several completely different processes. The decline in abundance of breeding populations of neotropical migrants in eastern North America may be accounted for by habitat loss *per se*, or by increasing habitat fragmentation and reduction of forest-interior habitat, or by increasing parasitism by cowbirds, or by increased predation, or by a loss of wintering habitat – all of these effects may produce the same general pattern. These represent alternative process hypotheses. By generating specific predictions and designing careful tests, one may be able to judge their contributions to the pattern. Aside from considerations of multiple pathways to the determination of latitudinal gradients in species diversity (e.g. Pianka 1966), little attention has been given to evaluating alternative process hypotheses in community ecology until quite recently.

### *Niche shifts and density compensation on islands*

The ways in which patterns may result from several causal pathways are especially clear when one attempts to interpret the patterns of niche shifts and density compensation revealed in island–mainland comparisons. On the one hand, the patterns (especially niche shifts) may be associated with differences in resource levels or predictability or in habitat characteristics between the island and mainland locations (Fig. 3.8). Island climates are often more benign than those on mainlands because of the ameliorating influences of the ocean, and environmental variations may be less dramatic there (although this is not always so; witness the dramatic rainfall variations on the Galápagos). If this leads to greater predictability

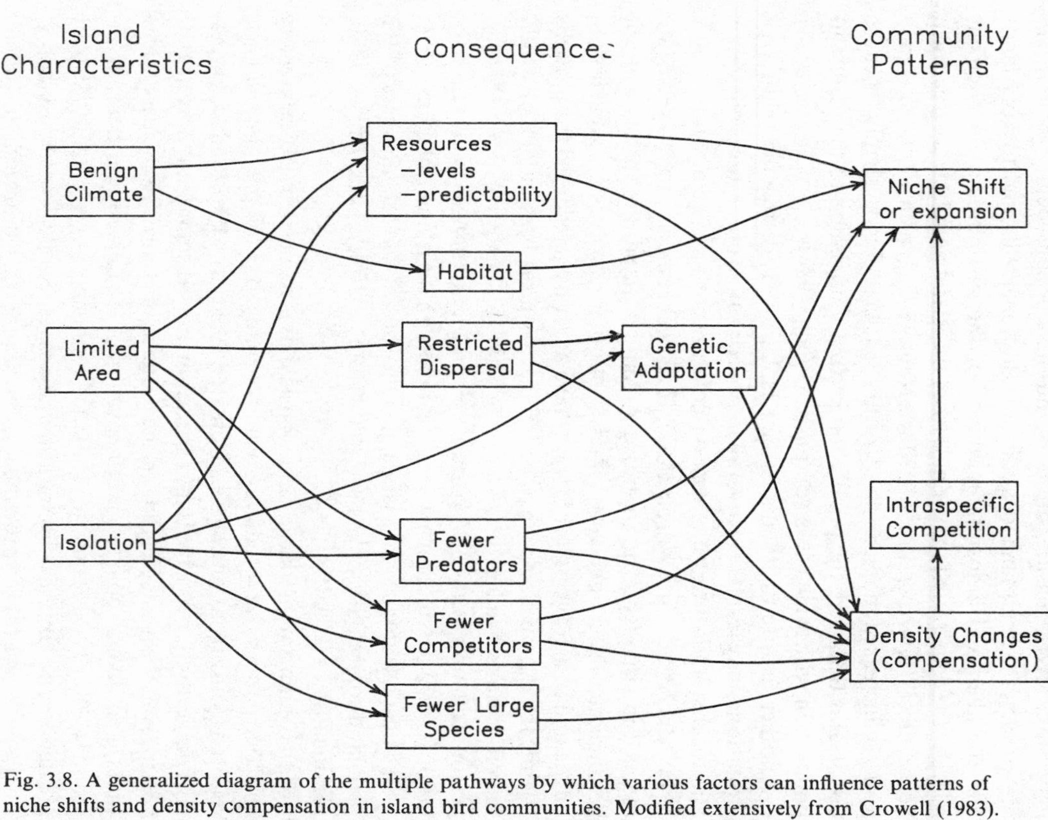


Fig. 3.8. A generalized diagram of the multiple pathways by which various factors can influence patterns of niche shifts and density compensation in island bird communities. Modified extensively from Crowell (1983).

in resource levels, densities on the islands may be consistently greater than in more variable mainland locations. In addition, the limited area of islands and the alien and inhospitable nature of the surrounding environment restricts the dispersal of individuals from the island populations, creating a 'fence effect' (Krebs *et al.* 1969, MacArthur 1972, Emlen 1979). This may lead to increased population densities on the islands simply because dispersal is thwarted; the population is closed in relation to the mainland population with which it is compared (Wiens 1986).

Another feature of islands is their isolation. This, together with area limitations, may result in an overall impoverishment of resources on islands due to area and distance effects on colonizations and extinctions (Lack 1976, MacArthur and Wilson 1967). If critical resources for a species are reduced in abundance or are missing from an island, the species will not become established there, even if levels of other resources are greater than on the mainland (Sherry 1985). Bees have never been recorded from Cocos Island, for example, and any species that requires bees in its diet would therefore have a hard time surviving should it disperse there (Sherry, personal communication). Isolation and area limitations may also contribute directly to a reduction in species numbers on the islands, of course (Fig. 3.8). Thus, the absence on islands of some species that are members of the mainland communities may itself result from multiple causes (Abbott 1980, Järvinen and Haila 1984).

If the number of predator species on islands is reduced, both the intensity and the patterns of predation will be changed, and prey populations released from predation may reach greater densities on the islands. Competing species (both other birds and species in more distant taxa) may also be absent from the island, leading to the patterns of competitive release, niche shifts, and density compensation envisioned in competition theory (Diamond 1978). The reduction in species numbers, however, may also be greater among large species than small ones, especially on small islands. If small species are over-represented on the islands, then more individuals can be supported on an equivalent resource base, and a pattern of density compensation may be evident (Fig. 3.8). Both the limited dispersal of individuals from island populations and the reduced flow of immigrants into these populations (due to island isolation) may contribute to the development of a greater degree of genetic adaptation of island populations to environmental conditions relative to their mainland counterparts, which are subjected to the diluting effects of gene flow from larger source areas (Emlen 1977, 1979). The increased resource-utilization efficiency of island populations may, in turn, lead to greater densities (Fig. 3.8). Finally, if island populations do occur at greater densities than those in mainland locations, intraspecific competition may be intensified, producing a pattern of niche expansion on the islands. It is clear that the observation of patterns of niche shifts or density compensation in itself says nothing about the cause of these patterns.

***Overlapping predictions***

This example illustrates the difficulty of devising and testing process hypotheses to explain community patterns in ecology. The problem of different processes or factors producing the same or quite similar patterns is not confined to this particular situation; virtually all community patterns may be produced through several pathways. In Table 3.4 I illustrate in a general way the conditions under which some of the community patterns discussed in Volume 1, Part II may be produced by the processes or factors discussed in this Section. Although this table appears formidable, it is still a vastly simplified summary of the relationships between patterns and their various causes. I include only 'main effects', the direct ways in which, for example, predation may influence niche breadth. Indirect effects involving interactions among several processes may also influence community patterns, of course, but to consider them here would lead to an impossibly complex table.

The entries in the cells of Table 3.4 are not hypotheses, in the strict sense, but simple statements of how or under what conditions a particular process or factor might produce a given pattern. The table serves as a guide to pattern-process relationships: if a given pattern is observed, one may find that pattern in the left column and then read across that row to see what processes or factors or conditions might have produced it. For each of these table cells, one may then frame a formal hypothesis that generates testable predictions that are specific to the observation. Some may be falsified immediately, but others may remain corroborated by the observation. By following such an approach, one is less likely to overlook possible explanations of the pattern than if one sets out to corroborate only a single process hypothesis.

Several general features of Table 3.4 merit comment. First, the statements relating interspecific competition to community patterns are generally simpler than those for most other processes. This does not mean that they are more likely to be correct, but it does suggest a reason for the widespread appeal of such explanations among ecologists (especially those who work with theory).

Second, many of the effects of intraspecific or interspecific interactions

Table 3.4. *Community patterns and the conditions in which various processes or factors might produce them.*

The patterns are described in detail in the appropriate chapters of Section II. The table cells indicate the circumstances relating a particular process (column) to a pattern (row), independently of any effects of other processes or factors (e.g. if habitats or resource distributions are patchy, the slope of the S/A regression will be high (because larger areas contain a greater diversity of patches, and thus of species); or, if environmental

| Predicted Pattern | Species-specific response | Intraspecific density/ competition[a] | Predation[b] | Parasitism |
|---|---|---|---|---|
| *Species-area:* | | | | |
| High value of z | Specialist species | Reduced, low densities | Intense | High loads reduce density |
| *Species-abundance distribution*[g]*:* | | | | |
| Geometric | Most species specialize on scarce resources; a few generalists | Little density-dependence permits dominance | Dominant species escape predation | Some species immune |
| Lognormal | Most species generalists (respond to many independent factors) | Intense, reduces abundances of many species | Density-dependent switching; specialization on dominants | Density-dependent |
| *Species diversity (S):* | | | | |
| Point and $\alpha$ diversity high | Generalist species | High ⇒ local spatial saturation | Low pressure ⇒ spatial saturation; *or* high ⇒ reduced dominance | Low |
| $\beta$ and $\gamma$ diversity high | Specialist species | Low ⇒ reduced habitat breadth/species | High; predation patterns differ between habitats | High; high parasite diversity |
| *Distribution:* | | | | |
| Parapatry | Species differ in physiological tolerances and/or habitat selection | | Intensity or prey/host preferences differ between areas | |
| Patchiness[h] | Species respond differently to undetected patch differences | Low ⇒ habitat not fully saturated | Intensity or prey/host preferences differ between patches | |
| *Species association in local samples:* | | | | |
| Negative[j] | Species respond to factors that covary negatively; species differ in physiological tolerances or habitat selection | Low ⇒ species occupy different preferred habitats | High; intensity and prey/host preferences differ among samples | |

disturbances cover large areas and are intense, generalist species may be favored, which are in turn likely to exhibit high niche overlap). The table is best used as a guide to the possible interpretations of observed patterns, by reading the row for a particular pattern (across the columns). The cell entries may then suggest formal, situation–specific hypotheses to be tested.

| Mutualism/ commensalism | Interspecific competition | Habitat/ resource variations | History[c] | Environmental disturbance[d] | Measurement/ sampling effects[e] |
|---|---|---|---|---|---|
| Many mutualist associations | Intense | Patchy | Considerable localized speciation (vicariance) | Frequent/small/ intense (⇒ patchiness) | Large areas undersampled[f] |
| | Dominant pre-empts niche space | Homogeneity permits dominance | | Low; *or* high & rare species sensitive to disturbance ⇒ absent | Rare species undersampled |
| | Resource over-exploitation ⇒ mutually adjusted competitors; balanced interference interactions (interspecific territoriality) | Patchy | | High ⇒ patchiness | |
| Many mutualist associations | Little ⇒ coexistence, spatial overlap; *or* intense ⇒ niche partitioning ⇒ coexistence | Complex habitat; many resource types | Long-term refugia; endemism | Intermediate ⇒ enhanced coexistence | Increased survey effort ⇒ more rare species recorded |
| | Intense ⇒ habitat and distributional exclusion | Patchiness | Vicariance; many biogeographic boundaries; greater age | High patchiness | |
| Nonoverlapping distribution of resources with which species are mutualists | Intense ⇒ exclusion | Change abruptly between areas | Dispersal barriers; low rate of expansion from refugia or centers of origin | Large-scale ⇒ habitat changes in one area | Inadequate sampling in zone of contact |
| Patchy distribution of mutualist resources | Intense ⇒ exclusion, combined with chance colonization of patches | Habitat or resource patchiness | Refugia; areas of endemism[i] | Recurrent disturbance of patches and lags in recolonization; habitat fragmentation | Inadequate sampling of distributions |
| Species are mutualists with resources that covary negatively | Intense (especially via interference) | Patchy, species differ in preferences | | Small-scale frequent; affect species differently | Oversampling of extreme habitats and undersampling of shared areas |

Table 3.4. (*cont.*)

| Predicted Pattern | Species-specific Response | Intraspecific density/ competition[a] | Predation[b] | Parasitism |
|---|---|---|---|---|
| Positive | Similar opportunistic response to resources; species aggregate socially[k]; species respond to factors that covary positively | High ⇒ overlaping habitat occupancy; *or* low but species prefer same habitats | High; affects prey/host species in same way[l]; intensity varies positively with variations in prey density | |
| *Morphology* | | | | |
| Size-ratio constancy[m] | Species differ in morphology[n] | | | |
| Character displacement | Geographical variation in resource use | Intensity varies between sympatry and allopatry | Size-dependent predation on one or both species in sympatry | |
| Character convergence | Species converge on use of abundant resources; extreme environment imposes same restrictions; mimicry to reduce aggression in mixed-species flocks | Intense where sympatric ⇒ greater niche breadth & overlap | Similar responses to predators present in areas of sympatry | |
| *Niche overlap*[o] | | | | |
| High | Generalists; species respond opportunistically to abundant resources; prefer similar habitats | Intense ⇒ increased niche breadths | Similar responses to high predation pressure[p] or parasitism; low predation pressure or parasitism ⇒ high densities | |
| Changes[q] | Species opportunistic when resources abundant, selective when resources scarce[r] *or* generalize when resources extremely limited | Intense when resources scarce ⇒ reduced overlap (specialization) *or* increased overlap (generalization) | Predation pressure varies with changes in availability of alternative prey | |
| *Niche complementarity:* | Species similar but respond to factors that vary inversely of one another | | | |
| *Niche breadth:* | | | | |
| Broad | Generalists; opportunistic use of abundant resources | Intense | Low predation pressure or parasitism; *or* pressure from a diverse set of predators | |

| Mutualism/ commensalism | Interspecific competition | Habitat/ resource variations | History[c] | Environmental disturbance[d] | Measurement/ sampling effects[e] |
|---|---|---|---|---|---|
| Species are co-mutualists | | Homogeneous, species have similar preferences; respond in same way to variations in limiting resources | Species share refugia/areas of endemism | Low intensity/ frequency; *or* large-scale; species respond similarly | Over-representation of shared areas |
| | Intense ⇒ limiting similarity | Discontinuities in availability of resource size classes | Divergence in size during phylogeny | | Inadequate sampling of intermediate-sized species |
| | Intense | Resources/habitats in sympatry differ from those in allopatry | Directional selection to reduce interbreeding in sympatry | | Inadequate sampling of clinical variation |
| Species share mutualistic associations in sympatry | Intense; convergence lessens spatial overlap (interspecific territoriality) | Resources more similar in sympatry | Species related phylogenetically | | |
| Species share mutualistic associations | Little; *or* spatial exclusion permits use of similar resources | Species respond opportunistically to abundant resources, habitat homogenous | Related species not ecologically differentiated; subjected to similar selective forces | Large-scale, intense ⇒ favor generalists | Resource categories too coarsely defined |
| | Intense with reduced resource levels ⇒ reduced overlap | Species specialize on resources or occupy different habitat patches when resource abundance low | | Reduce species' densities ⇒ smaller niches ⇒ reduced overlap | |
| Species are mutualists on independent resources that co-occur | Intense on some resource dimensions, absent on others | Habitat/resource features vary inversely of one another | | | Niche dimensions incorrectly defined |
| Little | Little; *or* intense competition from dominant forces niche expansion in subordinates | Broad resource spectrum; patchy or variable environment; resources abundant | Relaxed selection | Low disturbances ⇒ high densities; or high disturbance rate ⇒ ecological flexibility | Resource states excessively subdivided |

Table 3.4. (*cont.*)

| Predicted Pattern | Species-specific Response | Intraspecific density/ competition[a] | Predation[b] | Parasitism |
|---|---|---|---|---|
| *Niche shifts*[s]: | | | | |
| | Species flexible in response to different environmental situations | Intense + resource spectrum discontinuous or skewed[t] | Change in predation/parasitism pressure or prey choice by predator in different areas | |
| *Density compensation*[u]: | | | | |
| | Restricted dispersal; small species favored in "island" environment | | Reduced predation pressure or levels of parasitism | |
| *Interspecific Aggression:* | | | | |
| Dominance[v] | Species similar in appearance/ mistaken identity; individual distance; similar habitat choice ⇒ increased encounter probability | Intense ⇒ high densities increase probability of interspecific encounters | Low pressure permits greater aggressiveness | |
| Interspecific territoriality | Species similar in appearance; mistaken identity | Low ⇒ available habitat not saturated | | |

*Notes:*

[a] Assuming that intraspecific competition is density-dependent, and promotes expansion of species into additional habitats or resources at high densities (Svärdson 1949).

[b] 'Predation pressure' or intensity includes the effects of predator species diversity and functional and numerical responses within predator populations.

[c] Effects on an evolutionary or geological time scale.

[d] Unusual climatic changes or physical alterations of habitat or resource conditions, considered over short-term ecological time scales.

[e] Errors in sampling or measurement that may produce a systematic bias in community patterns; general 'sampling error' increases the variance of the patterns, but does not produce a consistent directional change in mean values.

[f] Because survey accuracy will decrease with the size of the area sampled, proportionately more species will be missed in large areas, depressing the rate of increase of $S$ with increasing $A$ at large values of $A$.

[g] Varies from community dominance by one or a few species (geometric) to low dominance and a high frequency of intermediate–abundance species (lognormal).

[h] e.g. 'checkerboard' distribution, in which similar habitat patches or islands support species A or species B but rarely contain both.

[i] Refugia or areas of endemism associated with past isolating events (e.g. Pleistocene glaciations).

| Mutualism/ commensalism | Interspecific competition | Habitat/ resource variations | History[c] | Environmental disturbance[d] | Measurement/ sampling effects[e] |
|---|---|---|---|---|---|
| Absence of resources for facultative mutualists | Absence or reduced densities of competitors | Changes in habitat or resouces between areas | Subspecific differentiation; past events different in areas; founder effects and genetic drift | Differences in disturbance regimes between areas; chance | Niche parameters measured differently in different locations |
| Greater abundance of mutualist resources | Absence or reduced density of competitors | Greater availability of resources or habitat or preferred habitat homogeneous on "island" | Reduced gene flow ⇒ enhanced adaptation to local conditions | Fragmentation or high disturbance rate increases densities of 'edge' or 'weedy' species | Individuals more conspicuous on 'island' |
| | Intense | Shared resources patchy ⇒ interspecific aggregations | Phylogenetic similarity ⇒ mistaken identity | | |
| | Intense ⇒ spatial exclusion | Patchy; species differ in habitat preferences | Phylogenetic similarity | | Territory boundaries inaccurately determined |

[j] Includes habitat segregation at a local scale; regional habitat segregation is expressed in parapatric or patchy distributions.
[k] e.g. mixed-species roosts or foraging flocks.
[l] e.g. predator generalizes to include both species in a single prey search image.
[m] Relationships are equivocal because a wide range of values may be consistent with the pattern.
[n] Differences alone will not produce ratio constancy, of course, but because the permissible variation about a 'constant' ratio is rarely specified, many ratios may appear to be generally similar.
[o] Includes patterns of limiting similarity and divergence in resource-use traits between species
[p] e.g. use of similar prey refuges, cover, etc.
[q] Changes in habitat or resource overlap between species associated with variations in resource abundance; low overlap is associated with reduced resource levels.
[r] In association with species-specific differences in utilization efficiencies.
[s] Here considered as changes in mean values of niche patterns of a species between locations.
[t] This is so because under such resource conditions the population cannot expand into niche space equally on both sides of the mean; the position of the mean will thus change.
[u] Here considered as an increased density of a species on an 'island' in comparison with a 'mainland' location.
[v] Not spatially fixed to a location such as a territory.

are mediated through changes in the densities of one or both species, which influence community patterns either directly (e.g. species-abundance distributions) or indirectly (e.g. niche breadth and overlap). Many of the relationships of patterns to habitat or resource features are mediated through the relative homogeneity or patchiness of the environment. The importance of population densities and habitat patchiness in these relationships suggests that they are of critical importance to most interpretations of community patterns.

Third, there are empty cells in the table. These occur because it is difficult to see how some processes could produce some patterns. There is no obvious direct relationship, for example, between mutualism and species-abundance distributions or between intraspecific competition and size-ratio constancy. In some cases the expectations for a certain pattern differ for different processes, but often the predictions converge toward the same patterns. The situation is complicated by the fact that the predictions in some instances are ambiguous; expectations regarding niche overlap or overlap changes are especially uncertain.

There are two additional complications to generating sets of predictions such as those of Table 3.4. First, the predictions may change as the spatial or temporal scale of their application changes. This is apparent in the contrast between patterns of regional diversity (beta and gamma diversity) and local diversity, but many of the patterns listed in Table 3.4 can be expressed on a variety of scales. As the scale changes, however, the contributions of various processes to the patterns may change. The effects of habitat variations on distributional patterns, for example, may differ depending on whether the variation is expressed as local patchiness or as regional landscape-level mosaics.

Second, all of these predictions are made in a *ceteris paribus* context – other factors are presumed to have little effect. Not only does this preclude the possibility that several processes might affect a pattern simultaneously, but it ignores the influences of disturbance. Although disturbance may have direct effects on some community patterns (e.g. diversity), more often it has a disrupting influence on the operation of other processes (Fig. 3.1). To the extent that disturbance is frequent and/or intense, the predictions of a given process may not be realized, even though that process may be operating and influencing community members. Its effects are diluted or overridden by the consequences of disturbance. In a sense, disturbance displaces a community from the trajectory that it might follow toward the pattern predicted by a given process in a *ceteris paribus* world (Fig. 3.9). If the disturbance is widespread, frequent, or of large magnitude, the community may be dis-

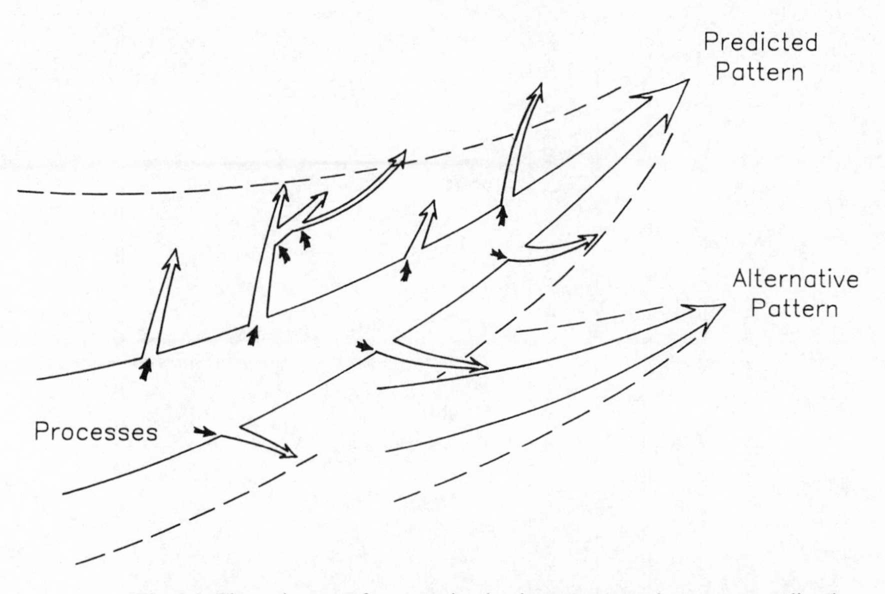


Fig. 3.9. The trajectory of community development toward a pattern predicted from particular processes or factors. Disturbances (solid arrows) displace the system from the trajectory; if these are moderate, they produce variance about the trajectory, but if they are frequent or intense, the system may be shifted beyond the boundaries of that trajectory (dashed lines) into the domain of trajectories toward different community patterns.

placed into a trajectory leading to an entirely different pattern. The record of such deviations is what makes history important in interpreting community patterns.

All of this is not to say that generating predictions and testing alternative hypotheses is an impossible task, although it does imply that the predictions probably must be more specific than the general statements given in Table 3.4. Alternative hypotheses have been tested in studies of the ecology of Galápagos finches (Abbott *et al.* 1977, Smith *et al.* 1978, Schluter and Grant 1982, Grant 1986b), diets of shrubsteppe birds (Wiens and Rotenberry 1979), distributions of Andean birds (Vuilleumier and Simberloff 1980), occurrences of species on islands in the Baltic Sea (Järvinen and Haila 1984) or in a Swedish lake (Nilsson 1977, Nilsson *et al.* 1985), the morphology of species on Tasmania (Abbott 1977), ecological patterns of species on the Faroe Islands (Bengtson and Bloch 1983), and distributions of species in tropical forests of Africa and South America (Endler 1982a, b), to name but a few. In all of these studies, the predictions were quite specific, and the tests were founded on a thorough knowledge of the natural history of the species and the communities.

Table 3.5. *Criteria that establish the occurrence of various processes influencing community patterns.*

Criteria 1 and 6 are necessary to document the unitary effect of a process on a pattern while the remaining criteria document its occurrence regardless of whether or not other processes also occur. Criteria 1–6 are from Table 1.4.

| | Process hypothesis | | | | | |
|---|---|---|---|---|---|---|
| Criterion | Interspecific competition | Individualistic | Intraspecific competition | Predation | Parasitism | Mutualism/ Commensalism |
| 1. Observed patterns consistent with predictions | X | X | X | X | X | X |
| 2. Species overlap in resource use | X | | | | | X (indirect) |
| 3. Intraspecific competition occurs (density-dependent resource limitation) | X | | X | | | |
| 4. Resource use by one species reduces availability to another species | X | | | | | |
| 5. One or more species is negatively affected | X | | | X | X | |
| 6. Alternative process hypotheses are not consistent with patterns | X | X | X | X | X | X |
| 7. One or more species benefits from the interaction | ? | | | X | X | X |

**Criteria and conditions for alternative processes**

In Chapter 1 I devoted considerable attention to the criteria required to document competition and the conditions under which it might be likely to occur. It is appropriate to do the same for the processes considered in this chapter. In order for any single process to explain a community pattern adequately, the patterns must be consistent with the predictions of that process (Table 3.4) and alternative process hypotheses must be rejected (criteria 1 and 6 of Table 3.5). If one accepts the conceptualization presented in Fig. 3.1, individualistic, species-specific characteristics are the basic factors determining community patterns, and none of the other criteria listed in Table 3.5 is required to document the occurrence of these characteristics. Species have these attributes as a consequence of their evolutionary history. To document them, however, one must consider autecological features such as morphology, behavior, or physiology. Documenting the occurrence of intraspecific competition requires further that one demonstrate density-dependent resource limitation and effects on the fecundity, survival, behavior, or habitat occupancy of individuals in the population.

Individualistic effects on community patterns are most likely to be expressed when they are not overridden by disturbance or by other processes involving species interactions. This may occur when resources are not directly limiting, as when resources are superabundant or in variable environments when climatic factors rather than resources limit distributions and abundances of species (Wiens 1977, Price 1984). It may also be expected in patchy environments, especially if disturbance is patchy and frequent. In general, individualistic effects should be more prominent in nonequilibrium than in equilibrium communities. Intraspecific competition, on the other hand, requires that resources be limiting to at least some individuals in a population. This is most likely to occur when population densities reach high levels in relation to resource availability ('carrying capacity'), which is traditionally associated with relatively stable environmental conditions in homogeneous habitats. It may also occur in variable environments during nadirs in resource fluctuations, or at any time when the resource demands of individuals exceed supplies. These are the same conditions that favor the occurrence of interspecific competition.

The other processes I have considered involve interspecific interactions, and documenting their occurrence requires that the interaction be of a specific sort. Predation and parasitism are both interactions in which individuals of one species benefit at the expense of individuals of another species (a +, − interaction). By definition, predation has direct effects on the survival of individuals. All that is required to demonstrate its occurrence is the observation (direct or indirect) that individuals of one species have killed individuals (or reproductive products) of another species; no determination of resource conditions is necessary. Parasitism, on the other hand, may not result in mortality but may lead to diminished performance among host individuals. Brood parasitism is documented by interspecific egg-laying, and the demonstration of internal parasitism is self-evident. In both cases, however, a diminishment in the performance of host individuals should also be demonstrated (Table 3.5) and this is not always easy. Mutualism by definition implies a (+, +) interaction, and commensalism is a (+,0) interaction between the species. No relations to resources are required, although each species is in a sense a resource to the other, especially if the mutualism is obligate. In indirect mutualism, in which two species share mutualistic relationships with a common species (e.g. two nectarivores adapted to pollinate the same flowers), resource overlap is also a criterion (Table 3.5).

Documenting that a particular process occurs, of course, is not sufficient to establish that it is important in producing a community pattern (Welden and Slausen 1986). To be important, a process must affect population as well as individual attributes. These effects are usually measured by changes in population densities, distributions, habitat occupancy, or resource use. Thus, if one wishes to determine whether or not predation (for example) might have contributed to an observed community pattern, it is necessary to document not only that predation occurs but also that it influences one or more of these population attributes. When a predator, parasite, or mutualist mediates the interactions among other species, the effects of interactions can become complex and lead to counterintuitive results (e.g. Price *et al.* 1986, Fig. 3.2). When the pathways of interactions are expanded to include even more species (e.g. alternative prey), the webs become still more complex. Unravelling these webs of direct and indirect interactions and determining their effects is one of the major challenges facing both theoretical and empirical community ecologists.

Under what conditions are these other processes likely to occur? Connell (1975) suggested that predation may be more important in tropical systems, where predator diversity is great. Higher trophic levels are likely to be both abundant and diverse when the resource base is relatively stable and productive. Predation may also be important where predators affect the dominant members of the community, as visualized in Paine's (1966) 'keystone predator' idea, or where predators are largely supported by a single prey species that varies in abundance, as in some high-altitude situations.

Brood parasitism is likely to have significant effects on communities where parasitic species are diverse and/or abundant. In North America, cowbirds frequent open habitats and forest edges, and communities in these situations may be influenced by parasitism more than those in large tracts of dense forest. Many internal parasites have an aquatic phase associated with intermediate hosts or require a humid microclimate, and birds frequenting these habitats may suffer higher levels of parasitism than those occupying arid environments. Many blood parasites are transmitted by aerial vectors that are more abundant in mesic than in xeric habitats (e.g. Fig. 3.5). It therefore seems more likely that parasitism may reach levels that affect birds sufficiently to have community consequences in damp than in dry environments.

Close mutualistic relationships among species are most likely to develop if there has been a relatively long and persistent association of the species. Mutualism would thus seem most probable where environments are relatively stable. If the interaction is related to resource use, mutualism should develop where the resource base is sufficiently productive to permit the species to become specialized. These conditions seem likely to be met in at least some tropical situations, although they should be anticipated wherever the association of species is evolutionarily stable.

## Conclusions

Although community investigations over the past several decades have focused on competition as the primary process determining community patterns, it is clear that other processes and factors may have important influences as well. One reason for this past emphasis on competition is its multi-species nature. Processes such as predation and parasitism have tended to be viewed as population phenomena and have been considered in discussions about population regulation rather than community dynamics. Mutualism has attracted attention primarily in a coevolutionary context. Species-specific features of behavior or physiology have been regarded as part of the domain of autecology. Because communities have also been viewed in an equilibrium framework, the effects of disturbance, history, and chance have also been minimized or ignored. Community ecology has followed pathways of investigation largely separate from those of population ecology or autecology. This is clearly a counterproductive approach – what happens in communities is ultimately a product of the behavior, physiology, and ecology of individuals of different species and of the factors that determine population sizes and distributions.

One can select a study situation either (a) to focus on one or more particular processes, such as competition and/or predation, or (b) to understand a particular slice of nature, such as shrubsteppe, tropical forest undergrowth, or foliage-feeding insectivorous birds. These objectives dictate different approaches, and the situations most appropriate for each will differ. If one wishes to study competition, an area in which all of the species are likely to be overwhelmingly influenced by such interactions is best for investigation. In most investigations, however, the study situation has been selected as if (b) were the objective but then interpreted as if (a) had been done. If one's objective is to understand community patterns, one is doing (b). This requires that one begin with an extensive knowlege of the attributes of the species in the assemblage, as these dictate the ecological patterns that may or may not be altered by more complex interactive processes or by disturbance or chance (Fig. 3.1). One must also consider all of the processes and factors that may influence the assemblage, not just those that are fashionable or that are suspected to be important. Ideally, these should be cast as testable alternative hypotheses, but, because there are multiple pathways to most community patterns (Table 3.4), this is not always possible. Moreover, documenting that a particular process *does* occur in a community does not assure that it is responsible for the patterns observed, as its influences may not extend to a population or multispecies level or may be overridden by other factors, especially if disturbance is frequent or intense. It is also essential to realize that contemporary communities are not static but bear the imprint of their history, over time scales ranging from years to geological epochs. Confronting these matters effectively will not be easy, but the potential effects of species attributes, predation, parasitism, mutualism, disturbance, history, and chance on communities are important enough that they cannot be disregarded.

# PART II

## Community variation in time and space

Many of the views of communities that developed during the 1960s and early 1970s portrayed a dream world of stability and homegeneity, in which spatial or temporal variations in environments were either nonexistent or were closely matched by responses of the community. To a large degree, these views were nutured by the mathematical constructs of community theory, which operated under the assumption that models reflecting a world in stable equilibrium and having no spatial dimension were reasonable representations of natural communities. Because natural communities were thought to tend strongly toward equilibrium, the effects of history disappeared, environmental perturbations had no lasting effects, spatial heterogeneity was of little consequence, and chance was limited to a small role in affecting the dispersal of species to a locality (Chesson and Case 1986).

But of course environments and communities *do* vary in time and space. They do so with varying amplitudes, periodicities, and degrees of stochasticity, on a wide range of scales. Communities vary in the presence and absence of species, in the density levels of these species, and in their densities relative to one another. Because the stability and predictability of other aspects of ecological relationships among the species hinge on the relative constancy of these patterns, such variability hinders attempts to detect and explain community patterns. This temporal and spatial variation of communities, although perhaps not entirely 'music to the ecologist' (Simberloff 1982a), at least can no longer be considered simply as uninteresting and unwanted 'noise', to be muffled by statistical analysis or simply disregarded altogether.

Fortunately, there is mounting evidence of an increasing awareness in both theoretical and empirical arenas of the importance of temporal and spatial variation (Hastings 1980, Pickett and White 1985, Chesson 1986). In the two chapters of this section, I focus on the patterns of temporal and spatial variability of environments and the bird communities that occupy them. Although they do not impugn the observations or conclusions I have developed in previous chapters, they do suggest some fundamental modifications in the ways we have viewed communities over the past few decades.

# 4

## Temporal variation of communities

Anyone who has watched birds for several years in a woodlot or other local habitat becomes aware of variations. Some species are present in some years, absent in others, and populations change in abundance between years. Other populations vary about a long-term average that seems to be relatively stable. The abundance of wintering Stonechats (*Saxicola torquata*) or tits in Europe, for example, changes from year to year, apparently in response to weather and food conditions during the previous winter that affect overwinter survival and the reproductive output of the birds during the summer. Long-term trends in abundance are not evident, however (Dhondt 1983, van Balen 1980, Klomp 1980). In other cases, variations are more episodic. The winter of 1962–63 was the harshest in southern England since 1740, and populations of the Wren (*Troglodytes troglodytes*) were reduced by 75%, the Song Thrush (*Turdus philomelos*) by 50% (Williamson 1975). A prolonged drought in the Sahel (sub–Saharan Africa) between 1968 and 1974 devastated populations of wintering palearctic migrants, leading to sharp reductions in their breeding numbers. Abundances of Whitethroats (*Sylvia communis*), for example, decreased by as much as 77% in some parts of England between 1968 and 1969, and similar reductions were recorded elsewhere in its breeding range (Winstanley *et al.* 1974, Berthold 1973). When the drought broke in 1975, Whitethroat abundances rebounded, increasing by 60% in 1976 (Batten and Marchant 1977).

Over a longer time period, populations may show systematic trends in abundance. Numbers of Ortolan Buntings and Wheatears decreased dramatically in some areas of Scandinavia between the 1920s and the 1970s, and populations of *Parus* and *Fringilla* species underwent substantial distributional shifts in Finland during the same period (Haila *et al.* 1979, Järvinen and Väisänen 1979). The spread and increase of the Cattle Egret (*Bubulcus ibis*) in North America since 1950 have been even more dramatic.

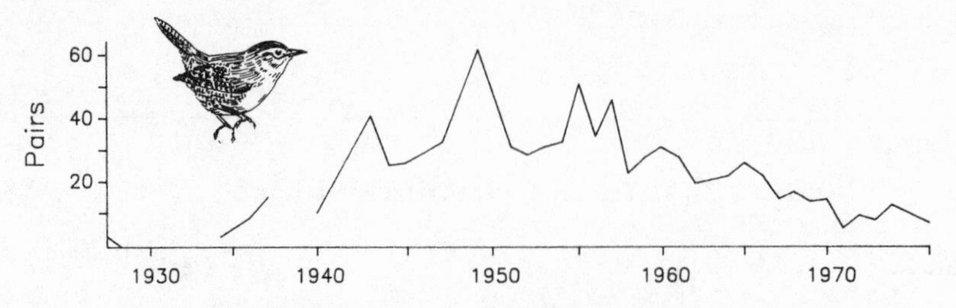


Fig. 4.1. Fluctuations in breeding populations of House Wrens in an Illinois woodland over a 50-yr period. From Kendeigh (1982).

In an Illinois woods censused by Kendeigh (1982) over a 50-yr period, on the other hand, House Wrens were scarce before 1934, increased to become the most abundant breeding species between 1942 and 1957, and then declined to complete absence by 1977 (Fig. 4.1). The period of greatest abundance coincided with the period of peak insect abundances in the woods (Kendeigh 1979), although wren densities increased before the build-up of insect populations and remained high for several years following the decline in insect levels. Kendeigh suggested that variations in wren densities were associated with winter temperatures in the Gulf States wintering areas.

**Views of community equilibrium**

These observations would seem to fly in the face of traditional community theory, which, in its simplest form, has held communities to be fully saturated with species and in a steady-state diversity. Giller (1984) considered the occurrence of such equilibria in communities to be 'generally accepted', and Terborgh and Winter (1980: 130) observed that 'a healthy species is normally at equilibrium with its environment, that is, overall its numbers are neither growing nor declining'. In this view of communities, disturbances are exceptional circumstances, and variation is considered to be a reflection of sampling inefficiencies or 'noise' that obscures what is really happening in the system (Karr and Freemark 1983, Chesson 1986).

Bird populations and communities clearly change through time, however, and these variations may represent real biological dynamics rather than sampling artifacts or noise. A second, alternative view considers these variations to be evidence of a dynamic equilibrium. Resource levels change through time, and the variations in populations and communities reflect close tracking of these resources through short-term behavioral, distributional, or demographic adjustments (Cody 1981).

A third view holds that populations and communities do not track resource variations at all closely but, instead, are strongly affected by episodic 'environmental crunches', periods of severe climatic or resource limitation interspersed with more benign periods in which resources are superabundant (Hutchinson 1961, Wiens 1974, 1977, 1986). The effects of severe winters or droughts noted above are examples of such crunches. In these situations, individuals or populations respond slowly in comparison with the rate and magnitude of the environmental variations. This produces time lags of varying lengths, a 'tracking inertia' (Wiens 1984b). The same lack of correspondence between resource or environmental variations and population levels may occur if the populations have 'density-vague' dynamics, a situation in which population growth parameters are directly affected by density-resource relationships only at very low or high densities and vary independently of density and resources over a broad range of intermediate densities (Strong 1984b, 1986). In either case, the lack of a close fit between populations and resources produces apparent nonequilibrium in community patterns.

A fourth, more extreme view holds that abundances of species are rarely, if ever, determined by resource levels and that communities are clearly nonequilibrial. Stochastic factors may have major effects on population variations, although nonequilibrium community patterns may also result from population dynamics that are largely deterministic but unrelated to resources, especially if the species vary independently of one another. Nonequilibrium does not necessarily imply a predominance of stochastic effects (Karr and Freemark 1983, *contra* Grossmann *et al.* 1982).

These views represent somewhat arbitrary positions along a spectrum from totally static equilibrium to chaotic nonequilibrium. Beyond specifying theoretical endpoints for community stability and resource coupling, it is unlikely that either extreme view can contribute much to our understanding of the dynamics of natural populations and communities. Most of the recent discussion and controversy have centered on the second and third of these views, especially about the closeness of resource tracking, the frequency of environmental crunches, and the importance of time lags (Cody 1981, Schoener 1982, 1983a, 1983b, Wiens 1983b, 1984a, 1986, Price 1984). There has also been disagreement, however, about the exact meanings of 'equilibrium', 'stability', 'nonequilibrium', and similar terms, despite a number of attempts to clarify this terminology (Holling 1973, Botkin and Sobel 1975, Connell and Sousa 1983, Chesson and Case 1986, Pimm 1986; Table 4.1). Stability connotes numerical constancy, but it does not necessarily equate with equilibrium, as populations tracking a variable resource may be in equilibrium with the resource but clearly unstable (i.e. nonconstant). To define nonequilibrium as a lack of constancy in species densities at a

Table 4.1. *Definitions of several elements of 'stability' in ecological systems.*

| Term | Definition |
|---|---|
| Stable (Units are nondimensional; either the system is stable or else it is not) | A system is deemed stable if and only if the variables all return to the initial equilibrium following their being perturbed from it. |
| Resilience (Units are those of time) | How fast the variables return to equilibrium following a perturbation. |
| Persistence (Units are those of time) | The time a variable lasts before it is changed to a new value. |
| Resistance (Units are nondimensional) | The degree to which a system is changed by a perturbation is the reciprocal of resistance. |
| Variability (Coefficient of variation is dimensionless) | The variance of population density (or similar measures like standard deviation or coefficient of variation). |

*Source:* From Pimm (1986)

location (Chesson and Case 1986) thus unrealistically restricts the meaning of equilibrium to static situations. Equilibrium (in the broad sense) has been defined as the tendency of a population or community to resist change when confronted with environmental variations (resistance), to return quickly to some stable point following a perturbation (resilience), or to vary within bounded limits without suffering extinction (persistence) (Table 4.1). For any situation, equilibrium or stability must be gauged with respect to the time scale of responsiveness of the populations (e.g. generation length) and the various characteristics of environmental variations or disturbance (Table 3.3).

In view of these terminological and operational difficulties, it may not be particularly rewarding to attempt to determine whether bird communities are 'stable' or in 'equilibrium' or not. It may be more worthwhile to examine how these assemblages vary in nature, to approach the patterns of variation as something of interest in their own right.

### General patterns of variation

Most investigations of bird communities have been conducted over short periods and thus provide a 'snapshot' view of patterns. In their reviews of experimental studies of competition, for example, Schoener (1983b) and Connell (1983) found relatively few (over all taxa) that were conducted long enough to assess whether or not there was temporal variation in the occurrence of competition. In view of this preoccupation with short-term studies, it is not easy to determine if there are any general

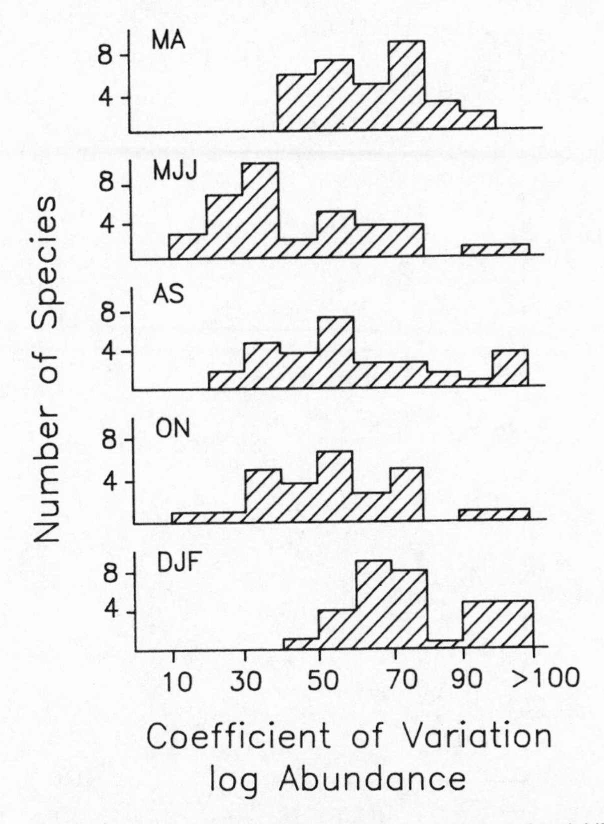


Fig. 4.2. Numbers of species showing various degrees of variability in abundance over a 6-yr period in riparian habitats on the lower Colorado River in southwestern United States. MA = March–April, MJJ = May–June–July, AS = August–September, ON = October–November, DJF = December–January–February. The entry for each species is the median CV across 10 or more transects. After Rice *et al.* (1983b).

patterns to the variation that occurs in bird communities inhabiting different environments. Moreover, the extent of variability in any particular location may change seasonally and differ among species or guilds (Carrascal *et al.* 1987), making comparisons sensitive to when and where they are made. Bird species occupying riparian habitats along the Lower Colorado River in southwestern USA, for example, vary substantially, but the extent of variation is generally less in late spring and early summer than in winter, and there are considerable differences among species in the degree of variability (Fig. 4.2). Wright and Hubbell (1983) examined such interspecific differences in variability (as measured by the coefficient of variation (CV) of abundances) for breeding species on the Farne Islands

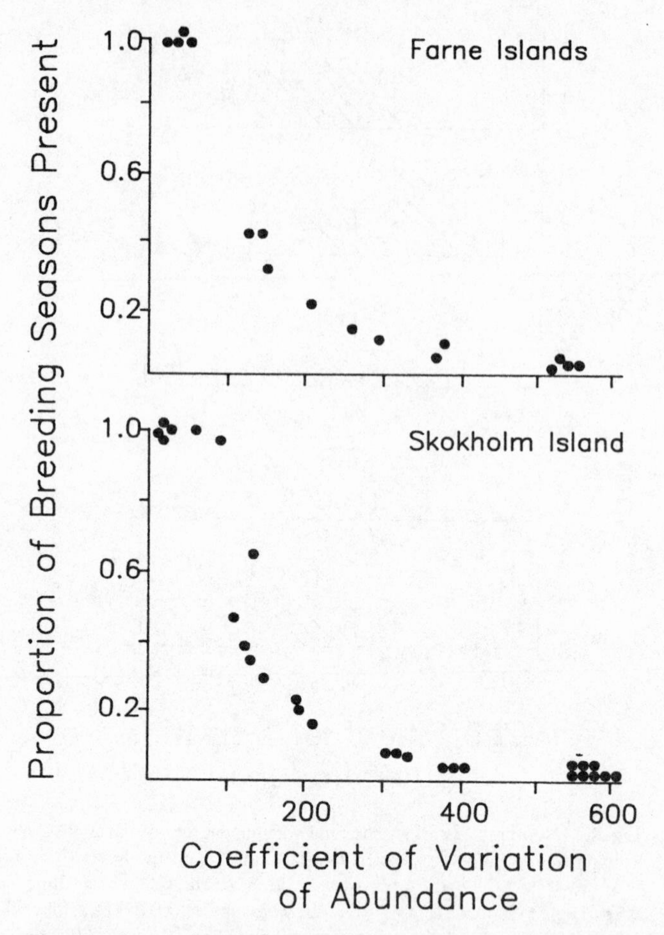


Fig. 4.3. The relationship between the proportion of years in which at least one breeding pair of a species was present on an island and the coefficient of variation in the abundance of the species. The data for the Farne Islands are for 16 species recorded between 1946 and 1974; for Skokholm, for 29 species recorded between 1928 and 1967. After Wright and Hubbell (1983).

and on Skokholm and found a close relationship to the persistence of the species on the islands – the more variable species were more irregular in their occurrence (Fig. 4.3).

To assess general patterns of variation, one must also determine which measures of variability are most appropriate. Most analyses have employed CVs of estimated population densities or of community parameters such as diversity (e.g. Brewer 1963, Järvinen 1979). Measures of CV are sensitive to high values of $N$, however, and Connell and Sousa (1983) and Williamson

(1984) suggested indexing variability instead by use of the standard deviation of log $N$ (which is sensitive to low values of $N$). In comparisons involving different sets of species at different places, however, this approach may not be valid, as the standard deviations would reflect not only differences in variation but differences in mean abundances. In such situations, Williamson (1984) advocated using Wolda's (1983) measure of annual variability:

$$\text{variance} [\log (n_{t+1}/n_t)].$$

If a population is changing following a trend (e.g. decreasing), this measure is better than the CV, which is calculated on the moving average and thus loses information (W. Rice, personal communication). Holmes *et al.* (1986) assessed population variability using Whittaker's (1975) Coefficient of Fluctuation,

$$\text{antilog} \sqrt{[\Sigma(\log n_t - \log \tilde{n})^2/(t-1)]}$$

where $n_t$ is the species' abundance in the year $t$ and $\tilde{n}$ is the geometric mean density. Because this measure is based on logarithms, it is less sensitive to differences in the absolute sizes of populations than are CVs. Järvinen (1979) and Noon *et al.* (1985) employed a variety of indices of variability, some based on CVs and others based on measures of compositional similarity between years.

All of these approaches share a potential bias due to temporal autocorrelation, as they measure variation between successive years in which population levels are not likely to be completely independent because of the persistence of resident individuals or the philopatric return of migrants (Järvinen 1979). For this reason, Connell and Sousa (1983) suggested that variability should be assessed over time scales longer than the turnover rates of populations (= generation length). However, one may argue that, although multi-year occupancy of a location by individuals violates the statistical requirement for independence of samples in variance analyses, it nonetheless represents the actual biology of the systems. If the communities in one location or habitat are less variable than those in another because of greater long-term persistence of individuals, this represents important information about the communities. To exclude it because of statistical provisions would miss the point of the analysis. Ideally, both local, short-term and global, long-term analyses of population variability should be conducted.

General patterns in the variability of bird populations and communities ('stability gradients') have been sought through comparisons among multi-year censuses conducted over a range of habitat types or latitude. Brewer (1963) compared CVs of population densities of breeding species surveyed for 7–18 years in five locations in the eastern and northcentral United States. Nor surprisingly, species differed from one another in the degree of population variability, but most species also differed in variability between the areas. Brewer suggested (perhaps a bit tautologically) that populations of a species varied least in the optimal section of its ecological range. In his studies in central Europe and Scandinavia, Järvinen (1979) found that populations tended to fluctuate least in the central portion of their geographical range, which is not quite the same thing (see also Brown 1984). Brewer found an overall tendency for CV values to increase from wet through mesic to xeric forests. Because his comparison included only five samples, however, this trend could not be confirmed statistically.

Järvinen (1979) and Noon *et al.* (1985) conducted broader analyses of variability patterns, using a variety of measures to index annual changes in populations and communities. Järvinen's chief axis of comparison was latitudinal, over a series of 15 locations from central Europe to northern Fennoscandia and Iceland. Northern communities were characterized by greater variation in total density ($\bar{x}$ CV = 22.2) than were those in southern Scandinavia and central Europe ($\bar{x}$ CV = 8.4), and they were somewhat more variable in species numbers as well ($\bar{x}$ CV = 14.2 vs 7.0). Overall, the northern communities had more variable species-abundance distributions and greater annual turnover in species composition (Fig. 4.4). Patterns among habitat types were not explicitly considered; because habitats changed latitudinally, however, there was a general trend to variability among habitats as well (Fig. 4.4). Järvinen attributed these differences to the greater degree of environmental unpredictability and harshness in the north. This interpretation is bolstered by A. Järvinen's (1983) finding that annual variation in the population densities and reproductive parameters of Pied Flycatchers, a southern species that has recently expanded into northern Fennoscandia, increases abruptly in the northern part of its range.

The analysis of Noon *et al.* (1985) was based on values of nine variability indices calculated for each of 174 study plots from North America that were censused over 3–22 yr. Direct pairwise comparisons of index values revealed few significant differences between habitats, and this general lack of clear patterns was reinforced by a Principal Components Analysis (PCA) conducted on the index-plot data matrix (with the effects of plot size controlled). Samples from the different habitat types were broadly spread over the first PCA axis, which was related to indices of annual turnover of species

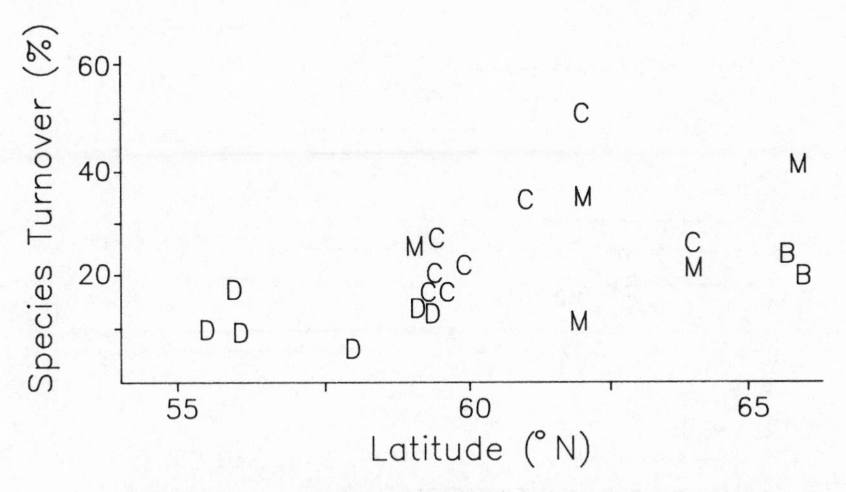


Fig. 4.4. Species turnover as a function of latitude for 21 forest census areas in Sweden. B = mountain birch, C = coniferous, D = deciduous, M = mixed deciduous-coniferous. Turnover is defined as the percentage of immigrating and disappearing species of the total number of species present in an area. After Järvinen (1980b).

and individuals (Fig. 4.5). Grassland habitats exhibited greater variation in diversity index values than did plots in most other habitat types (Axis II), and there was an indication that populations in coniferous forests may vary more between years than those in other habitats (Axis III, Fig. 4.5). Noon *et al.* also found some evidence that annual turnover of species increases as one moves north, but there were no clear geographical trends in any of the other variability measures. This is in accord with Järvinen's findings in Scandinavia (Fig. 4.4), but Noon *et al.* failed to find the clear differences in density variation that characterized Järvinen's latitudinal gradient. If increasing environmental unpredictability and harshness are associated with increasing variation in bird communities, as Järvinen suggests, the absence of clear latitudinal patterns in variability in North America may reflect the fact that climatic conditions there are less directly associated with latitude alone than in Scandinavia, due to the longitudinal influences of the Rocky Mountain chain.

Overall, the North American analysis was distinguished by the general absence of any systematic differences in population or community variability among habitat types or on geographical gradients, confirming the findings of a similar unpublished analysis conducted by Myers, Mumme, and Pitelka (Wiens 1984a). This finding contrasts with the results of individual studies that have suggested that communities in forested habitats

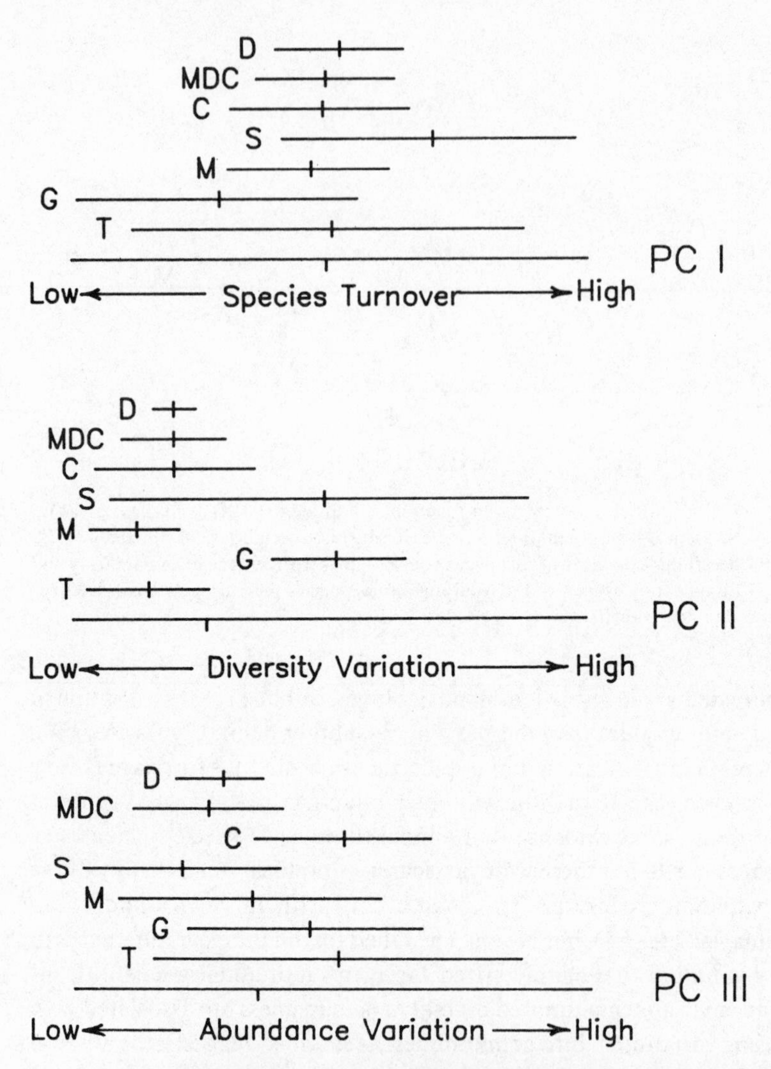


Fig. 4.5. Ordination of North American breeding bird census plots on the first three axes of a PCA of measures of population and community variability. Plots are grouped by major habitat types (D = deciduous forests, MDC = mixed deciduous-coniferous, C = coniferous forests, S = scrub habitats, M = mixed scrub-grassland, G = grassland, T = tundra), and for each type the mean factor score (vertical line) and 95% confidence interval (horizontal line) are indicated. Axis PC I represents a gradient in species turnover, PC II a gradient in diversity variation, and PC III a gradient in abundance variation. After Noon *et al.* (1985).

may be less variable than those in shrubsteppe and grassland (Kendeigh 1982, Winternitz 1976 vs Wiens and Rotenberry 1980, 1981a, Rotenberry and Wiens 1978, 1980b). This contradiction may occur because the local plots included in broad comparisons are each subjected to a wide variety of factors that effect annual variations and thereby obscure any general patterns that might emerge. The spread of plots on the PCA axes derived by Noon *et al.* (Fig. 4.5) certainly indicates substantial between-plot differences in variability, whatever the causes. In any event, we clearly are not yet able to predict the variability of communities or populations on the basis of habitat or geography alone.

## Responses to environmental variations

### *Environmental tracking*

It is unrealistic to expect populations and communities to persist in steady-state stability, given the variations in environmental conditions that occur from year to year. Optimal foraging theory, however, leads one to expect a close matching between resource-exploitation patterns and changing resource levels, and competition-based community theory translates this into population and community effects. Other things being equal, we might expect variations in environmental conditions to be closely tracked.

Cody especially has championed this view, noting (1981: 107) that 'recent complaints that bird community structure appears variable and unpredictable from year to year . . . may be more a reflection on our failure to measure varying resource levels tracked by a changing set of consumers from year to year rather than a failure in the consumer community to match resources that we wrongly assume to be constant.' Cody supported this argument with observations from pine-oak woodlands in Arizona that he monitored over several nonconsecutive years. Yearly rainfall in the area varied, and Cody related these variations to fluctuations in the abundance of insects, which he measured using sticky boards. Insect abundance was severely reduced during a drought year, and there was considerable change in bird community composition. In subsequent years, rainfall was average, and the bird community returned to its former state. Unfortunately, the census data for the critical drought year were lost, and Cody based his conclusions on qualitative recollections of population levels. Censuses for the years immediately preceding and following the drought year were not available, and the measures of insect abundance bear an unknown relation to actual food availability to the birds (Smith 1982, Wiens 1983a). Cody's conclusion that 'the community tracks the varying food supply by adjusting the identity, number, and density of its members' (1981: 112) scarcely seems justified, although the pattern is suggestive.

There is nonetheless clear evidence of close tracking of environmental and resource variations in other situations. Nectarivores, which are energetically closely linked with floral nectar supplies, may undergo extremely rapid changes in abundance if nectar availability changes (e.g. Gass and Lertzman 1980, Heinemann 1984; see Chapter 2) and raptors that specialize on voles may sometimes track variations in their prey populations closely (Korpimäki 1985). Wintering populations of species that feed on oak mast in North America, such as Red-headed Woodpeckers or Blue Jays (*Cyanocitta cristata*), are more variable between years than are insectivores and frugivores, and the covariation of the woodpeckers and jays suggests that their population fluctuations may be tied to variations in the mast crop (Smith 1986a, b). Insectivores such as warblers or vireos, however, may vary in accordance with changes in insect abundance during the breeding season (MacArthur 1958, Morse 1980, Kendeigh 1982). Birds inhabiting undergrowth in tropical forests may shift their selection of habitat and microhabitat in response to diurnal, seasonal, and yearly variations in climatic conditions. Because the species track these variations largely independently of one another, however, patterns at the community level appear to be quite nonequilibrial (Karr and Freemark 1983). In peatland habitats in northern Finland, several rare species may breed in years in which spring temperatures are warm, but they are absent when temperatures are moderate or cold; apparently the birds respond to variations in food supplies (Väisänen and Järvinen 1977*a*). Densities of Redpolls (*Carduelis flammea*) vary dramatically in northern Sweden, and on a local scale these variations can be closely related to variations in the abundance of their favored food, birch seeds (Fig. 4.6; Enemar *et al.* 1984). In a similar fashion, variations in the abundance of frugivorous Phainopeplas (*Phainopepla nitens*) and Northern Mockingbirds (*Mimus polyglottos*) in riparian habitats of the southwestern United States closely matched annual changes in the abundance of mistletoe berries caused by unusually cold weather (Anderson and Ohmart unpublished).

Not all members of a local community necessarily track climatic or resource variations in the same way. In the southwestern riparian habitats, for example, American Robins (*Turdus migratorius*), Western Bluebirds (*Sialia mexicana*), Cedar Waxwings (*Bombycilla cedrorum*), and Sage Thrashers (*Oreoscoptes montanus*) also fed on mistletoe berries, but densities of these species varied independently of mistletoe crops (Anderson and Ohmart unpublished). Faaborg *et al.* (1984) examined patterns of popula-

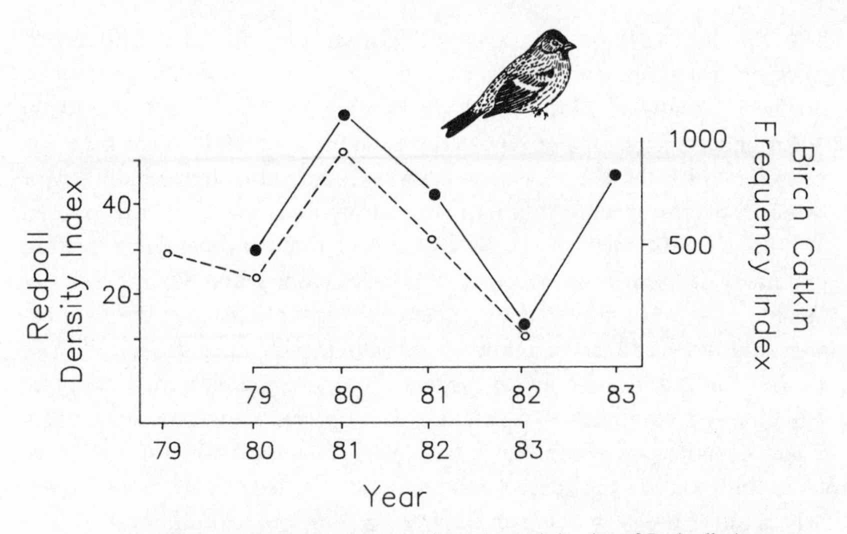


Fig. 4.6. The relationship between the estimated density of Redpolls (open circles) and the supply of birch seeds (expressed as the catkin frequency in the preceding year, solid circles) in Swedish Lapland. From Enemar *et al.* (1984).

tion fluctuations and rainfall over 9 yr in a dry forest in Puerto Rico and found an association between total community density levels and variations in spring rainfall during the previous year. Different guilds of species varied in different ways, however, frugivores fluctuating considerably more than insectivores. Only insectivore numbers showed a significant relationship to rainfall variations.

Fluctuations in precipitation are an obvious sort of environmental variation that may bear a close relationship with food abundance, and most attempts to document environmental tracking have related bird abundances to rainfall variations. Folse (1982), for example, measured changes in bird densities, arthropod abundances, and vegetation structure between wet and dry seasons over a single year at several locations in the Serengeti. Variations in abundances of Wing-snapping (*Cisticola ayresii*) and Zitting (*C. juncidis*) cisticolas closely followed arthropod levels, but the other species did not track these variations (although some apparently responded to seasonal changes in vegetation structure). In South Africa, records of the sizes of hunting bags of Orange River Partridge (*Francolinus levaillantoides*) from a single estate over a 76-yr period were positively correlated with rainfall amounts during the breeding season (Berry and Crowe 1985). Dunning and Brown (1982) found that total densities of wintering sparrows were significantly correlated with precipitation during the previous summer at four of five locations in southern Arizona (see also Pulliam 1986, Pulliam and Dunning 1987), but Laurance and Yensen (1985) found no evidence of tracking of rainfall variations by wintering sparrows in a similar analysis of northern Great Basin locations.

Rotenberry and I have repeatedly examined our data from breeding communities in North American grassland and shrubsteppe habitats for evidence of close tracking of rainfall variations and have found rather little. We surveyed 14 shrubsteppe locations over three consecutive years, an extremely dry year followed by two unusually wet years (Rotenberry and Wiens 1980b). Habitat structure changed substantially over the sites between the dry year and the following wet year, producing an overall increase in the height and coverage of vegetation and a decrease in horizontal patchiness. Coverages of different shrub species remained unchanged. There was considerable annual variation in local population densities of the birds, but none of the species exhibited any consistent pattern of change across all of the sites between the dry and wet years. Variations in bird densities on any single site were largely independent of variations on other, nearby sites, and there were no clear indications of any close tracking. When we extended these analyses for five of the locations over an additional 4 yr, the patterns remained much the same: considerable annual variation in vegetation that was consistent over the sites, little change in shrub species coverages, and substantial variation in bird densities that was neither consistent over the sites nor correlated with the variations in habitat structure. An analysis of annual variations at a shortgrass prairie location (Rotenberry and Wiens 1980a) indicated a similar lack of concordance between variations in bird densities and environmental variations.

The shrubsteppe analyses have two weaknesses. First, we began our studies in a drought year and therefore lacked critical information about bird densities for the previous year. In his studies of drought effects on bird populations in several montane habitats, Smith (1982) found the greatest differences between predrought and drought years. The sudden drop in annual precipitation was closely tracked, but the subsequent return of precipitation to normal levels was not. Smith also found clear differences between habitats in the impact of the drought – communities in deciduous forests were strongly affected, whereas those in alpine meadows were relatively unaffected.

Second, our annual comparisons were regional, in that we related changes in densities of the birds on local plots to regional precipitation variations. Maurer (1985) has justifiably criticized this approach for possibly obscuring tracking that might occur at a local scale, a problem we have also acknowledged (Wiens 1986). When we look at rainfall variations and bird densities at this local scale, however, we still find that variations in abundances of most species are unrelated to precipitation, although there are indications of possible tracking by some species. At a shrubsteppe site in central Oregon, for example, annual changes in abundances of Sage Sparrows (*Amphispiza belli*) were generally parallel in nearby census plots and tracked variations in rainfall during the winter–spring period preceding the *previous* year, although there was no relationship to rainfall during this period in the same year (Wiens, Rotenberry, and Van Horne 1986). At a shortgrass prairie location in Texas, we recorded annual changes in the densities of three breeding species that closely paralleled local rainfall variations in a way one would expect from the more general habitat relationships of the species, although such patterns were not evident at other grassland sites (Wiens 1974). In short, we do find occasional evidence of tracking of environmental variations in these systems by some species at some localities, but the general pattern is one of apparently independent variation in bird densities.

In order to document true tracking rather than general correlations, one must have not only density estimates over a series of consecutive years but measures of environmental conditions and resource levels as well. This information is necessary to determine whether or not the 'tracking' is anything more than coincidence and to establish what is being tracked, with what precision. Measures of resource levels are especially important. Resources may display a variety of temporal dynamics (Price 1984), each of which offers a different potential for close tracking. Simply recording general environmental variation in features such as rainfall offers no assurances that the important resources vary in the same way. Howe (1983), for example, documented how unusually heavy rains on Barro Colorado Island, Panama produced a fourfold variation in fruit production and dramatic changes in fruiting phenology among most plants. These changes had effects on many of the frugivorous animals in the forest, but one group of species remained unchanged. These species fed primarily on the fruits of *Virola surinamensis*, which remained a stable resource despite the variations in climate and in the other plants.

None of the investigations of tracking discussed above really meets the requirements of simultaneous measurement of birds, resources, and environment over several years. Some of the long-term investigations that I discuss later in this chapter, although not directly focused on tracking, provide some important insights. The most careful documentation of tracking of environmental and resource variations by bird populations comes from the studies of Galápagos finches conducted by Grant and his colleagues. As this example is especially relevant to an evaluation of the 'ecological crunch' view, I will consider it after some aspects of that model are first explored.

***Environmental crunches***

Environments often vary episodically, sudden, high-magnitude fluctuations in temperature, rainfall, tidal extremes, sea temperature and the like occurring unpredictability or on cycles longer than the generation lengths of organisms (e.g. Bleakney 1972, Quinn and Neal 1982, Davis 1986, Pickett and White 1985). Kerr (1984), for example, suggested that drought cycles in the North American Great Plains and patterns of seasonal monsoons in India are influenced by both an 18-yr lunar cycle and an 11- or 22-yr sunspot cycle; when the cycles coincide, the effects may be especially severe. Fluctuations in oceanic conditions associated with El Niño produce well-documented changes in weather over a large area (Chavez and Barber 1984) and may have dramatic effects on breeding seabird populations (Duffy & Merlen 1986, Schreiber and Schreiber 1984). When environmental variations are pulsed and extreme, changes in resource levels may be so large and sudden that close tracking is difficult. Such environmental crunches may reduce population levels or community membership through resource limitation, and competition at such times may be intense. If the resource levels recover more rapidly than population sizes after an amelioration of environmental conditions, resources may be superabundant. This may remove competitive constraints, leaving individuals and populations free to wander some distance from theoretically optimal forms of resource use. Populations may vary at such times due to a variety of density-independent or chance effects, following 'density-vague' dynamics (Strong 1984b, 1986). If the environmental crunches are severe but occasional, resources may be superabundant and community structure only coarsely tuned a good deal of the time.

This is the basic structure of a verbal model I proposed several years ago (Wiens 1974, 1977) to explain why bird communities sometimes might not be as tightly structured and equilibrial as the community theory in vogue at that time suggested they should be. I questioned the assumptions that (1) populations and communities were at resource-determined equilibria, (2) selection on resource-utilization traits was thus continuous and intense, (3) interspecific competition was the major selective force influencing community patterns, and (4) the theoretically optimal state was available to a population in its evolutionary development. I concluded that (1) competition may be less pervasive in nature than had been presumed, (2) what we witness in nature may at times only coarsely fit the optimal states expected from theory, and (3) documenting competition in nature may thus be difficult, given its sporadic occurrence. These notions were not entirely new or original. Andrewartha and Birch had offered a somewhat similar perspective on population dynamics two decades earlier (1954), and MacArthur (1972) noted that food might be scarce enough to cause severe competition only 1 year in 20, although the effects of that competition might persist for some time. Cody (1981) recognized (but did not discuss further) the possibility that food resources might be superabundant in some years and some habitats and that community structure might degenerate under these conditions.

This is not the place to discuss the details of this view (see Wiens 1977, 1986; Grant 1986b presents a detailed evaluation of the arguments). The ideas were prompted by our findings that in North American grassland and shrubsteppe breeding bird communities evidence of the patterns expected in competitively-structured communities was scarce (see Volume 1, Chapter 10) and food supplies were seemingly superabundant (Wiens and Rotenberry 1979, 1987, Wiens 1977, Rotenberry 1980). Other studies have also provided evidence suggesting an intermittency of resource limitation and competition associated with environmental variations. The effects of the abnormally cold winter in England or of the Sahel drought described at the beginning of this chapter are clear examples. Abbott and Grant (1976) noted that the number of landbird species on islands in the Australia–New Zealand region is not in equilibrium, and they attributed this to irregular climatic fluctuations. On Trinidad and Tobago, variations in the food supplies of nectarivores undergo substantial seasonal variations, and populations may encounter scarce resources only during the 3-month dry period (Feinsinger and Swarm 1982). Wintering sparrows in southern Arizona may encounter limiting quantities of seeds only following summers of exceptionally low rainfall (Pulliam and Parker 1979, Dunning and Brown 1982, Pulliam and Dunning 1987), and frugivorous and nectarivorous birds in Hawaiian native forests may experience resource-based competition only infrequently (Mountainspring and Scott 1985). In Costa Rica, Freed (1987) found that Rufous and White Wrens (*Thryothorus rufalbus*) killed nestlings of syntopic House Wrens only during one 8-week period of food shortage over the 4-yr study. In hardwoods forests in New Hampshire, on the other hand, outbreaks of lepidopteran larvae produce superabundant food levels at irregular periods separated by intervals of as long as a decade. In this situation, food resources are apparently often limiting and competition may be important (Holmes *et al.* 1986). Periods of resource superabundance do occur, but the 'crunch' conditions persist for longer periods.

In the aspen parkland-prairie region of Canada there is a spectrum of increasing temporal variability in the habitats of breeding waterfowl from those of parkland diver species through parkland puddle ducks and prairie divers to prairie puddle ducks. Nudds (1983) found clear patterns of community structure only among the parkland divers. He related the absence of such patterns in the other assemblages to the greater environmental variation of their habitats and the frequent absence of resource limitation. In montane habitats of Utah and Idaho, bird densities were reduced and community structure altered by a severe drought (Smith 1982). The population declines coincided with dramatic reductions in insect abundance, especially in aspen woodlands; although insects returned to predrought levels in the following year of normal precipitation, bird densities remained low. Variability in the degree of apparent resource limitation has also been recorded in experimental studies. Eriksson (1979) recorded yearly variations in the responses of ducks to experimental removals of potential fish competitors, for example, and Källander (1981) found that Great Tits responded to food supplementation manipulations during a severe winter but not during a mild winter.

We normally think of the likelihood of resource limitation in variable environments being greatest when environmental conditions are most severe. This may often be the case, but it is not necessarily so. Resource limitation, after all, is a consequence of the relationship between resource availability and the demands placed on the resources by the populations. 'Crunches' occur when the ratio of supply to demand is quite low. Anderson and Ohmart (unpublished), for example, used bioenergetic/activity budget calculations to determine the food demands of the species feeding on mistletoe berries in Arizona and related these demands to estimates of berry supplies per territory. By these calculations, food supplies appeared to be limiting during only 3 months in 1 of the 10 years of the study. This occurred not when berry abundance was lowest but during a winter of moderate berry crop when wintering bird densities were unusually high.

One of the most thoroughly documented examples of responses of populations to variations in environment and resource levels is provided by the Galápagos finches. Over the decade or so that Grant and his colleagues studied these finches, finch biomass over a series of islands was closely related to resource levels (seed biomass) during the dry seasons but not in the wet seasons (Grant and Grant 1980, Grant 1986a). On some islands and in some years, rains were extensive and food was not limiting even in the dry

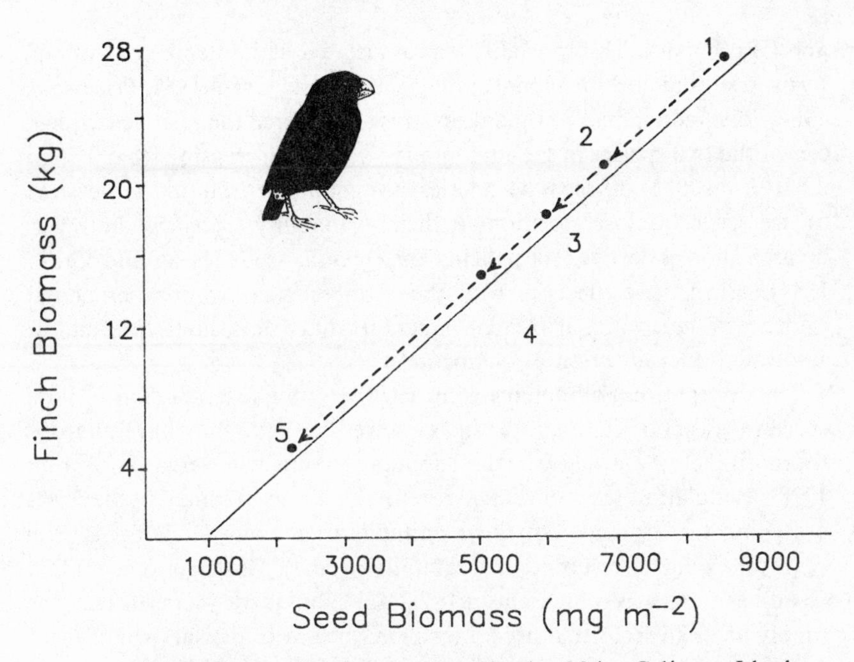


Fig. 4.7. Biomass of *Geospiza* species on Daphne Major, Galápagos Islands, through an extended dry period in relation to seed biomass on the island. The observed values are indicated by the numerals (1 = May 1976, 2 = December 1976, 3 = March 1977, 4 = June 1977, 5 = December 1977). The least squares regression is shown by the solid line; the dashed arrows indicate the progressive decline in finch and seed biomasses. From Grant (1986b).

season. Resource limitation is thus intermittent and clearly associated with dry periods. On Daphne Major, for example, 1973 was abnormally wet, food supplies were apparently abundant throughout the year, survival and reproduction of the finches were high, and no interspecific competition was in evidence (Grant and Grant 1980, Smith *et al.* 1978; the same conditions obtained during the El Niño period in 1982–83 – Gibbs and Grant 1987b, c, Grant and Grant 1987). There was a severe drought in 1977, however. This drought almost completely eliminated the small populations of *Geospiza fuliginosa* and *G. magnirostris* on the island and sharply reduced the abundance of the numerically dominant species, *G. fortis*. The pattern of progressive decline in total finch biomass as the drought on Daphne intensified tracked rather closely the decline in estimated seed biomass (Fig. 4.7). The two resident finch species on Daphne, *fortis* and *scandens*, initiate breeding at different rainfall thresholds, and only *scandens* bred during the drought, with low success. Mortality of both species was high, and by 1978 the population of *scandens* had decreased by 66%, that of *fortis* by 85% (Boag and Grant 1984). This mortality was associated with intense selection on *fortis*, resulting in shifts in morphology (Boag and Grant 1981, Price *et al.* 1984). The reductions in population sizes also altered the relative frequencies of the two species in the community, *scandens* increasing from 0.25 to 0.50 (Grant 1985). As seeds become less abundant, *fortis* shifted its diet so as to use seeds in closer relation to their availability; *scandens*, however, became more selective, specializing on *Opuntia* seeds (Boag and Grant 1984). All of these effects provide the strongest sort of nonexperimental evidence of the severity of an environmental crunch, food limitation, and an associated intensification of competition.

The drought on Daphne broke in 1978, and the reproduction of both species was good, although the male-skewed sex ratio may have retarded the reproductive output of the total population somewhat (Boag and Grant 1978). Population sizes of *scandens* returned to approximately the levels preceding the drought, but *fortis* populations remained low for several years following the resumption of rainfall (Fig. 4.8). This might suggest that resources were superabundant in 1978 and subsequent years. In fact, the supply of small seeds that are critically important to the survival of *fortis* during the dry season remained low during this period despite the resumption of normal rains (Fig. 4.8). This is because many seed-bearing plants died during the severe drought of 1977 and were not immediately replaced. One of the major producers of small seeds, *Heliotropium angiospermum*, became common again only in the extraordinarily wet El Niño year of 1983 (Grant 1986a).

Thus, despite appearances, the Galápagos finches do not fit the ecological crunch model very well. Dry periods are frequent enough and recovery of some of the food resources slow enough that a finch probably experiences severe limitation at least once during its lifetime. Although periods of resource superabundance and relaxed competition do occur, they apparently are not a common feature of life on the Galápagos.

Why do the finches on the Galápagos appear to track resource variations rather closely whereas birds in the shrubsteppe, which also experience episodic environmental crunches, appear not to? One reason clearly has to do with the spatial constraints of the systems. Finch populations on the islands in the Galápagos are largely closed, and most individuals spend their entire lives within a small area. Moving elsewhere to avoid severe environmental conditions is not an available option. The shrubsteppe birds, on the other hand, are all migrants, wintering in areas well to the south. The local communities are quite open. This means that they may not be constrained to follow local environmental variations closely and that the year-to-year

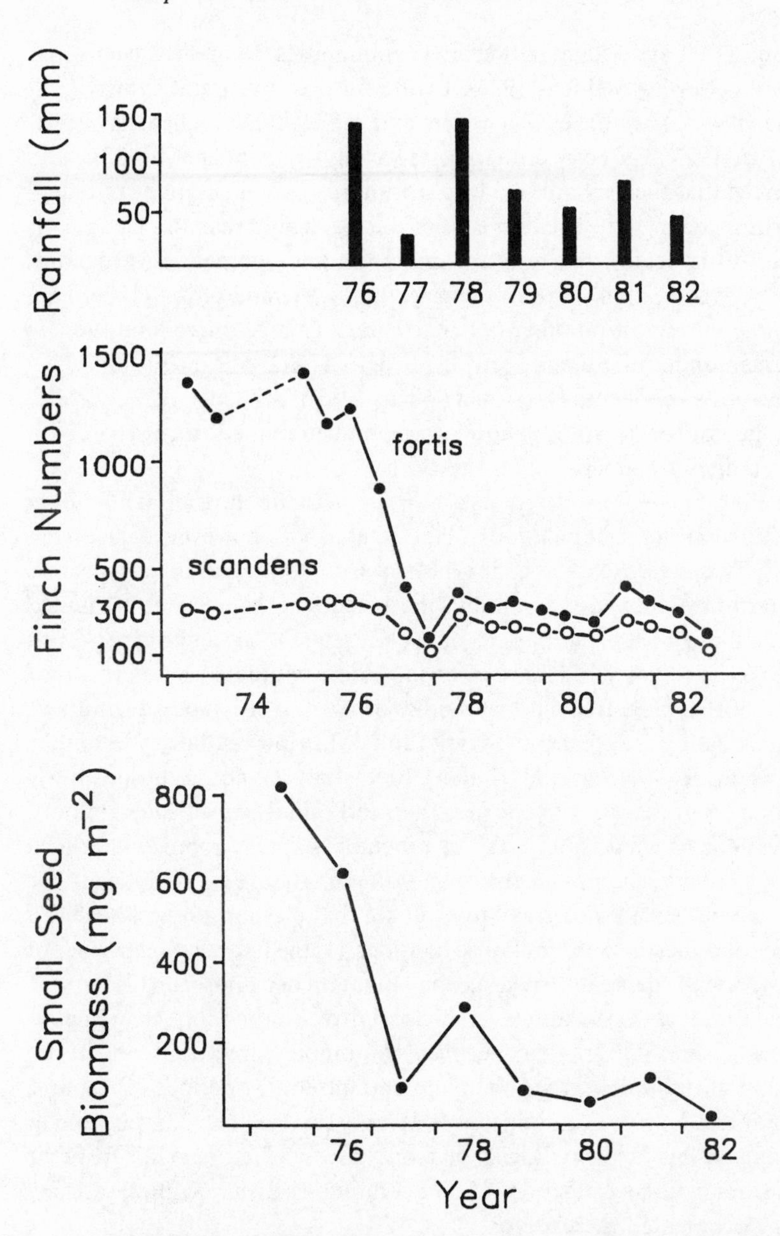


Fig. 4.8. Rainfall (top), population sizes of *Geospiza fortis* and *G. scandens* (middle), and biomass of small seeds present at the beginning of the dry season (June–July, bottom) over a sequence of years on Daphne Major, Galápagos Islands. From Grant (1986b).

dynamics that are evident in breeding communities may reflect events that occurred elsewhere (Wiens 1986). I will return to this point shortly.

The view of variable environments and ecological crunches that I proposed in 1977 has been criticized, especially by Schoener (1982), who suggested that the available evidence (primarily documentations of reductions in niche overlap between species during 'lean' times; see Chapter 1, Table 1.6) indicates that resource limitation and competition occur frequently, even in highly variable environments. He and Grant (1986b) are probably correct in pointing out that I portrayed resource limitation as occurring quite infrequently, but Schoener's (1982) interpretation of my arguments was not entirely correct (Wiens 1983b, Schoener 1983a). In any case, the critical question is how often populations find themselves with either limited or superabundant resources.

Put another way, we may ask how unusual limiting or nonlimiting conditions are to a population. What is 'unusual', however, depends on one's perspective. Weatherhead (1986) surveyed 380 papers in four journals to determine under what conditions the authors of these papers attributed their findings at least in part to unusual events. One might expect such events to be encountered more often in the longer studies, but in fact they were reported more frequently in studies of short to intermediate duration. It seems that we may perceive a year of somewhat low rainfall, for example, as unusual in a 2- or 3-yr study, but in a 10-yr study it is not so conspicuously different from the other years. Weatherhead's analysis indicates that our impressions of what is 'unusual' or potentially an environmental crunch may not be very reliable and that quantitative analyses of variation patterns and magnitudes are necessary to evaluate the question posed above. To answer this question, one must also consider: (1) the frequency of periods of severe resource depression relative to population demands, and (2) the time lags that prevent populations from closely tracking rapidly changing resources. There is a gradient between continuous limitation and precise tracking of resources on the one hand and infrequent periods of resource limitation interspersed among times of plenty on the other. The position of a system on this gradient is dependent on these two factors. Aside from the finch studies, no investigations of bird community dynamics have satisfactorily documented these factors.

### *Time lags in responses to environmental changes*

Individuals may not respond immediately to changes in environmental conditions or resource levels, especially if those changes are relatively small. Moreover, alterations of individual behavior with changing resource levels may be expressed as changes in niche measures (e.g. diet, foraging tactics or locations) but not immediately as changes in demography. Time lags in demographic features may either stabilize or destabilize population dynamics and equilibrium densities, at least in theory (e.g. May 1973, Maynard Smith 1974, Nunney 1983, 1985). Regardless of the effects on theoretical population stability, however, time lags erode the match between population levels and proximate environmental and resource conditions. This makes it much more difficult to associate community patterns with presumed causal factors, especially in short-term investigations.

Time lags are evident in many aspects of bird biology, especially at the population level. Even in the Galápagos finches, which appear to track environmental variations rather closely, populations may not respond to increasing resource levels by immediate density increases (Grant 1986b). Immigration is restricted and breeding is seasonal, so that population growth may lag at least several months behind resource changes. Because the reproductive capacity of individuals is rather low, recruitment to reach densities at which resources become limiting may take several years, even if survivorship is high. If the sex ratio is skewed, additional lags in population growth may occur. In other situations the elements contributing to response lags may be less clear, but the lags are nonetheless evident. Holmes and Sturges (1975), for example, noted how abundances of foliage-gleaning insectivores increased in a hardwoods forest in association with an outbreak of the moth *Heterocampa guttivitta.* The abundance of caterpillars and adult moths peaked in 1970 and declined sharply thereafter, but foliage-gleaner densities continued to increase through 1972, presumably as a consequence of the high recruitment and survival rates in 1969 and 1970.

Populations may also lag in their responses to habitat changes. During our studies in the North American shrubsteppe, for example, one of our sites, Guano Valley, was subjected to 'range improvement' alterations by land management agencies (Wiens and Rotenberry 1985). A large area of sagebrush-dominated habitat was aerially sprayed with a herbicide, and the following autumn the dead shrubs were broken down and removed and an exotic grass, *Agropyron cristatum*, was planted. The treatment had dramatic effects on sagebrush and grass coverage and on several other features of habitat structure (Fig. 4.9). We expected that such massive changes in the habitat would have immediate effects on the densities of breeding birds, but they did not. Sage Sparrows, which generally exhibit a close numerical association with sagebrush coverage (Wiens and Rotenberry 1981a), did not decrease in abundance in the following years, and Horned Larks

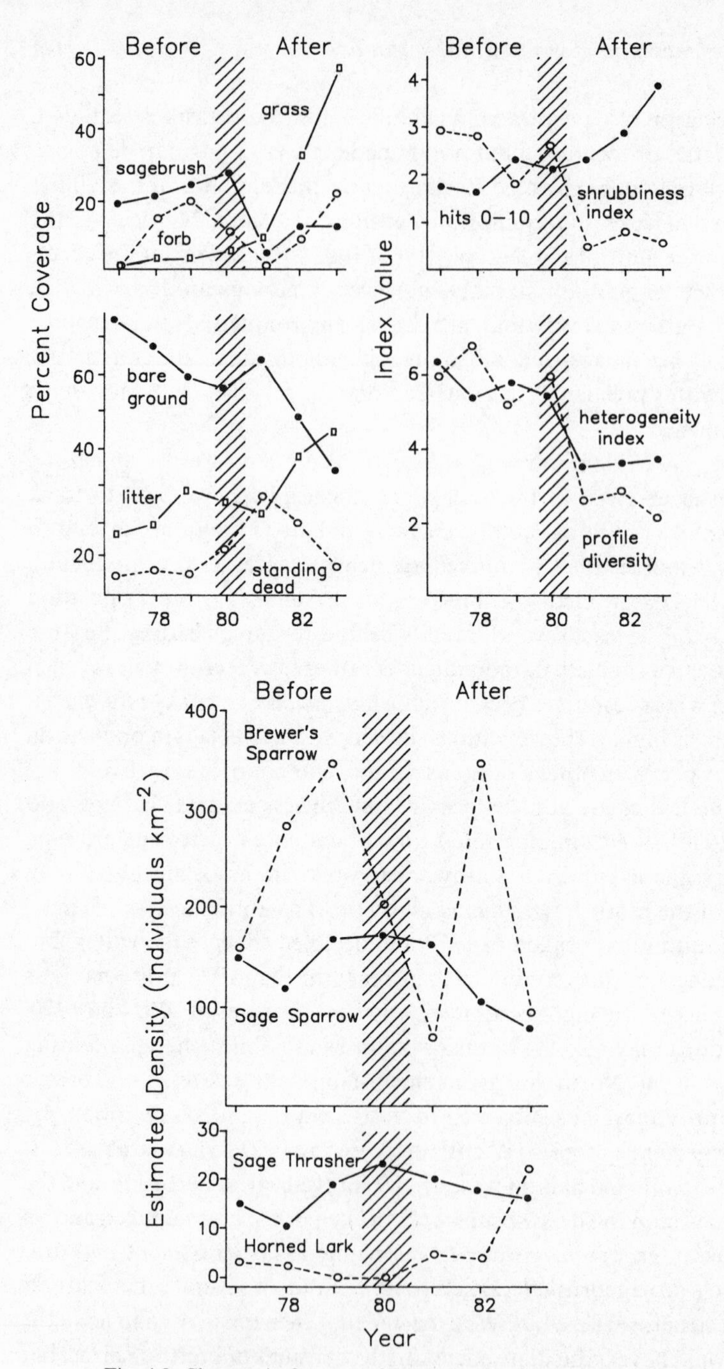


Fig. 4.9. Changes in features of vegetation coverage and habitat structure (top) and in the abundances of breeding bird species (bottom) in shrubsteppe habitat at Guano Valley, Oregon that was subjected to herbicidal treatment, shrub removal, and planting of exotic grasses in 1980 (hatched). After Wiens and Rotenberry (1985).

(*Eremophila alpestris*), which are generally more closely associated with higher grass coverage, did not increase in density. After two or three years, however, populations of each of these species began to change in the anticipated manners (Fig. 14.9). We documented a similar time lag in the responses of Sage Sparrows to a more natural reduction of sagebrush coverage by a rangefire in another location (Rotenberry and Wiens 1978, Wiens 1985b). Lags in responses of Brewer's Sparrows (*Spizella breweri*) and Field Sparrows (*Spizella pusilla*) to sudden changes in habitat structure have also been reported (Schroeder and Sturges 1975, Best 1979).

Adult birds that have previously bred successfully in an area often return to the same location to breed in subsequent years, and we interpreted the time lags in responses to the shrubsteppe habitat alterations as consequences of such philopatry or site tenacity (Wiens and Rotenberry 1985, Wiens 1986). Many of the Sage Sparrows that bred at Guano Valley, for example, might have returned there after the range treatment and attempted to breed despite the habitat changes. If such site tenacity were strong, a population response to the alterations might begin to be evident only after such individuals died or moved elsewhere. Some Sage Sparrows live at least 7 yr (Wiens, Rotenberry, and Van Horne 1986), so the time lag introduced by such behavior could be considerable. O'Connor (1985) also drew attention to the ways in which site tenacity distorted the patterns of distribution of Chaffinches among optimal and suboptimal habitats in England from those that would be expected on the basis of densities. Chaffinch abundance was reduced substantially after the severe winter of 1961–62. A portion of the survivors continued to occupy the suboptimal habitats with which they were familiar in the 1962 breeding season, despite the availability of more suitable habitat. Their offspring settled in the more suitable habitat, but it was only after mortality over the next several years reduced the numbers of 1961 survivors occupying suboptimal habitats that the habitat distribution of the population began to approximate the pre-1961 pattern.

What factors influence the magnitude or importance of time lags in population responses to environmental variations? Site tenacity is clearly one such factor, and its importance is related to both the degree of expression of tenacity in individuals and the longevity of the tenacious individuals in a location. Bird species differ in average and maximum longevity and individual turnover rates in populations (e.g. Jedraszko-Dabrowska 1979), and the potential for considerable variation among members of a community in time lags is therefore great.

Beyond this, however, there may be lags introduced in the steps necessary

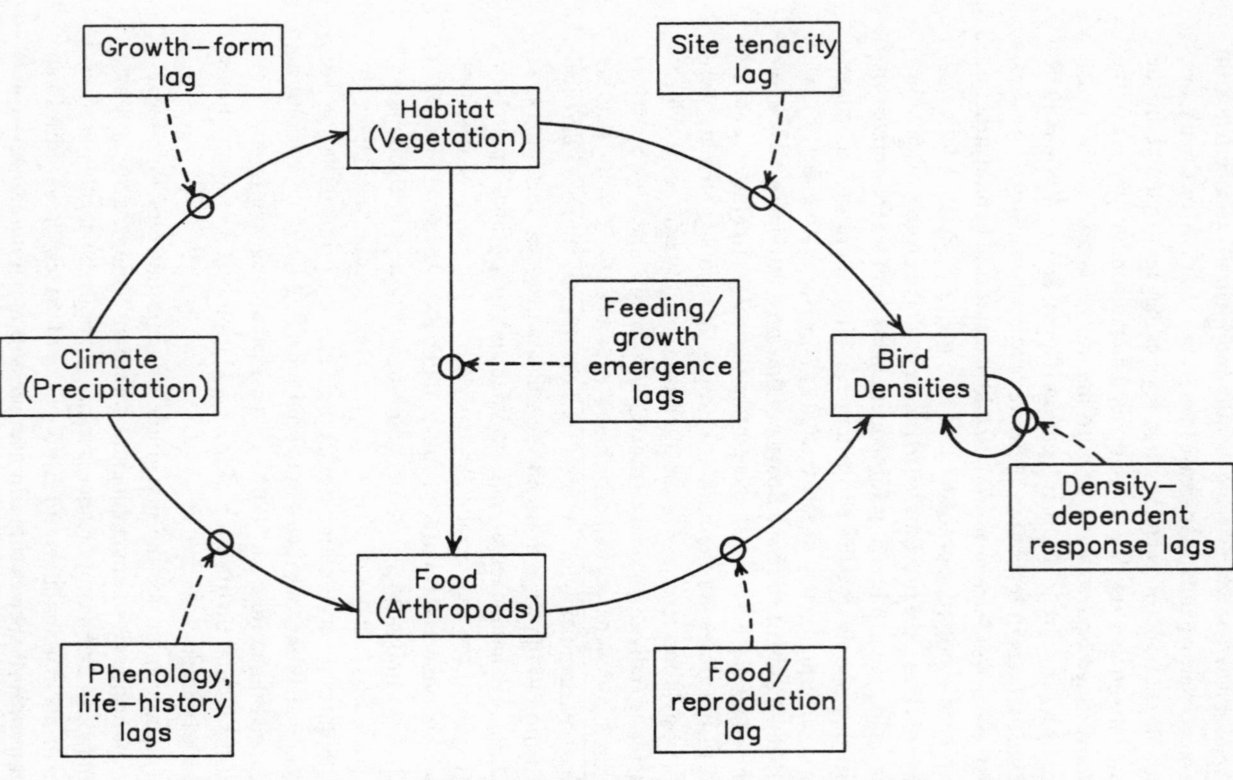


Fig. 4.10. Potential influences of variations in climatic factors (precipitation, in this case) on bird population densities, showing the avenues of effects and sources of time lags in population responses to climatic variations. After Wiens (1986).

to translate an environmental change into a population response. Variations in a climatic factor such as precipitation, for example, may influence bird populations through effects on habitat features and/or food supplies, but such effects are not likely to be direct (Wiens 1986; Fig. 4.10). Depending on the season of rainfall and the growth form of the plants, there may be a variable time lag in the response of the vegetation through germination or plant growth. There will be further lags before these changes are reflected in changes in seed availability. Arthropod populations may also respond to the changes in moisture or herbaceous food supplies with lags associated with their life-history traits and physiology. Converting these changes in resource levels into changes in bird densities may involve additional time lags, some associated with site tenacity, others with reproductive phenology or potential or the operation of density-dependent population processes. Bird populations that have a high potential growth rate, for example, may respond more rapidly to environmental changes than will low-growth populations; by avoiding the mismatches with resource levels produced by time lags, they may achieve a greater degree of long-term stability in numbers (Pimm 1984), at least in the absence of large-magnitude environmental fluctuations. Other factors, such as the periodicity or severity of environmental fluctuations, the degree of closure or openness of the community, or the availability of alternative resources in the system, may also influence the operation of time lags or the closeness of tracking of variations. Because these factors and lags differ among resource types and among species in a community, there is a large and bewildering array of possible temporal dynamics in the community as a whole (Price 1984, Wiens 1986). Given this, it seems likely that both the close tracking and environmental crunch views of community variation, as well as other intermediate or even more extreme views, can contribute to increasing our understanding of community dynamics.

**Spatial aspects of temporal variation**

It is a commonplace observation that population of a species do not vary in the same way in different locations or habitats. Community dynamics also vary from place to place, even in quite similar habitats with the same species composition. In grassland or shrubsteppe locations, for example, populations of the same species in study plots within 100 m of each other in the same habitat type may vary between years in quite different ways, sometimes even changing in opposite directions (Wiens and Dyer 1975b, Rotenberry and Wiens 1980b). We have often recorded substantial independence in annual changes of populations and communities between local plots or among plots at various locations within the grassland or shrubsteppe. When we combine samples and compare yearly dynamics averaged over a regional scale, however, we find substantially less variation in population densities or community composition (Rotenberry 1986, Wiens 1986). Superimposed on the variability within and among plots at a local scale is a regional-scale equilibrium of sorts.

Such spatial differences in population and community variation are not unique to arid and semiarid environments. Bird assemblages in tropical forest understorey habitats, for example, vary considerably at the scale of local sampling stations or study plots but are much more stable when viewed over a collection of study plots (Karr and Freemark 1983, 1985). A similar scale-dependence of variation occurs in northeastern hardwoods forests in North America (Holmes *et al.* 1986) and among small islands in Finnish archipelagos (Haila and Järvinen 1981). Noon *et al.* (1985) attributed their failure to find systematic differences in community stability between different habitat types in North America to the substantial variation among widely scattered samples within each habitat type.

Two general hypotheses may be offered to explain the relationships between population and community dynamics at the local and regional scales (Väisänen *et al.* 1986). First, changes at the regional scale may simply result from a summation or averaging of variations that occur at more local scales within the region. If the temporal dynamics of populations at a number of local sites are influenced by a variety of factors that operate more or less independently among the sites, then variation viewed at the regional scale will inevitably be less than that at any given local site as a consequence of the Central Limit Theorem. The variations among local sites are especially likely to be somewhat independent of one another if community dynamics at each site are influenced by stochastic factors.

One way such factors may affect local dynamics is through 'sampling error' of one form or another. Thus, some of the local variation may be a methodological consequence of inaccuracies in censusing methods or variations in techniques or observers from year to year or site to site. There may be biological sources of such 'sampling' variation as well, however. Rotenberry and I have described this problem using the analogy of a checkerboard (Rotenberry and Wiens 1980b, Wiens 1981a). We represent the territories of breeding individuals by checkers and scatter some number of checkers over the board. We then conduct a 'census' of this population by recording the number of checkers and portions of checkers contained in some part of the board, say a 9-square block (Fig. 4.11). This provides an estimate of population density for a given year. We then simulate the process

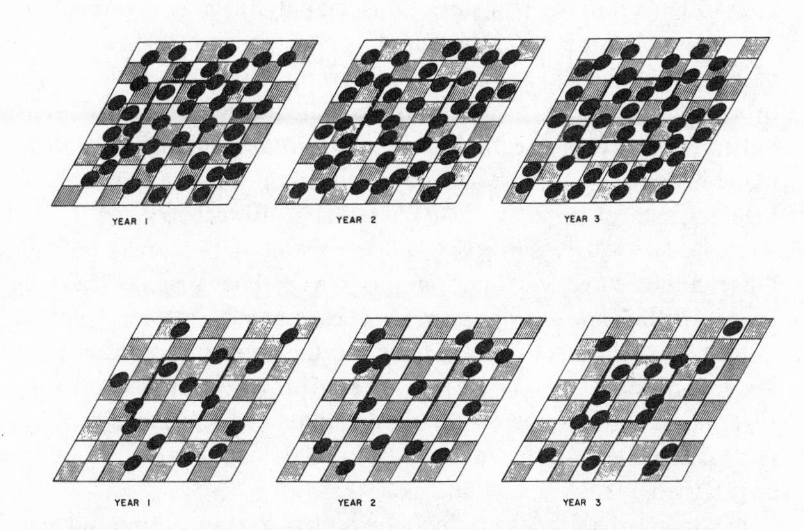


Fig. 4.11. A 'checkerboard model' of local population variation. Part A represents the distribution of territories of a species (checkers) in a nearly saturated habitat (densely packed board), and Part B depicts the dispersion of territories in a sparsely packed habitat. The habitat is assumed to be uniform, and the squares on the board therefore do not represent habitat patches of different types. The solid line encompasses a census plot established in the area. In A, the number of territories included in the plot (and thus the estimate of density) varies little between years, but in B there is substantial yearly variation in the estimates, despite the fact that the total number of checkers on the board as a whole remains constant. From Wiens (1981a).

of annual migratory departure from the plot and the return and resettlement of individuals in the following year by shaking the board to redistribute the checkers (having placed a barrier around the edges of the board to keep checkers from falling off). We then 'census' the 9-square block again, to estimate densities in the second year, and repeat this procedure for a number of years to chart the annual dynamics of the species in this local plot. Note that we have not added or removed checkers from the board – the population occupying the entire board remains constant. Nonetheless, we are likely to record year-to-year changes in the density estimates derived from the 9-square block. These variations will be especially great if there are few checkers on the board (i.e. the habitat is not saturated with breeding individuals); in fact, the local censuses will accurately reflect the dynamics of the 'true' population (constancy in numbers) only when the available habitat is completely packed with individuals (Fig. 4.11).

Translated into the real world, this analogy suggests that much of the variation we record in local census plots may represent chance reshufflings of individuals (as well as mortality and recruitment) that are less apparent over a larger, regional area. This is especially likely if the populations are not at resource-defined equilibria and the habitat is not fully saturated with individuals, as we believe occurs much of the time in communities of breeding shrubsteppe birds (Wiens and Rotenberry 1979, 1981a, Rotenberry 1980, Wiens 1977, 1981a, 1986). If two areas or habitats differ in the degree of saturation, this 'checkerboard' model leads one to expect greater annual variation in densities estimated from local plot counts in the less saturated environment, even if the populations at a larger scale are equally stable (or variable). The checkerboard analogy also suggests that the variations we record in populations at a local scale may not necessarily reflect the dynamics of the 'true' populations, which occur on a larger spatial scale.

Expanding the size of local survey areas would reduce this mismatch, but there are logistical constraints on the size of area that can be censused accurately. By increasing the area surveyed, one may trade increased resolution of the true dynamics of the population for increased variance due to sampling error in the traditional sense. Had we sampled a larger area (say, a 25-square block) in our checkerboard exercise, we would have recorded less yearly variation, even in the unsaturated environment. This indicates that the magnitude of temporal variability may be influenced by the size of the census area (a factor that has rarely been considered in analyses of the temporal dynamics of bird communities) and, at the very least, that comparisons of temporal patterns in areas of different sizes surveyed at different times are likely to be invalid.

A second hypothesis linking local to regional temporal dynamics states that the variations at a local scale are influenced by changes at the regional scale; the local dynamics cannot be fully understood without reference to the regional context (Väisänen *et al.* 1986). If, for example, there are gradual regional changes in climate, one would expect some degree of conformity among the long-term changes at local sites, even though they might be more variable from year to year. There are a number of ways in which regional events might influence local dynamics, depending on the spatial scale to which 'regional' applies. Broadly defined, events on distant wintering areas may influence year-to-year dynamics at local breeding sites, for example. Influences at this scale will be considered later in this chapter.

At a smaller 'regional' scale, it is apparent that the surrounding environmental context of a given local census plot may have profound influences on the population and community dynamics one records in that plot. In forest habitats in eastern North America, for example, variations in annual turnover rates of species in mature stands may be affected by proximity to younger stands (Yahner 1987) and the abundance of certain guilds may be associated with the distance of forest patches to other areas of forest or to suburban habitats (Askins and Philbrick 1987).

Patches of habitat are arrayed in mosaics over a landscape, and different patches may differ in suitability or 'quality' to individuals of a species. Densities may or may not be greater in the higher-quality habitat patches (see Volume 1, Chapter 9), but *variability* in density is likely to be less in suitable than in marginal habitat patches, at least if habitat-occupancy patterns are strongly influenced by density (Fretwell and Lucas 1969, O'Connor 1986). Local populations of Willow Warblers in favored deciduous woodlands in southern Finland, for example, vary less between years (CV = 0.05) than do those in nearby spruce-dominated habitat patches (CV = 0.28) (Tiainen 1983). This greater stability in the more suitable habitat patches is consistent with the notion that such habitats are usually saturated or nearly so and that additional individuals are forced into the less suitable patches. Variations in reproduction or survivorship between years may have rather little effect on densities in the suitable habitats but are reflected in differing magnitudes of 'overflow' into other habitat patches. At the extreme, a regional landscape mosaic may represent a series of 'source' habitat patches, in which population densities are relatively stable and reproduction and survival are high, and 'sink' patches, in which breeding performance and/or survival are poor. Populations in sink patches may persist largely through recruitment from source patches and may vary considerably (Fig. 4.12). Their dynamics are thus closely linked to events in the source patches (Lidicker 1975, Wiens and Rotenberry 1981b), but density variations in the source patches may likewise be influenced by the availability of nearby sink patches to absorb emigrants. If the regional landscape mosaic changes, as through fragmentation of forests or changing human land use, the relationships of source and sink patches may be changed, altering temporal dynamics in the sink patches and perhaps the source patches as well (Helle and Järvinen 1986) (Fig. 4.12). The distinction between source and sink patches obscures the fact that habitat patches in a landscape mosaic represent a gradient in suitability, but it serves to draw attention to the ways in which local population dynamics and variations may be linked at a larger scale.

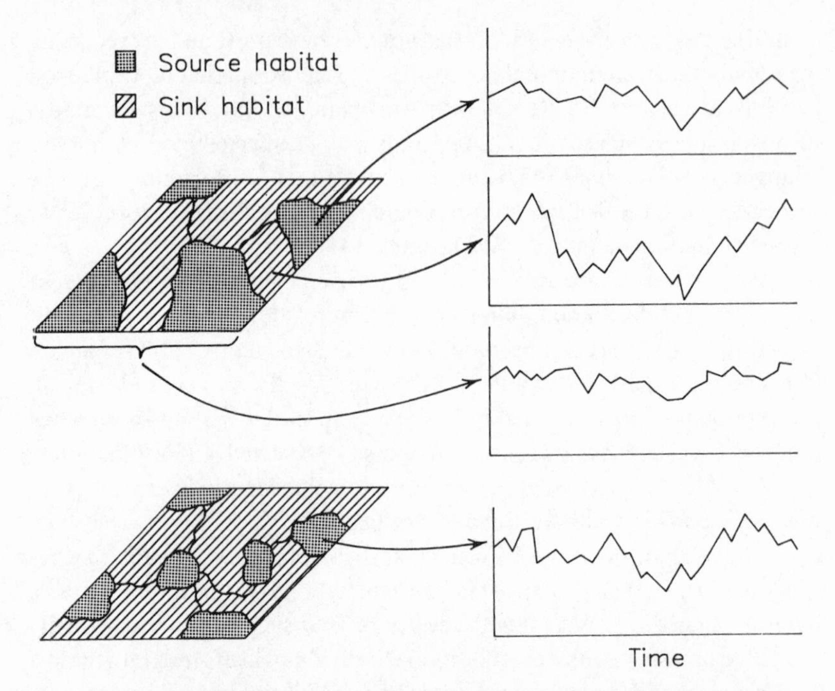


Fig. 4.12. Diagrammatic representation of the dynamics of populations in patches of a landscape mosaic. Populations in 'source' habitat patches may vary less than those in 'sink' patches (above), which may fluctuate in and out of existence at a local scale. At a regional scale summed over patches that are following independent dynamics, the population may be more stable. If the structure of the habitat mosaic changes in a way that reduces the size and increases the isolation of the 'source' patches, within-patch dynamics may become more variable (below).

## Long-term changes in populations and communities

Studies conducted over a few years give some indication of the sorts of temporal variation that characterize populations and communities. However, a long-term perspective in which changes over decades are considered may allow one to see patterns that would be obscured by the 'noise' of year-to-year variation in local plots. Long-term studies may thus provide insights about the overall trends of the maintenance of general equilibria in species composition or numbers that are unavailable even after several years of study.

### *Species turnover in communities*

One focus of long-term studies has been on changes in the presence or absence of species in communities – species turnover. Populations in marginal patches that are either less suitable ecologically or at the edges of the geographic distribution of a species are often small and are subject to a variety of stochastic effects. The probability of extinction of such local populations may be relatively great. In fact, local populations at the edges of a species' range are often visualized as fluctuating between extinction and re-establishment through waves of immigration from more central locations during good years (Andrewartha and Birch 1954, 1984, Mayr 1963). Because there are time lags in the recolonization of such patches, a species may be absent from some seemingly suitable habitat patches as a result of chance local extinction in the past. Fritz (1980), for example, documented such extinction-related local distributional patchiness at the edges of Spruce Grouse (*Dendragapus canadensis*) range in northern New York. The same sort of patch-related extinction can produce a turnover of populations in patches of differing ecological suitability well within the main distribution of a species.

The composition of a local community is therefore not fixed but undergoes turnover as local populations suffer extinction and the same or other species immigrate. On the small British island of the Calf of Man, for example, abundances of Wheatears, Stonechats, and Skylarks (*Alauda arvensis*) fluctuated over a 16-yr period, and populations of the first two species underwent several episodes of extinction and re-establishment during this time (Fig. 4.13). Meadow Pipits (*Anthus pratensis*) were present throughout the period, gradually increasing in abundance (Diamond 1979). Such faunal turnover indicates that community assembly is an ongoing process.

Species turnover is usually considered in the context of island biogeography theory (MacArthur and Wilson 1967). According to this view, the faunal composition of an island (or, in principle, any patch of habitat) is determined by an interplay of colonization and extinction. Early in community development, colonizations outnumber extinctions and species accumulate, but a point is reached at which additions and losses of species are roughly balanced. Extinctions and colonizations continue to occur, but at equilibrium rates. The area and isolation of an island affect the probabilities of colonization and extinction and thus the number of species that occur at equilibrium, but species turnover occurs under all conditions (see Abbott 1980, Williamson 1981, Brown and Gibson 1983, Rey 1984 for additional detail). With reference to species turnover, the MacArthur–Wilson model thus proposes that, at equilibrium: (1) the species composition of a community is not static; (2) turnover is balanced, extinctions being compensated for by colonizations; (3) the community is thus saturated, in

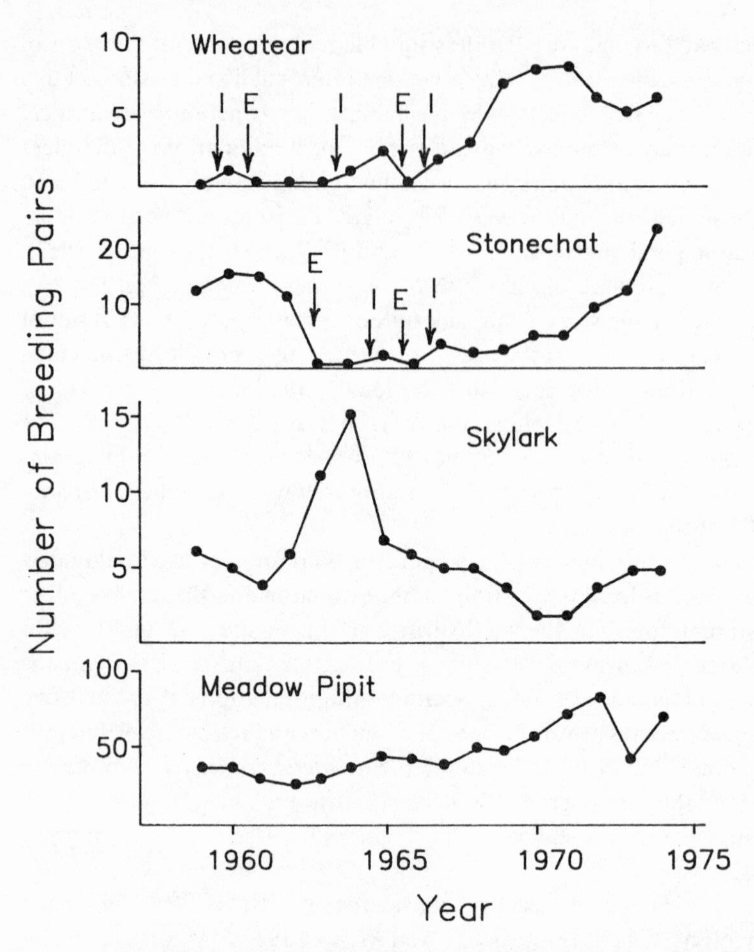


Fig. 4.13. The number of breeding pairs of Wheatears, Stonechats, Skylarks, and Meadow Pipits on Calf of Man, a small island in the Irish Sea, from 1959 to 1974. E = extinction of a local population, I = immigration (re-establishment) of a local population. After Diamond (1979).

that the number of species present remains relatively constant; and (4) the turnover is not related to systematic environmental changes but is due to the largely stochastic extinction and colonization events (Lynch and Johnson 1974, McCoy 1982). If a community is not in equilibrium, our expectations regarding turnover obviously change, and the applicability of the MacArthur–Wilson theory to a given situation may be invalidated if any of the above criteria is not met (Simberloff 1976).

Species turnover in natural communities has most often been calculated by comparing species lists recorded for an area at widely separated times.

Diamond (1969, 1971) calculated turnover on the Channel Islands of California by comparing a summary of distributional records published in 1917 with the results of surveys conducted in 1967–68. Although there were a fair number of changes in species composition, the number of species recorded on each island (*S*) was similar between the two listings, suggesting avifaunal equilibrium in accordance with the MacArthur–Wilson theory. Turnover rates were greater on the small islands (5.6% per year) than on larger islands (0.9% per year). Turnover of wader species on the Krunnit archipelago of northern Finland also seems to support the equilibrium model (Väisänen and Järvinen 1977), as do data for the Farne Islands of Great Britain (Diamond and May 1977) and small islands off the coast of Maine (Morse 1977), although turnover rates for these archipelagos are considerably greater than those Diamond calculated for the Channel Islands (see below).

In other situations, however, turnover does not appear to be balanced. On 12 of 14 islands off Australia and New Zealand, for example, there was a net increase in *S* between censuses separated by periods of 30–132 yr (Abbott and Grant 1976), and Reed (1980) reported a general increase in *S* on 26 islands off the coast of Britain during this century. Haila *et al.* (1979) found that two species became extinct on the Åland Islands between Palmgren's (1930, 1936) surveys in the 1920s and censuses conducted in 1975–77 (0.04 species per yr), whereas 8–18 species colonized the archipelago during the same interval (0.16–0.36 species per yr). Moulton and Pimm's (1983) analysis of the fates of introduced species on the Hawaiian Islands indicated that turnover was equilibrial on some islands but not on others. Calculations of turnover in continental rather than island settings have also provided evidence of a long-term balance in *S* in some situations but clear nonequilibrial turnover in others (e.g. Lynch and Whitcomb 1978, Głowacinski and Järvinen 1975, Johnston personal communication).

In all of these situations there is clearly a considerable amount of species turnover going on, although the degree of long-term balance in turnover and the magnitude of turnover rates vary. Part of the difficulty in assessing such results in relation to the MacArthur–Wilson model relates to how one defines 'equilibrium'. Diamond and May (1977), for example, considered a fauna to be in equilibrium if the CV of species number was $<0.20$, whereas Abbott and Black (1980) required a CV of $<0.05$ for equilibrium. Simberloff (1983c) noted that the Farne Island bird fauna, which Diamond and May suggested was equilibrial, actually varied in *S* by more then 100% (extremes of 4 and 9 species over a 29-yr period; $\mathrm{CV}=20.2$). Järvinen and his colleagues also found little evidence of balanced relationships between losses and additions of species during turnover in northern European communities (Järvinen and Ulfstrand 1980, Järvinen and Väisänen 1979).

The notion that competitive interactions may produce at least some of the species turnover observed on islands is clear in the writings of MacArthur (1972, MacArthur and Wilson 1967) and Diamond (1975a, 1982). According to this view, the extinction of one species from an island and its replacement by an ecologically similar species (preserving an equilibrium value of *S*) might occur either through the second species occupying a place in the community vacated by the first or by the second actively contributing to the decline and extinction of the first. Diamond (1975a) described several situations in the New Guinea region that are suggestive of such interactions. An indication that such replacements may not always be what they seem is provided by two turnover episodes on Socorro Island, Mexico (Jehl and Parkes 1983). There, the endemic Socorro Dove (*Zenaida graysoni*) became extinct between 1958 and 1978 and the endemic mockingbird *Mimodes graysoni* was rare and approaching extinction by 1981. Coincident with these losses, the mainland Mourning Dove (*Z. macroura*) became established on the island between 1971 and 1978, and the Northern Mockingbird colonized between 1978 and 1981. This would appear to be a textbook example of species turnover involving replacements by close relatives. In fact, the elimination of the endemic species was probably due to predation by feral cats that were introduced to the island in 1957 or shortly thereafter. The mainland species had previously been recorded as vagrants on the island but had not become established. They were apparently able to do so in the 1970s because of the provisioning of free water associated with the establishment of a human settlement (Jehl and Parkes 1983). The turnovers were real, but they were linked only through environmental changes produced by human activities on the island, not by competitive interactions.

Species turnover in communities may be of several sorts, only some of which bear directly on the MacArthur–Wilson equilibrium model. In fact, testing the model requires data sets in which the community surveys are accurate, environmental conditions are known not to have changed during the period considered, and the species recorded are known to be *bona fide* members of the community. If these requirements are not met, the turnover data are simply not relevant to the equilibrium model, although they may still represent valid documentations of turnover, rather than 'artifacts' as Wright (1985) suggests.

Census accuracy is of obvious importance, as it is to any community investigation. The data that Terborgh and Faaborg (1973) used to calculate species turnover on Mona Island in the Caribbean, for example, were based on surveys conducted during 6 days in winter in several years and probably do not permit an accurate quantitative determination of turnovers. It is especially important that surveys be intensive and careful enough to document that a species present in a previous survey but not observed in the current survey is actually absent rather than simply overlooked. If species not recorded in previous surveys were included in the species list because they were expected to be there (e.g. Cody 1975), one cannot interpret their documented absence in a subsequent census. If the surveys compared to determine turnovers were conducted by different individuals using different survey procedures, the resulting values must be interpreted with considerable caution (Abbott 1980).

Changes in habitats or other environmental features between surveys can produce turnovers as species respond differently to the changes. Järvinen and Ulfstrand (1980) concluded that most of the avian extinctions that occurred in northern Europe between 1850 and 1970 were due to human persecution or habitat changes, and at least 50 of the 88 colonizations were facilitated by human activities. Many of the changes that Abbott and Grant (1976) recorded for islands off Australia and New Zealand were also attributable to habitat changes or human-produced extinctions or introductions. On Rottnest Island off Western Australia, for example, three species went extinct and 10 species were recorded as colonists between 1904 and 1983, a turnover rate of 0.36% per year (Saunders and de Rebeira 1985). All of the extinctions and all but three of the colonizations were consequences of human activities, however, so 'natural' turnover was really quite low (0.12% per year). In addition to changes like these on islands, changes in the composition of the source pool of colonists, whether natural or human-induced, may also affect turnover patterns in the area (Jones and Diamond 1976, Williamson 1983).

To be considered as a true member of a community, a species should have established at least the foundation of a self-perpetuating population. This requires evidence of breeding activity (Lynch and Johnson 1974). If such evidence is not obtained, transient or vagrant individuals recorded in surveys may be included in the species list for the community, and their absence at a later time will be recorded as an extinction and will contribute to turnover calculations. This 'pseudoturnover' is most likely to occur in continental regions, where habitat 'islands' are not surrounded by inhospitable habitat and the incidence of vagrants may be high (McCoy 1982), but the potential for it to contribute to errors in turnover calculations on islands is also great. As an extreme example, Lynch and Johnson (1974) pointed to Southeast Farallon Island off California. Only one landbird species, the

Rock Wren (*Salpinctes obsoletus*) has been recorded breeding on this island. During banding studies in June 1969, however, 307 individuals of 50 landbird species were recorded on the island. Of 109 sightings of vagrants on Rottnest Island between 1905 and 1983, only one included a sufficient number of individuals to establish a breeding population (Saunders and de Rebeira 1985).

The criticisms that Lynch and Johnson (1974) leveled against Diamond's (1969, 1971) analysis of turnovers for the Channel Islands involved all of these factors. They questioned the adequacy of the survey data that provided the baseline for the calculations, pointed out that over half of the extinctions could be attributed directly to human activities on the islands, and suggested that another 26% of the extinctions involved pseudoturnover. Jones and Diamond (1976) responded to these criticisms in detail, reaffirming the adequacy of the baseline surveys as an indication of the presence of breeding species on the island while admitting that perhaps a third of the turnovers they tallied were due to human activities. McCoy's (1982) criticisms of the analyses of turnover in a forested area in Maryland conducted by Whitcomb *et al.* (1977) were based on similar points. He calculated that 40% of the colonizations and 60% of the extinctions that Whitcomb *et al.* recorded represented pseudoturnover and that perhaps half of the remaining turnovers were consequences of habitat changes or inadequate sampling.

Determining turnover rates is clearly no easy matter, but there are yet additional difficulties. Pseudoturnover tends to produce overestimates in turnover rates, perhaps by as much as threefold (Nilsson and Nilsson 1983). Turnover calculations are usually based on surveys taken at times separated by intervals of varying lengths, however, and long intervals may produce a systematic undersampling of true turnover (Diamond and May 1977). This is because some species may colonize an area during one year and become extinct a year or two later. With an increasing interval between surveys, more and more of these 'in-and-out' turnovers are not recorded, and true turnover may be underestimated (Fig. 4.14), perhaps by as much as an order of magnitude (Jones and Diamond 1976, Diamond and May 1977, Lynch and Whitcomb 1978). The evidence of such an effect is inconsistent, however. Abbott (1978) and Saunders and de Rebeira (1985) found a positive relationship between calculated turnover rate and the interval between surveys for Australian islands, and Väisänen and Järvinen (1977b) reported the same relationship for several Finnish islands. Svensson *et al.* (1984) found that turnover between successive years in alpine heath and mire habitats in Swedish Lapland was less than that between surveys several

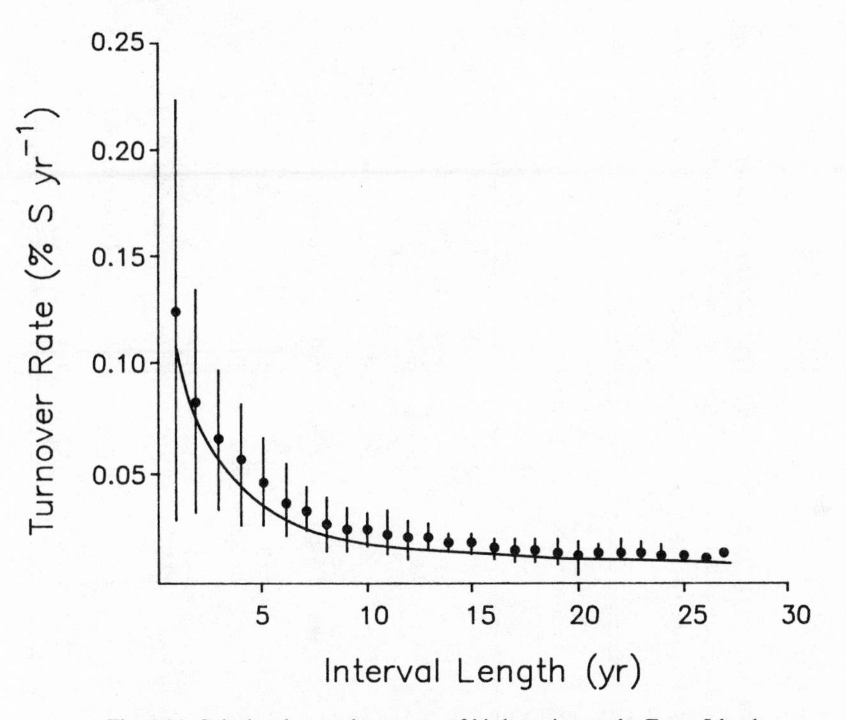


Fig. 4.14. Calculated annual turnover of bird species on the Farne Islands, Great Britain, determined from censuses separated by time intervals of varying lengths. Solid circles indicate the mean observed turnover rate, the vertical lines 1 standard deviation on each side of the mean. The solid curve indicates the predicted mean. From Diamond and May (1977).

years apart; they attributed this to time lags in community change produced by factors such as site tenacity. Turnover rate values did not begin to decrease with increasing interval until the interval exceeded 15 yr.

There has also been disagreement about how turnover rates should be calculated. Most investigators have used some form of relative turnover rate, such as:

$$\mathrm{RT} = (I + E)/0.5(S_1 + S_2)t,$$

where $I$ = immigrations, $E$ = extinctions, $S_1$ = number of species recorded in the first survey, $S_2$ = species recorded in the second survey, and $t$ = the interval between surveys. This measure has the advantage of standardizing turnover values for differences in the species richness of communities (Schoener 1983c). Others (e.g. Wright 1985, Simberloff 1976, Williamson 1978) have advocated the use of absolute turnover rates, calculated as:

$$T = (I + E)/2,$$

Table 4.2. *Species turnover values derived from several continuous long-term studies in different habitats. Absolute turnover is calculated as* $(I+E)/2$, *where* I = *number of species immigrating per year, and* E = *extinctions per year. Relative turnover is calculated as* $(I+E)/0.5(S_1+S_2)$, *where* $S_1$ = *number of species present in year* t, *and* $S_2$ = *number of species recorded in year* t + *1*.

| Habitat | Years | Number of species | | Absolute turnover | | Relative turnover | | Reference |
|---|---|---|---|---|---|---|---|---|
| | | $\bar{X}$ | $S$ | $\bar{X}$ | $S$ | $\bar{X}$ | $S$ | |
| Hardwoods forest, NE USA | 16 | 21.1 | 2.83 | 1.6 | 0.81 | 0.15 | 0.08 | Holmes *et al.* 1986 |
| Spruce forest, E USA | 22 | 9.7 | 1.61 | 2.1 | 0.87 | 0.41 | 0.18 | Hall 1984a |
| Oak woodland, England | 25 | 31.7 | 2.44 | 3.0 | 0.88 | 0.19 | 0.06 | Williamson 1981 |
| Mixed woodland, S Sweden | 10 | 17.8 | 1.93 | 1.8 | 0.79 | 0.21 | 0.09 | Enemar 1966 |
| Birch woodland, N Sweden | 20 | 17.8 | 1.88 | 2.7 | 0.90 | 0.30 | 0.09 | Enemar *et al.* 1984 |
| Alpine heath-mire (low elevation) N Sweden | 20 | 11.6 | 1.64 | 1.6 | 0.51 | 0.27 | 0.08 | Svensson *et al.* 1984 |
| Alpine heath-mire (high elevation) N Sweden | 20 | 8.0 | 1.15 | 1.5 | 0.80 | 0.37 | 0.18 | Svensson *et al.* 1984 |
| Shrubsteppe, Oregon | 8 | 4.9 | 1.46 | 1.1 | 0.45 | 0.42 | 0.19 | Wiens & Rotenberry, unpublished |
| Skokholm Island | 46 | 11.3 | 2.44 | 0.8 | 0.92 | 0.14 | 0.13 | Williamson 1983 |
| Farne Islands | 29 | 5.9 | 1.19 | 0.7 | 0.60 | 0.25 | 0.19 | Diamond & May 1977 |
| Deciduous forest, Illinois (breeding) | 44 | 27.3 | 6.81 | 3.9 | 1.42 | 0.29 | 0.11 | Kendeigh 1982 |
| Deciduous forest, Illinois (wintering) | 48 | 20.2 | 6.22 | 3.8 | 1.29 | 0.37 | 0.11 | Kendeigh 1982 |

which represents the actual change in species number over some time period. Each measure portrays something a bit different about turnover; I have used both equations to calculate values from several studies in which community composition was surveyed on an annual basis (i.e. $t = 1$) in order to illustrate some turnover rates that are not subject to the interval-length bias Diamond and May (1977) have noted and to show the different patterns one obtains from absolute versus relative turnover values (Table 4.2).

The species in a community differ in population variability (e.g. Figs. 4.2, 4.4). Those that undergo little variation in population sizes are more likely to persist in a community than are highly variable species, and thus will contribute less to community turnover. The species that have suffered extinction on Barro Colorado Island following its creation in the Panama Canal have usually been those that evidence high variability in population sizes on the adjacent mainland rather than those that are rare but stable (Karr 1982b). In general, species with high densities tend to be more variable than those with low densities (Taylor and Woiwod 1980, 1982, Anderson *et al.* 1982), and this pattern seems to hold in at least some bird communities (e.g. Holmes *et al.* 1986, Fig. 4.15). Although variance in densities may be greater in large populations, small populations may nonetheless be the most likely to suffer local extinction if their populations do vary. Most of the turnover in the studies listed in Table 4.2, in fact, involved species that were never common and that fluctuated in and out of the communities as they suffered local extinction and then recolonized the areas. On the island of Skokholm, for example, 29 species have been recorded, but 9 of these have been recorded only once or twice (always as single pairs) and 8 more have never exceeded two pairs. The remaining 12 'common' species have contributed little to the species turnover (Williamson 1983; Table 4.2).

Because a high proportion of the turnover in a community may involve the uncommon, 'in-and-out' species, a preoccupation with calculating turnover rates may lead to a neglect of other important long-term community patterns. Lynch and Whitcomb (1978), for example, recorded an approximate equilibrium in turnover (40 extinctions and 34 colonizations) in a deciduous forest tract in eastern United States between 1950 and 1976. During this period, the mean number of breeding species was reduced by 25%. Some of the turnover thus represented long-term declines and disappearances of formerly abundant species in the community, but these turnover events were lost among the many in-and-out changes that occurred. In Williamson's (1981, 1983) view, much of the species turnover in

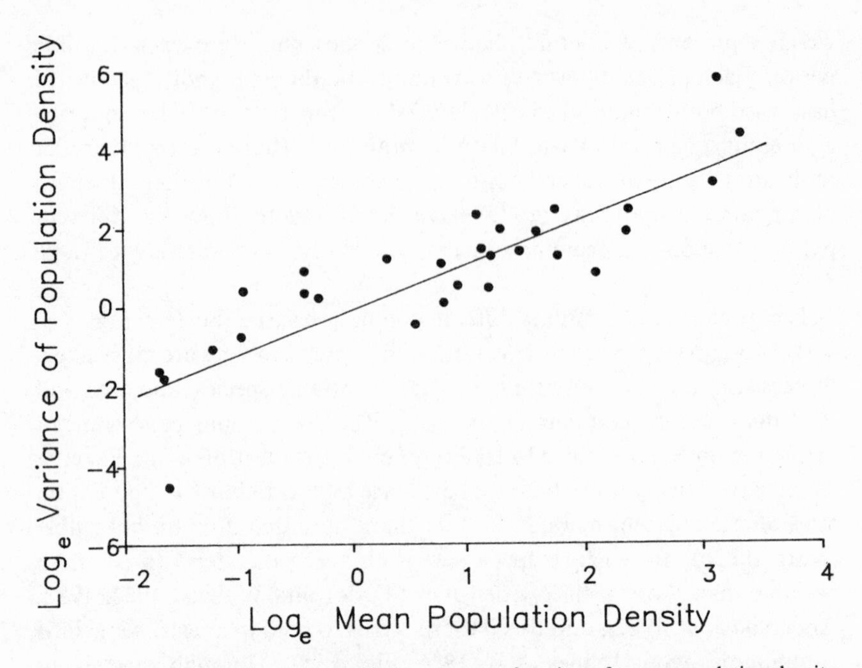


Fig. 4.15. Variance in population size as a function of mean population density for 29 bird species breeding in a mixed-hardwoods forest in New Hampshire. After Holmes *et al.* (1986).

communities is thus 'ecologically trivial'. Changes in total species number in a community may be of little importance unless they reflect turnover affecting the 'core' of a community, some common species declining to long-term absence from an area and other species invading and becoming abundant. These long-term changes clearly may have a greater impact on community composition and the patterns of guild structuring and species interrelationships in the community than the turnover produced by several casual, in-and-out species.

***Long-term density changes***

The focus of turnover studies is on the presence or absence of species in an area, and the basic data are species lists recorded at different times. Greater insight into the long-term dynamics of communities and their constituent populations can be obtained from monitoring changes in abundances as well as occurrences through time. This requires density estimates obtained over long intervals by means of comparable survey methods, and there are relatively few such data sets available.

Some of the most valuable and impressive long-term studies have been made in Scandinavia, where the careful, quantitative censuses of bird populations conducted by several ornithologists in the 1910s and 1920s (e.g. Merikallio 1929, Palmgren 1930, Kalela 1938) provide a solid foundation for long-term comparisons. Using these and other studies, von Haartman (1973) concluded that 34% of the 233 breeding species recorded in Finland had increased in distribution and abundance during the previous century, whereas 25% decreased in abundance, 22% fluctuated, and only 20% remained fairly stable. Some of these changes were attributed to alterations in human land-use practices or shifts in the behavioral adaptations of the birds, but, following earlier Finnish workers (e.g. Kalela 1949), von Haartman also emphasized the possible role of climatic amelioration since the 1870s in promoting these changes in bird abundances. Järvinen and Väisänen (1977a, b) conducted a similar but more restricted analysis of population changes in Finland from 1945 to 1975, basing their work on line-transect surveys made in 11 latitudinal zones and restricting their attention to species that were common (> 20 000 pairs) in Finland in 1945. Of the 86 species they considered, 43% increased by 1% or more per year, 28% were stable, and 29% decreased by at least 1% per year. Ten species increased by 4–8%/yr on average, and four species decreased by > 8%/yr. Contrary to the pattern reported for many systems (e.g. Fig. 4.15), the stable species were often abundant, and the two most abundant species (Willow Warbler and Chaffinch) varied rather little over the 30 years. Despite the overall stability of these two species, however, their abundances increased dramatically in forested areas during this period (Järvinen *et al.* 1977).

This pattern suggests that habitat changes may have contributed to the population shifts of many of the species. Indeed, during this 30-yr period, forested habitats in Finland underwent extensive modifications as a consequence of changing forestry practices. Clearcutting was extensive, reducing the area of old forests in northern Finland, creating smaller forested plots (and thus more edge habitat) and increasing the proportion of spruce in southern Finland, and shifting the age-structure of the forests toward younger trees. Artificial fertilization of forests began in the 1960s, increasing the density of herb and brush layers. Large areas of peatland were drained for forestry. Finally, many agricultural areas were converted to forest, and cattle grazing in the forests was abandoned in the 1950s.

Because these practices and their effects on habitats differed in different parts of the country, their influences on bird populations were not the same everywhere. Several foliage-gleaning species overwintering in coniferous forests declined dramatically in total density in northern Finland but maintained stable populations in more southern areas (Fig. 4.16; Järvinen

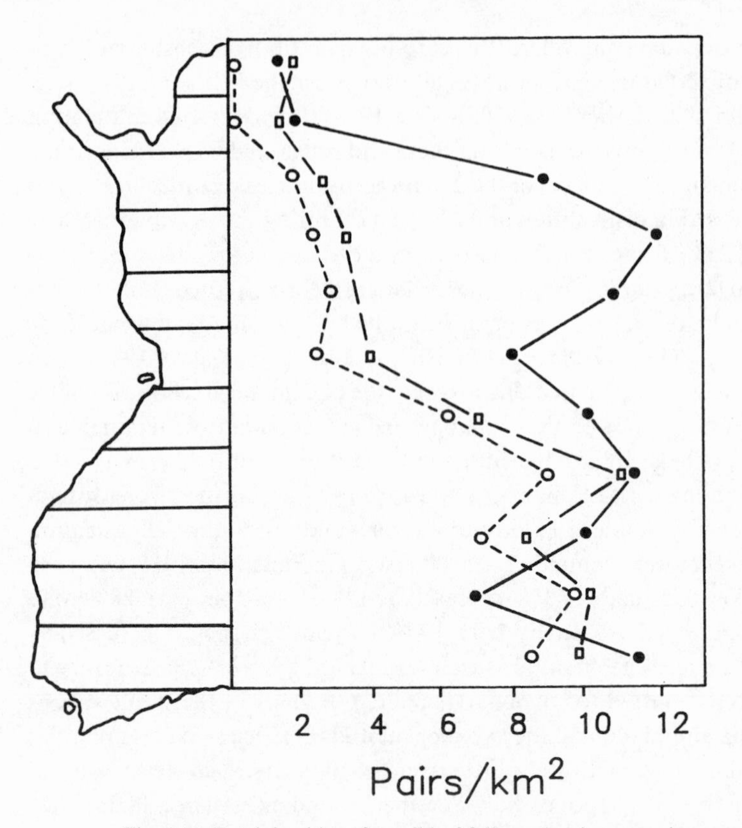


Fig. 4.16. Total densities of a guild of foliage-gleaning passerines overwintering mainly in coniferous forests, as recorded in censuses in several latitudinal sections of Finland in 1945 (solid circles), 1955 (open squares), and 1975 (open circles), Members of the guild were *Parus montanus, P. cristatus, P. ater, P. cinctus, Perisoreus infaustus,* and *Certhia familiaris*. From Järvinen *et al.* (1977).

*et al.* 1977, Väisänen *et al.* 1986). Chaffinches and Willow Warblers increased in southern Finland, probably due to forest fragmentation and increases in edge habitat, but only the warblers also increased in the north. There, Chaffinches are a habitat specialist occurring mainly in river valleys, and its populations were apparently little affected by the changes in forest composition and area (Järvinen and Väisänen 1978a). In southern Finland, the breeding densities of 22 of 40 species (55%) steadily increased over a 50-yr period beginning in the mid-1920s, largely in association with increasing coverage of young forest and spruce and the increasing amount of edge habitat produced by forest fragmentation (Järvinen and Väisänen 1978b). Seven species exhibited irregular fluctuations (some associated with severe winters), and only four species steadily decreased. Most of the systematic changes in the abundances of Finnish birds over the last half-century or so may thus reflect these habitat changes rather than a more general climatic amelioration (Järvinen and Väisänen 1979, Helle and Järvinen 1986).

These long-term studies in Finland also provide insights into the interplay of local and regional determinants of population changes. Väisänen *et al.* (1986), for example, compared population trends over a 30-yr period in one area of mature forest and peatland that was drastically changed by clearcutting and draining with those in another area of spruce forest that remained intact and relatively unchanged. The patterns of population changes in the managed area were similar to those seen regionally, although there were more declines. These declines indicate the effects of the local habitat changes, which were especially severe. Population changes in the spruce forest also paralleled regional trends, despite the stability of that local habitat. Järvinen (1978) also found that the avifauna of an extremely stable peatland reserve changed in accordance with regional patterns.

In another analysis, Helle and Järvinen (1986) used independent data to derive estimates of the habitat preferences of several abundant breeding species in northern Finland and then related these to the patterns of long-term population changes of these species. Collectively, variables expressing habitat-edge preferences of the species and their responses to forest age-structure accounted for 67% of the variation in population trends of species with northern affinities, but there was no relationship between the habitat-selection measure and population variations for species derived from southern Finland (Fig. 4.17). The dynamics of these species depended instead on overall population trends in the southern part of the country. Of the northern species, those with low values of the habitat selection index (preferring large, old-growth stands) generally decreased in abundance and those with higher index values (> 100, indicating a preference for younger, fragmented stands with more edge) generally increased (Fig. 4.17). Because the values for this analysis were derived regionally rather than locally, these results suggest that the long-term changes represented adjustments to alterations of landscape mosaics on a regional scale, and that changes in local communities may have had little to do with local habitat changes (Järvinen 1980, Helle and Järvinen 1986).

The studies of Järvinen and his colleagues have all involved comparisons of quantitative surveys taken some years apart, and in some instances (e.g. Järvinen and Väisänen 1978a) substantial data adjustments were required to compare the censuses. The investigations of Enemar *et al.* (1984), although not nearly so extensive as the Finnish studies, have the advantage of charting population levels over a continuous 20-yr period. Enemar and

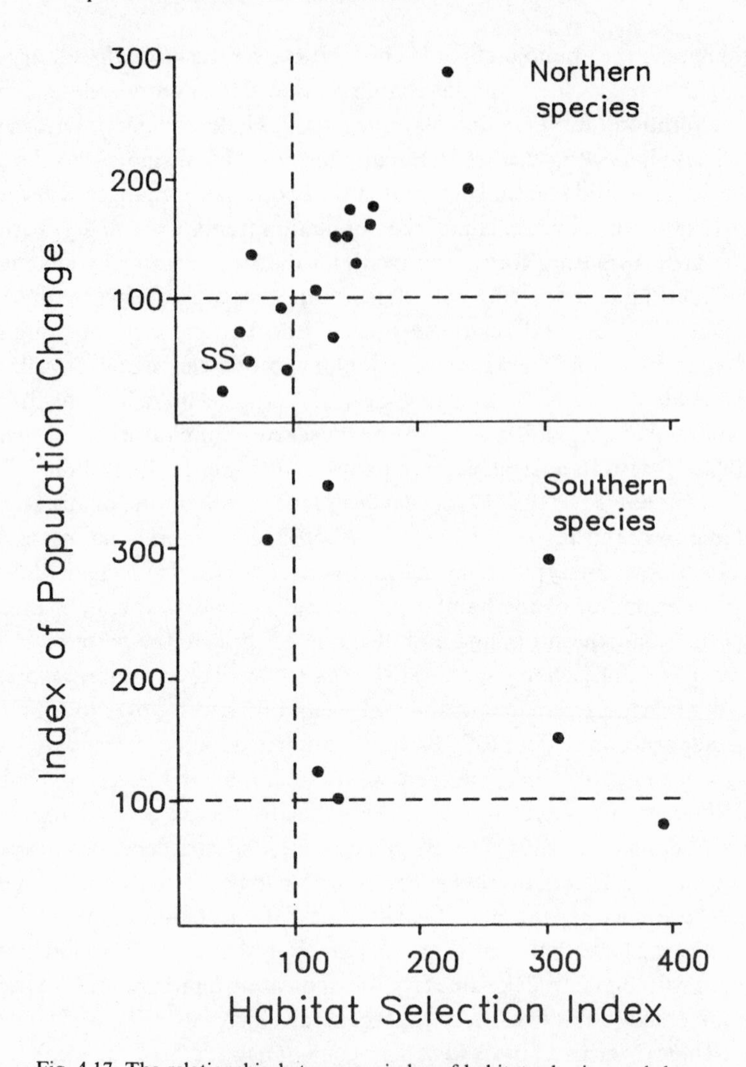


Fig. 4.17. The relationship between an index of habitat selection and the magnitude of population change for 22 bird species breeding in forested habitats in northern Finland. Values of the habitat-selection index below 100 indicate a preference for large, mature forest stands, whereas higher values are associated with occupancy of younger, more fragmented, brushy stands. A population-change index less than 100 indicates a decreasing population, whereas a greater value denotes population increase. Species located in the lower left sector of the graphs occupy old-growth stands and have decreased in abundance, whereas those in the upper right sector occur in younger stands with more edge, and have increased; SS = sedentary species of old forests. The top graph shows the patterns for species of northern affinities, the bottom graph, those for species of southern affinities. After Helle and Järvinen (1986).

his associates used territory mapping and line-transect procedures to estimate the breeding densities of species in a subalpine birch forest in Swedish Lapland from 1963 through 1982. The number of species recorded fluctuated about a mean of 37 (CV = 9.5) and showed a slight increase over the 20-yr period. Total community density, on the other hand, peaked in 1973 in association with an outbreak of *Epirrita autumnata* caterpillars and declined steadily for several years thereafter (even though caterpillars were still abundant) (Fig. 4.19). This decline was due largely to a long-term decrease in the abundance of Willow Warblers, which had numerically dominated the community before 1973 (Fig. 4.18). The three *Turdus* species in the community also contributed to this overall decline, their numbers decreasing following the superabundant birch seed crop of 1968 (Fig. 4.18). *Fringilla montifringilla* was generally common in the community but increased dramatically during the initial years of the *Epirrita* outbreak in the early 1970s. Other species, such as *Luscinia svecica, Prunella modularis*, and *Emberiza schoeniclus*, showed no systematic density changes over the period, although they fluctuated considerably (Fig. 4.18).

A scanning of Fig. 4.18 suggests that several of the species fluctuated in similar ways. Enemar *et al.* examined such patterns through cluster analysis based on correlation coefficients of the density variations of the species. The resulting dendrogram (Fig. 4.19) defined three clear groupings. With the exception of one species (*Turdus philomelos*), all of the species in group A are long-distance migrants that overwinter in tropical Africa. All of the species in group B except one (*T. iliacus*) intermittently show irruptive migratory behavior. The species in group C are all nonirruptive, short-distance migrants. Overall, densities of group A species consistently declined over the 20-yr period; those of group C remained relatively stable. Much of the variation in total community density was due to variations in group B, the irruptive species (Fig. 4.19). The dynamics of group A species suggest that conditions outside the breeding area (e.g. on the tropical wintering grounds) may contribute to the community patterns, but the decline in abundances of short-distance, irruptive species over the last decade of the study indicates that conditions on the breeding ground (e.g. reduction in *Epirrita* availability, harsh weather) may also be involved. Enemar and his colleagues recorded no evidence of compensatory fluctuations between species or (aside from *Carduelis* and *Fringilla*) of close tracking of resource variations, and they concluded that the community was not ecologically saturated.

Several North American studies provide a similarly detailed long-term perspective on local community dynamics (Kendeigh 1982, Hall 1984a,

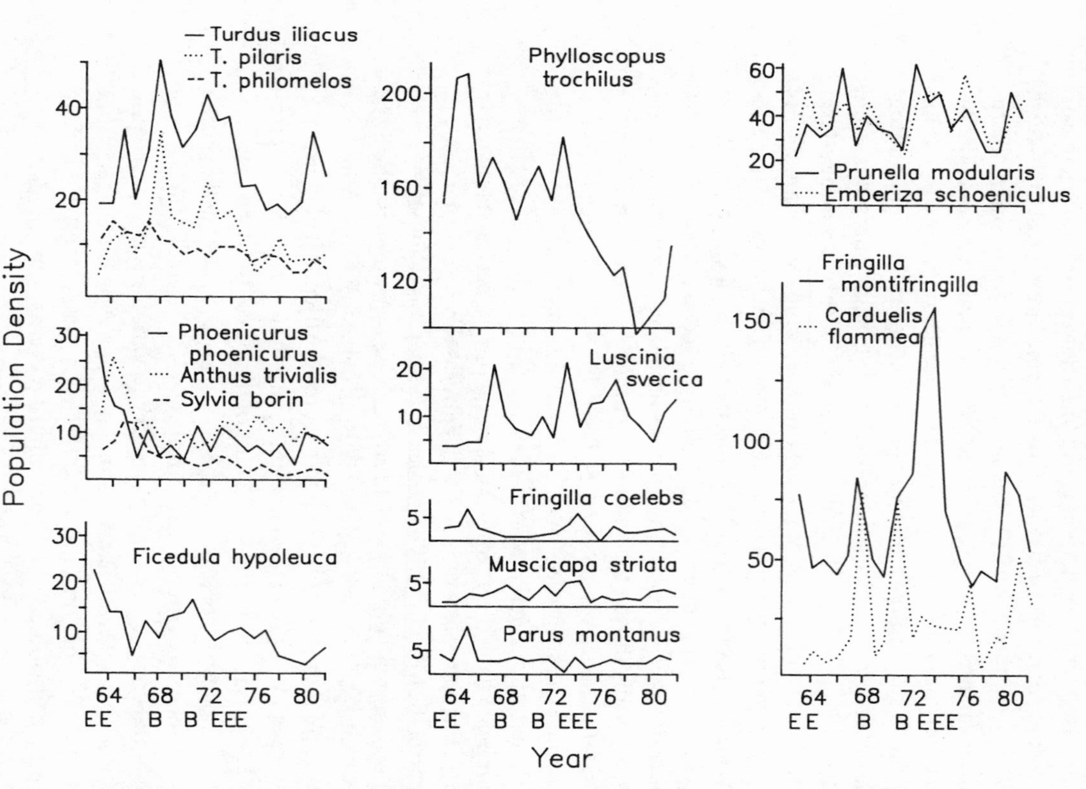


Fig. 4.18. Population fluctuations of 16 small passerine species regularly breeding in a subalpine birch forest area of northern Sweden, 1963–82. B or E on the abscissa indicate years of superabundant birch seed crop or outbreaks of caterpillars of *Epirrita autumnata*, respectively. The density index corresponds approximately to territories km$^{-2}$. Modified from Enemar *et al.* (1984).

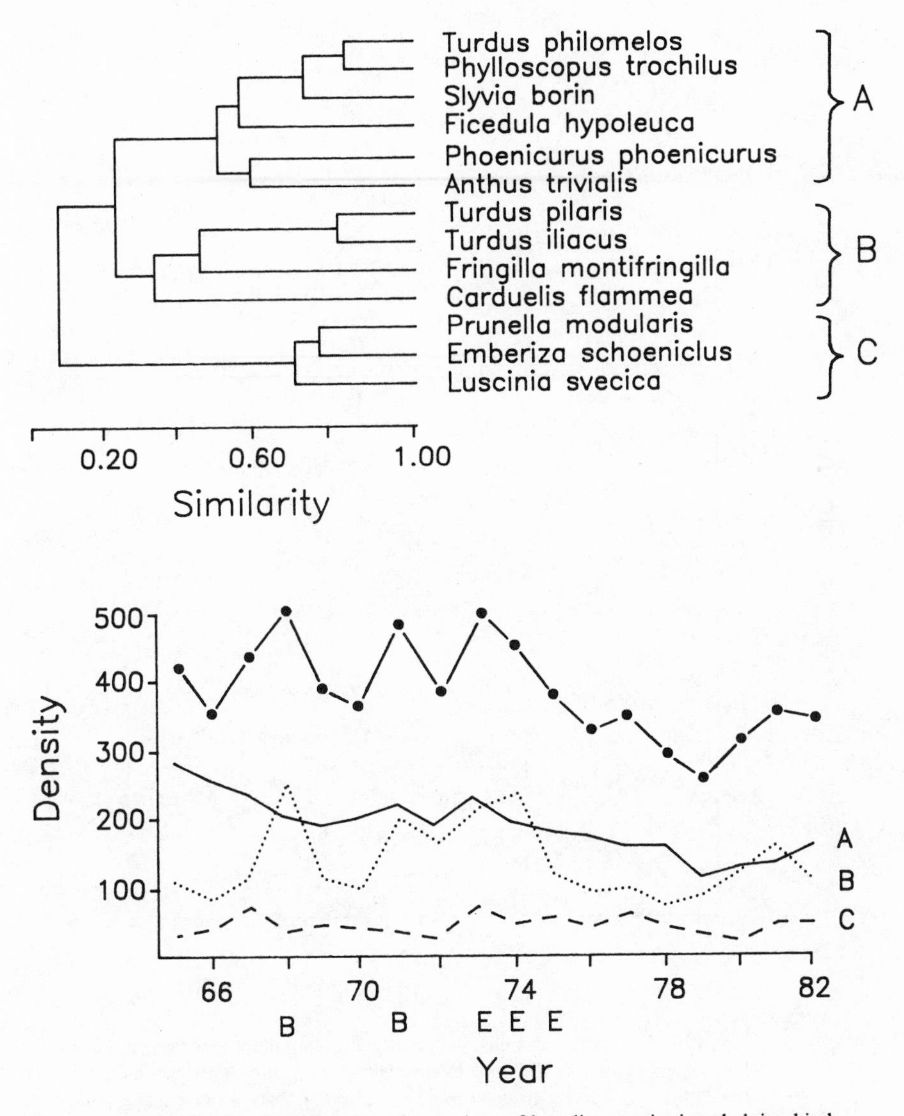


Fig. 4.19. Top: Dendrogram of groupings of breeding species in subalpine birch forests in northern Sweden, clustered according to the degree of similarity in density fluctuations, 1965–82. Group A includes mostly long-distance migrants that winter in tropical Africa, group B species that are short-distance migrants that are often irruptive, and group C short-distance, nonirruptive species. Bottom: Fluctuations in total densities of the species groups shown at the top and in total community density (solid circles). After Enemar *et al.* (1984).

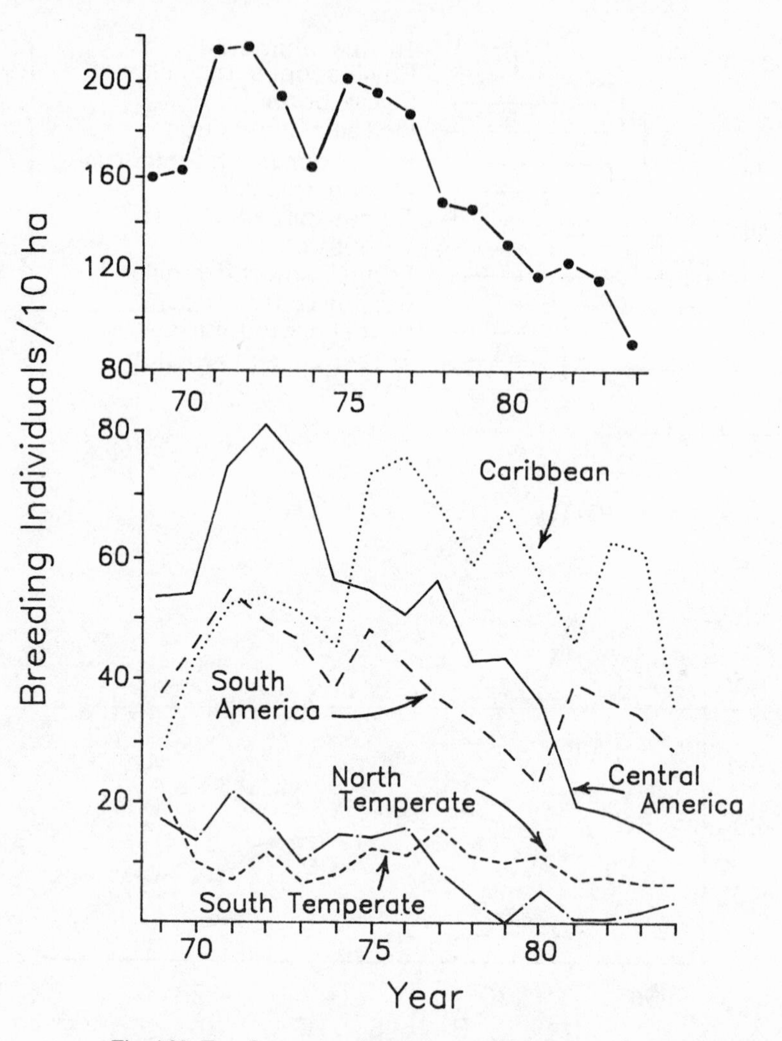


Fig. 4.20. Top: Long-term changes in total bird community density in a mixed-hardwoods habitat at Hubbard Brook, New Hampshire. Bottom: Changes in the densities of species grouped by their wintering areas. After Holmes *et al.* (1986).

Raphael *et al.* 1987, Askins and Philbrick 1987). Those of Holmes and his colleagues in a mixed hardwoods forest at Hubbard Brook, New Hampshire (Holmes *et al.* 1986) parallel the studies of Enemar *et al.* in several ways and provide an interesting comparison. Bird populations were censused from 1969 through 1984 on a 10-ha plot within a large tract of undisturbed, unfragmented forest. During this period, overall community density declined dramatically (Fig. 4.20); roughly 70% of the species contributed to this drop in abundance. Least Flycatchers (*Empidonax minimus*) numerically dominated the community at the beginning of the study but declined sharply in 1974 and thereafter (Fig. 4.21), perhaps as a consequence of local changes in the vertical distribution of foliage. American Redstarts (*Setophaga ruticilla*) increased in the first two years, coincident with an irruption of *Heterocampa guttivitta* caterpillars, and then abruptly increased again in 1974 (Fig. 4.21). This increase may have been due to a release from competition with Least Flycatchers, which harassed other species (particularly redstarts) when they were abundant (Sherry 1979). Other evidence (some experimental) also indicates that the flycatchers may influence the local distribution patterns and abundance of redstarts (Sherry and Holmes 1988). The decline of the redstarts in later years was associated with other factors, perhaps variations in local reproductive success due to weather or predation (Holmes *et al.* 1986). Levels of *Heterocampa* have generally been low since 1971, and this may also have contributed to the long-term declines of redstarts and several other species.

Abundances of Scarlet Tanagers (*Piranga olivacea*) also declined sharply in 1974, but this change was related to a period of unseasonably cold, wet weather in late May that caused high adult mortality (Zumeta and Holmes 1978). Numbers of Hermit Thrushes (*Catharus guttatus*) and Dark-eyed Juncos dropped precipitously in 1973 and again in 1977 and 1978 (Fig. 4.21). These changes were most likely associated with severe cold weather on the wintering grounds in southeastern United States during the previous winters, which had large-scale effects on populations of these species. Some other species also declined, but at different times, and other species fluctuated in abundance with no evident long-term trends. Overall, most species fluctuated independently of one another, suggesting that their variations were influenced by different combinations of factors. Similar patterns of independence in species' fluctuations have been reported from long-term studies in riparian habitats in Arizona (Rice *et al.* 1983b), alpine heath/mire habitats in Sweden (Svensson *et al.* 1984), and North American shrubsteppe (Wiens and Rotenberry, unpublished).

Because events on the wintering grounds were implicated in the population variations of several of the species at Hubbard Brook, Holmes and his colleagues considered the overall breeding-season dynamics of species grouped by wintering areas. Except for the group wintering in the Caribbean region, all of the groups showed trends of declining abundance over the 16-yr period (Fig. 4.20). The reductions in abundances of the groups wintering in Central America and in south-temperate locations were especially large. To determine whether the species in these groups followed generally similar patterns of fluctuations, Holmes *et al.* conducted a cluster

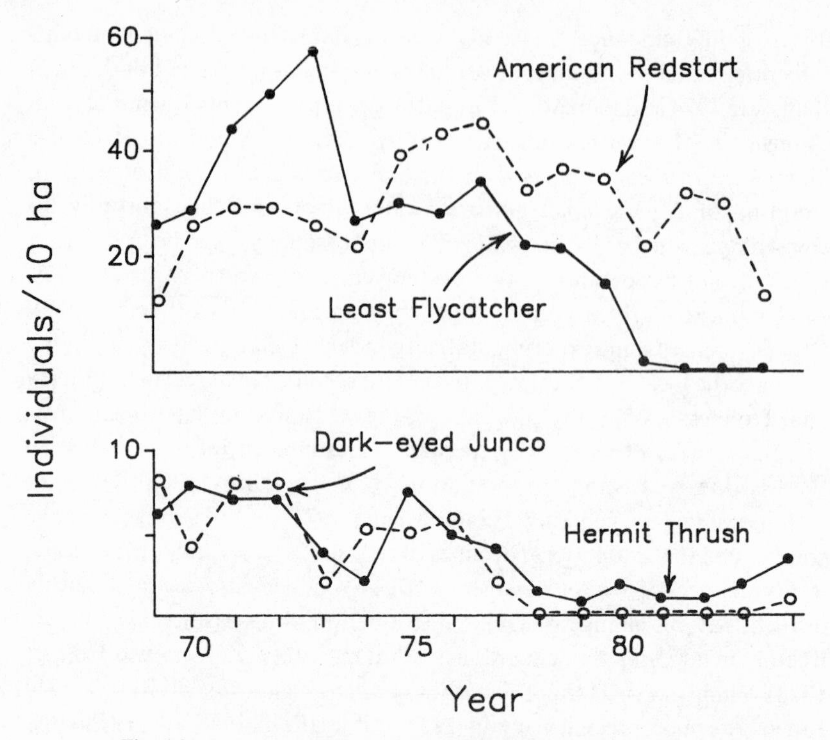


Fig. 4.21. Long-term changes in breeding densities of Least Flycatchers, American Redstarts, Hermit Thrushes, and Dark-eyed Juncos in mixed-hardwoods habitat at Hubbard Brook, New Hampshire. After Holmes *et al.* (1986).

analysis similar to that of Enemar *et al.* (1984). Black-throated Blue Warblers (*Dendroica caerulescens*), Ovenbirds (*Seiurus aurocapillus*), and American Redstarts, which winter in the Caribbean, clustered together, as did the thrush and junco noted above. In contrast to the pattern in the Swedish woodland, however, most of the other species clustered without regard to wintering areas.

These observations bear on the proposal that recent declines in breeding abundances of many neotropical migrants in forested habitats of eastern North America are related to events on their wintering grounds, particularly the widespread destruction of tropical forests (Wilcove and Whitcomb 1983, Wilcove and Terborgh 1984, Morse 1980). Because of the shapes of the continents, North American species that migrate to the tropics are concentrated into a much smaller wintering area (Moreau 1966). The destruction of 1 ha of forest in Mexico, for example, may be equivalent to the clearing of 5–8 ha in northeastern North America (Terborgh 1980). Sharp decreases in neotropical migrants (especially warblers and vireos) have clearly occurred in areas of Maryland (Lynch and Whitcomb 1978), Illinois (Kendeigh 1982), and West Virginia (Hall 1984a, b). In a deciduous forest in Virginia and a mixed conifer-hardwoods forest in Connecticut, on the other hand, the declines of several species were associated with local habitat changes, and long-term changes in total community density followed no consistent trends (Johnston personal communication, Askins and Philbrick 1987).

Interpretation of these changes, even where they are dramatic, is difficult. The clear decline in overall abundance of Central American migrants recorded by Holmes *et al.* (1986) (Fig. 4.20), for example, was due almost entirely to the loss of Least Flycatchers from the forest. This was a strictly local change associated with habitat succession. The species remained abundant in nearby areas, and the change had no obvious relationship to events on the wintering area. There are also other reasons to question the generality of the tropical-forest destruction explanation. Many North American migrants to the tropics do not occupy tropical forest but frequent patches of second-growth or edge habitats instead; these habitats increase in availability with forest clearing. Moreover, although the magnitude of recent logging activities in the tropics is great, these habitats were probably extensively altered by natives before the arrival of Europeans (Morse 1980). The fragmentation of forests in the North American breeding areas has also increased over the past several decades (Burgess and Sharpe 1981), and a separation of the effects of this fragmentation or of spring weather or local habitat change from those of tropical deforestation or severe weather in the wintering or migration areas is difficult. Several factors are involved in the long-term dynamics of breeding populations, and to attribute these variations to a single environmental change is premature (Holmes *et al.* 1986). There are plenty of reasons to be alarmed about the accelerating rate of destruction of tropical forests, but a loss of birds from forests in eastern North America is not unequivocally one of them.

A related aspect of these linkages between temperate and tropical areas merits brief mention. Climatic conditions in the tropics are generally less extreme than in higher latitudes, fostering a more continuous production and availability of food. The number of resident species may therefore be considerably greater than in temperate locations. During winter, these communities are enriched by the arrival of temperate-zone migrants. Opinions vary on how this seasonal influx affects the structure of the tropical bird assemblages. Terborgh and Faaborg (1980) stated that the distributions of several resident warblers in the Greater Antilles are 'drastically truncated' by the activities of large overwintering populations of migrants, and Keast (1980) noted the same effect in Central America and Mexico. This would suggest that the tropical communities vary seasonally in species packing and that residents may show niche shifts when the migrants arrive or leave. Emlen (1980), however, found that resident species in Florida and the Bahamas did not expand to fill the available niche space created by the departure of migrants, and he concluded that the winter assemblages of residents plus migrants represent 'full, integrated ecological communities'. Stiles (1985) reached the same conclusion for Costa Rican communities, and Gochfeld (1985) found no evidence of an inhibitory effect of migrants on resident species in three habitats on Jamaica. In Kenya, Rabøl (1987) found that both local resident species and palearctic migrants changed in behavior and ecology with the influx of large numbers of migrants, but he concluded that the migrants should best be considered as full members of the tropical communities that happen to be absent for part of the year rather than 'foreign intruders'. Whether migrants are superimposed on established tropical assemblages and are forced to occupy peripheral ecological positions or combine with residents to form integrated communities that become depauperate when the migrants leave, it is apparent that these movements produce considerable temporal change in community patterns (see Rice *et al.* 1983b). Incorporated into these patterns are the effects of events occurring elsewhere in the ranges of the migrants. The communities are not closed.

**Conclusions**

When one monitors the abundances of populations in communities over several years, a substantial amount of variation is usually evident, especially at a local scale. Many factors may contribute to this variation: habitat changes; changes in habitat selection behavior; population variations in nearby 'source' habitat patches; fluctuations of food resources; site tenacity; weather conditions during winter, migration, or breeding; variations in predation or parasitism; a lack of saturation of suitable habitats; and so on. To some extent these factors have direct, deterministic effects on individuals and hence on local populations and communities. Some of the variation is therefore interpretable through careful study at this local scale, and understanding the sources of this variation may provide a key to developing better insights into community processes and richer, more realistic community theory.

Because there is also an element of chance in the operation of many of these factors, and because different species respond to such factors in different ways, the temporal dynamics of local communities may sometimes give the appearance of being hopelessly chaotic. This is particularly likely if the communities are composed of migrants that spend portions of their lives in widely separated areas. Such assemblages are open to influences occurring well beyond the boundaries of the breeding or wintering 'community' that an investigator arbitrarily defines, and an interpretation of local dynamics through the measurement of local 'causes' is likely to be difficult. The openness of local communities and the complexity of variations they display may lead one to examine community dynamics at a broader, regional scale. At this level, the dynamics of communities may indeed reflect long-term trends that operate at a regional (or larger) scale. There is also little question that such a regional perspective may enable one to assess the degree to which particular local situations seem idiosyncratic. Regional patterns are detected by comparing among a particular set of local samples, however, and they are thus sensitive to exactly which samples are included in the comparison. This raises the possibility that these patterns are epiphenomena, artifacts of averaging over a particular set of local samples, rather than consequences of real biological processes (Wiens 1981b, 1984a, 1986). Clearly, both local and regional perspectives may be necessary in order to begin to understand why bird populations and communities vary as they do.

Because bird populations and communities vary from year to year, no single year is 'average' or 'typical'. A short-term approach based on single-year samples at one or several sites will therefore provide only a misleading snapshot of a changing community frozen at a moment in time. To see the true dynamics requires long-term studies, in which population abundances, resource levels, and other environmental factors are recorded annually. The value of long-term data sets is evident in the insights obtained from the comparisons that Järvinen and Väisänen made in Finland or from the studies of Enemar, Holmes, and Grant and their associates. Long-term studies provide the only way to determine whether populations *do* track resource variations closely or with considerable time lags or to assess how frequently populations are exposed to environmental crunches or conditions of resource plenty (Wiens 1984c). Long-term studies are not easy, nor are they encouraged by contemporary views of scientific productivity or research funding policies. Nonetheless, they are essential if we are to progress toward an understanding of how communities really behave in nature.

It is apparent from the studies reviewed in this chapter that populations and communities vary in a wide variety of ways. A simple categorization of communities as 'equilibrial' or 'nonequilibrial' is therefore unrealistic and unproductive. An assumption of strict equilibrium may simplify community theory (Real 1975, Tilman 1982), but it may also make much of the theory irrelevant to a substantial portion of nature that may not match the assumption (DeAngelis and Waterhouse 1987). Charles Elton noted this problem in a somewhat different context over half a century ago, and his comments merit quoting at some length:

> It is further suggested that if we knew enough about the ecological relations of the animals we could predict the effect of any interference, just as a clockmaker can work out the ultimate effect of the twirling of one wheel upon the rate of revolution of any of the others. At the same time it is assumed that an undisturbed animal community lives in a certain harmony, referred to as 'the balance of nature,' and that although rhythmical changes may take place in this balance, yet that these are regular and essentially predictable and, above all, nicely fitted into the environmental stresses. . . . The picture has the advantage of being an intelligible and apparently logical result of natural selection in producing the best possible world for each species. It has the disadvantage of being untrue. 'The balance of nature' does not exist and perhaps never has existed. The number of wild animals are constantly varying to a greater or less extent, and the variations are usually irregular in period and always irregular in amplitude. Each variation in the numbers of one species causes direct and indirect repercussions on the numbers of others, and since many of the latter are themselves independently varying in numbers the resultant confusion is very remarkable. The simile of the clockwork mechanism is only true if we imagine that a large proportion of the cogwheels have their own mainsprings, which do not unwind at a constant speed. There is also the difficulty that each wheel retains the right to arise and migrate and settle down in another clock, only to set up further trouble in its new home
>
> (Elton 1930: 16–17).

Natural communities are arrayed along a spectrum between the extremes of static equilibrium and chaotic nonequilibrium (Wiens 1984a), with few natural situations matching either extreme. Some populations may fit a dynamic equilibrium in which varying resource levels are closely tracked, but there is no evidence that such tracking is a community-wide phenomenon, except perhaps in small, closed assemblages such as the Galápagos finches. Most often, the tracking that might produce a dynamic equilibrium is distorted by time lags, is nullified by resource superabundance, or is overridden by the effects of other environmental factors. The spectrum of variability expressed in populations and communities indicates that no single theoretical model is likely to apply very generally. We should approach the study of communities using a multiplicity of views, including theories and hypotheses based on traditional equilibrium, dynamic resource tracking, 'ecological crunch' scenarios, and stochastic processes (Wiens 1986, Chesson and Case 1986).

# 5

## Spatial patchiness and scale

Viewed at any scale of resolution, from an individual's territory to a biogeographic region or a continent, environments are mosaics of patches. Individuals and populations repond to this spatial heterogeneity in various ways, some spending long periods of time within specific patches, some using certain patches for feeding and others for reproduction, still others moving over a mosaic in an apparently aimless fashion but distinguishing between different mosaics. Because the patch structure of an environment changes through time, the responses of individuals and populations to spatial patchiness are dynamic. The structure and dynamics of communities, in turn, are strongly influenced by this spatial variation.

Kareiva (1986) observed that ecologists have had difficulty making sense of insect communities because environmental heterogeneity has generally not been included in theories of species interactions and because spatial dimensions of variation have usually been ignored in field studies. The same could be said of many investigations of bird communities. Much of the theory that has guided community studies has, for simplicity, assumed nature to be spatially homogeneous. Often these theories fail to generate realistic predictions when spatial patchiness is introduced (Schluter 1981, Tilman 1982, Chesson 1986). Likewise, field studies have often been restricted to small sites selected on the basis of their apparent internal homogeneity. In recent years, however, efforts to incorporate the influences of heterogeneity into population or community models have increased (e.g. Roff 1974, Okubo 1980, Paine and Levin 1981, Nisbet and Gurney 1982, Chesson 1981, 1985, 1986, Levin *et al.* 1984), and the importance of 'patch dynamics' has become more widely appreciated in field studies (Pickett and White 1985). Spatial heterogeneity is more often explicitly included as a factor in the design of studies.

Spatial variation can influence community patterns in numerous ways. In this chapter I focus on the consequences of alterations in the spatial configuration of habitats, particularly those associated with the fragmentation of landscapes. I also consider the importance of spatial scale and the degree to which community patterns may be dependent on the scale on which they are viewed.

## Changes in habitat patchiness

In the absence of disturbances, the vegetation of an area should develop within the limitations imposed by underlying edaphic factors. Areas of uniform soil composition and nutrient levels should thus be occupied by relatively homogeneous habitats, other things being equal (Wiens *et al.* 1985). Disturbances, however, are a fact of life in all environments, and they fragment landscapes into complex mosaics of patches of different habitat types. Variations in the frequency, extent, and magnitude of disturbance (Table 3.3) produce complex patterns in the composition, age-structure, and size-distribution of patches in these mosaics (White 1979, Sousa 1984, Brokaw 1985). Populations and communities are confronted not only with habitats that are usually heterogeneously distributed but that are also continuously changing mosaics. It is little wonder that ecologists have often chosen to disregard this complexity.

The alteration of the spatial configuration of habitats can take a variety of forms, but we can distinguish two general types (Fig. 5.1). *Gap formation* occurs when an area of habitat is internally disturbed, creating a patch of different characteristics within a larger patch. *Fragmentation*, on the other hand, involves external disturbance that alters the large patch so as to create isolated or tenuously connected patches of the original habitat that are now interspersed with an extensive mosaic of other habitat types. As with most dichotomies, these extremes grade into each other: a gap that is large relative to the patch in which it occurs may fragment that patch into isolated units (Fig. 5.1). At the extremes, however, the effects of these habitat disruptions on birds may be rather different.

### *The effects of gap formation*

The consequences of gap formation have been studied primarily in forests, where treefalls create small patches of altered light availability, soil characteristics, and microclimate that foster the development of different vegetation than occurs in the surrounding forest (Shugart 1984, Brokaw 1985). Treefalls are frequent in lowland forests in Panama, and some bird species respond selectively to the habitat patches created by the gaps; others avoid them. As a result, the assemblages of understorey bird species occupying gap patches are quite distinct from those in the adjacent forests

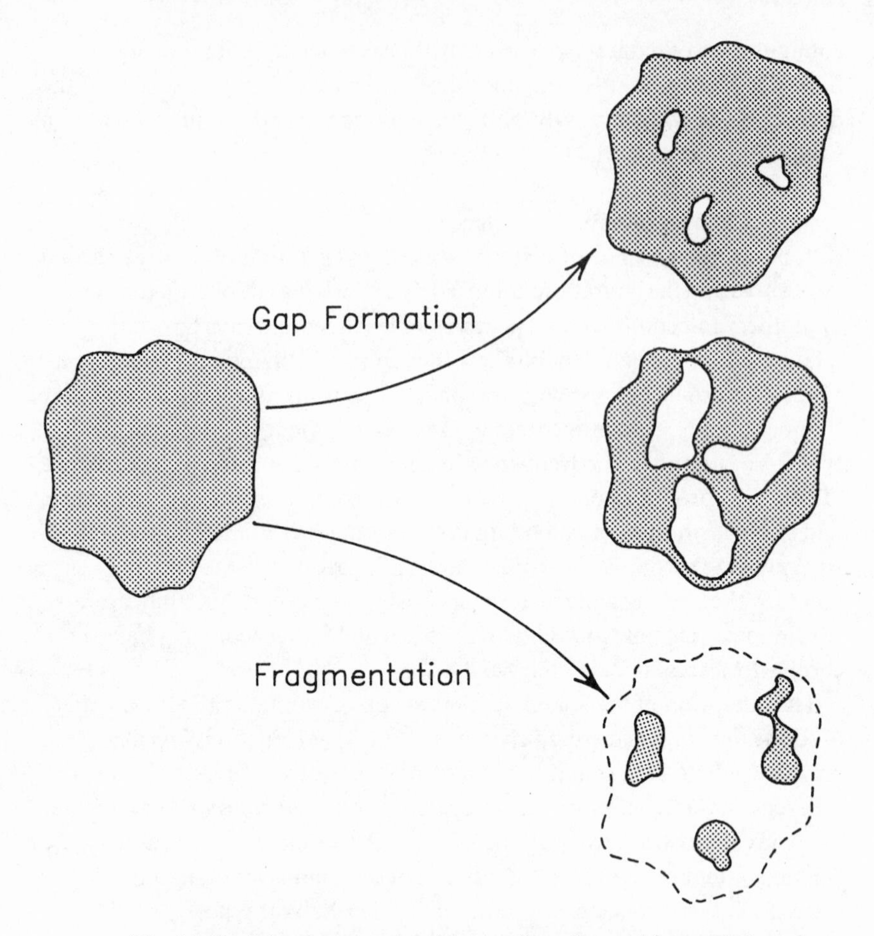


Fig. 5.1. Spatial patterns produced by gap formation and fragmentation of an area of previously homogeneous habitat. The intermediate situation (formation of large gaps) is also shown.

(Schemske and Brokaw 1981). In mist-net surveys, more species are recorded in the gaps than in the forests (Fig. 5.2). The species richness of the area as a whole is thus enhanced by the heterogeneity produced by the gaps.

Similar patterns have been reported in temperate forests in Illinois, where foliage-gleaning insectivores used gaps more frequently than forest during migration; other insectivores, however, did not differentiate in their use of gaps and forest (Martin and Karr 1986, Blake and Hoppes 1986). Frugivores concentrated their activity in the treefall gaps during late fall, although in spring, when they are primarily insectivorous, they used gaps and forest areas equally. Understorey foliage (and thus foliage-feeding

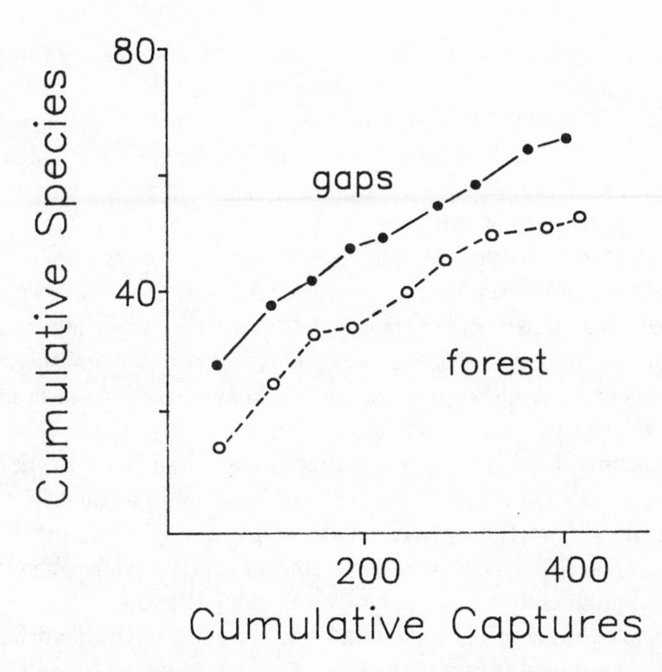


Fig. 5.2. Cumulative number of bird species recorded in treefall gaps and in adjacent areas of undisturbed tropical forest in relation to the cumulative number of captures of individuals in mist-net samples. From Schemske and Brokaw (1981).

insects) and fruits are more abundant in the gaps than in the forest understorey, and the patterns may therefore be related to differences in resource availability and cover from predators (Blake and Hoppes 1986). As in Panama, more individuals and species were captured in mist nets in the gaps than in the surrounding forests. It is worth noting that mist nets may be an especially suitable way to survey the avifauna of small treefall gaps in forests (which are not easily censused in other ways), although they are likely to produce inflated estimates of the number of individuals and species present in the gaps because they capture forest birds moving across the small openings. Some of the difference between gaps and forests shown in Fig. 5.2 may reflect this bias.

**Habitat fragmentation**

Fragmentation of habitats has received far more attention than gap formation, especially in suburban or agricultural areas where natural vegetation is broken into small, isolated patches as land is converted to human uses. Ecologists have focused especially on the fragmentation of forests (e.g. Burgess and Sharpe 1981, Harris 1984, Saunders *et al.* 1987), perhaps because the loss of forest habitat is so apparent and the recovery time so long, perhaps because we associate forests with a more nearly equilibrial (climax) state of nature. Any habitat type can suffer fragmentation, however. The effects of breaking a large area of prairie into a patchwork of prairie remnants interspersed with cornfields or housing developments are no different, qualitatively, from those accompanying the fragmentation of forests. Fragmentation of habitats can occur as a consequence of natural disturbances (e.g. floods in riparian areas, landslides) as well as human activities, of course, and it is a mistake to think of fragmentation as solely an artificial process. Also, human alterations of landscape may increase the size of homogeneous habitats as well as fragment habitats. Changing agricultural practices and mechanization in much of the world over the past several decades, for example, have led to a consolidation of habitats: small parcels have been amalgamated into larger farm holdings, hedgerows and shelterbelts have been removed, and larger fields have been planted in monocultures, reducing overall avian diversity.

When a large area of a given habitat type is fragmented by disturbance, a complex chain-reaction of events ensues. The overall area of the habitat and of habitat patches is reduced, patches are isolated from one another, the amount of edge habitat relative to interior habitat is increased, and the spatial relationships among the patches are altered (Fig. 5.3). The responses of bird species to these changes are determined by species-specific habitat or space requirements or life-history attributes and by changes in interactions with other species. These factors may lead to the extinction of local populations of some species and to increases in the distribution and abundance of other species. In either case, community composition will be altered.

### *Species' responses to fragmentation*

Different species have characteristic individual spatial requirements and predilections for edge versus interior locations in habitat patches, and these attributes determine how they will respond to habitat fragmentation. Differences in individual area requirements are perhaps most obvious. In Japan, the woodpecker *Dendrocopus kizuki* is rarely found in forest patches of less than 100 ha, but a bulbul (*Hypsipetes amourotis*) occurs in fragments as small as 0.1 ha (Higuchi *et al.* 1982). Moore and Hooper (1975) found that most species occurred at greatest frequencies in large woodlots in Britain. However, some species, such as Redstarts and Nuthatches, were mostly restricted to the larger patches whereas others, such as Wrens and Blackbirds (*Turdus merula*), were frequently present in

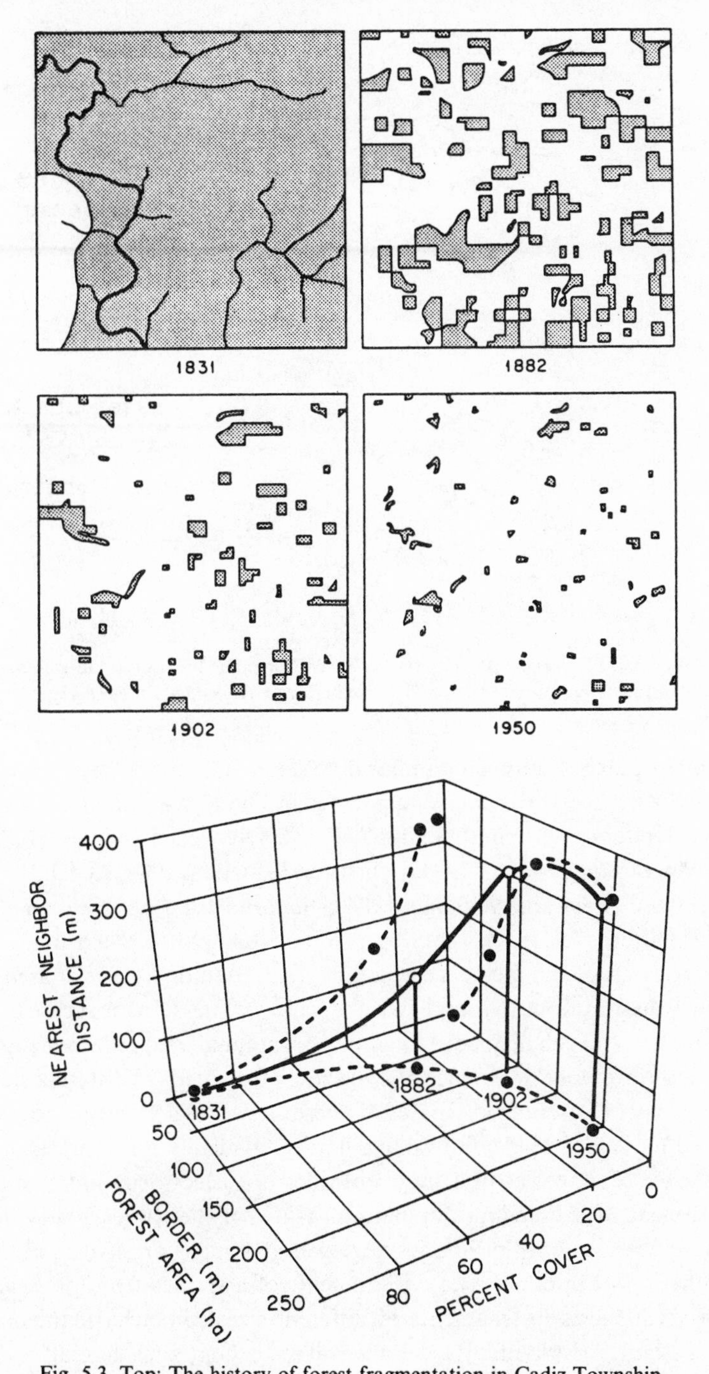


Fig. 5.3. Top: The history of forest fragmentation in Cadiz Township, Wisconsin, from settlement in 1831 to 1950. From Curtis (1959). Bottom: Trajectory of changes in total forest area, edge area, and mean distance between forest patches for the same location. From Urban (1986), after Sharpe *et al.* (1981).

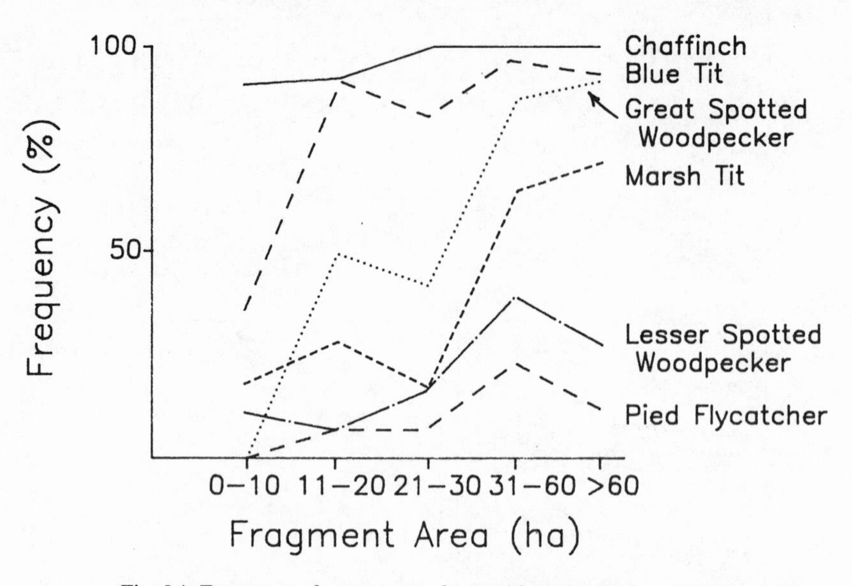


Fig. 5.4. Frequency of occurrence of several breeding bird species in woodlot fragments of different sizes in The Netherlands. After Opdam *et al.* (1985).

the smallest patches sampled. Similar differences in minimal-area requirements among species have been documented in The Netherlands (Opdam *et al.* 1985, Opdam and Schotman 1987; Fig. 5.4), Sweden (Nilsson 1986), North America (e.g. Martin 1981, Lynch and Whigham 1984, Lynch 1987), New Zealand (East and Williams 1984, Diamond 1984), and elsewhere.

These patterns of species-specific response to habitat-patch area are another version of the incidence functions that Diamond (1975a) used to relate bird distributions to island area or species richness. In most cases, the determination of such incidence functions or minimal area requirements is based on the presence or absence of a species in fragments of different sizes. A species may be absent from a patch of a certain size for a variety of reasons unrelated to its area (e.g. a failure to colonize, inadequate resources), and the presence of a species in a fragment does not necessarily indicate the establishment of a breeding population. Many of the same factors that complicate assessments of species turnover (Chapter 4) also make the determination of minimal area requirements of a species from presence–absence frequencies in fragments of different sizes somewhat equivocal. Still, it is clear that individuals generally will not occupy sites that fall below some threshold defining adequate area or territorial space, and the position of this threshold varies considerably among species (Calder 1984a).

The affinity of a species for edge versus interior habitats also influences how it will respond to fragmentation. As the size of a habitat patch is reduced by fragmentation, the proportion of the fragment that adjoins other habitat types (edge) increases (Levenson 1981, Forman and Godron 1986). The same change occurs if the shape of a patch changes from circular to oblong or rectangular without any change in its size. Some species characteristically locate breeding territories on the edge of patches, whereas others occupy only interior situations (Fig. 5.5). A reduction in fragment area or a change in patch shape is therefore likely to lead to a loss of interior species and an increase in edge species (e.g. Butcher *et al.* 1981, Whitcomb *et al.* 1981, Lynch and Whigham 1984, Freemark and Merriam 1986, Temple 1986). Other species associated primarily with the adjacent habitat types may also occur more frequently in the smaller habitat fragments, so the overall diversity of the community may be increased by fragmentation, at least to a point (Noss 1983, Haila *et al.* 1987). In North America, many long-distance (neotropical) migrants breed primarily in extensive stands of mature, floristically diverse forest, whereas permanent resident species and short-distance migrants are less closely associated with such forests or may even exhibit opposite distributional patterns (Whitcomb *et al.* 1981, Lynch and Whigham 1984). Variation in the abundances of the forest-interior/migrant/canopy species are closely related to variations in fragment area (Fig. 5.6), and these species are therefore particularly sensitive to forest fragmentation. Permanent residents are less strongly affected by patch area, and short-distance migrant/edge species are more closely related to habitat heterogeneity (Fig. 5.6), and therefore may benefit from such habitat changes. The long-distance migrants are often characterized by life-history attributes such as small clutch size, single-broodedness, use of open nests placed on or near the ground, and large-scale dispersal of young following fledging that collectively inhibit their capacity for rapid population growth, thereby amplifying their sensitivity to habitat loss.

Interior and edge species may respond to changes in the availability of suitable habitat independently of one another, but interactions between these sets of species have also been implicated as contributing factors in the decline of interior species with fragmentation. In Wisconsin forest fragments, Ambuel and Temple (1983) found that interior-dwelling long-distance migrants decreased as patch size decreased, and they suggested that this was due at least in part to the active exclusion of the interior species by forest-edge and farmland species, which increased. Ambuel and Temple felt, in fact, that this exclusion was a more important determinant of overall community composition in small forest fragments than were area-dependent changes in habitat or the degree of patch isolation. Their argument was

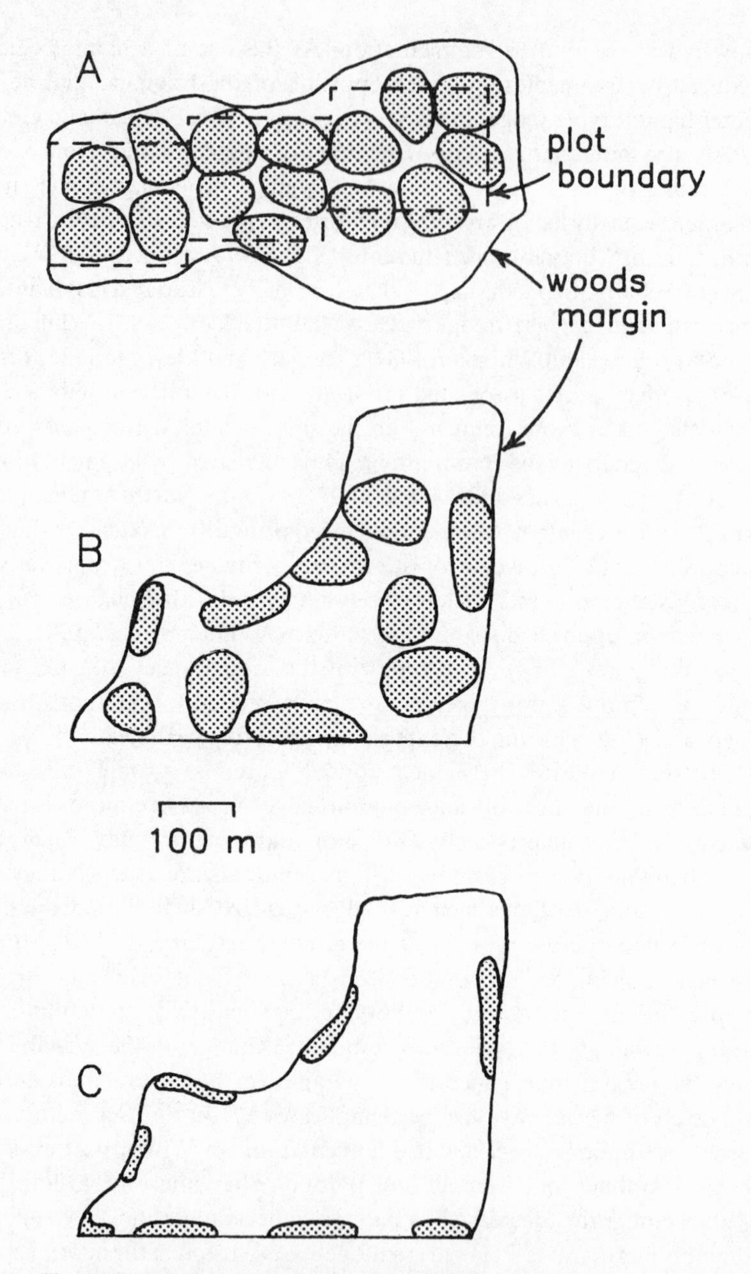


Fig. 5.5. Territory placement of (A) Hooded Warblers (*Wilsonia citrina*), (B) Northern Cardinals (*Cardinalis cardinalis*), and (C) Indigo Buntings (*Passerina cyanea*) in tulip-tree/oak forest fragments in Maryland. The warbler is confined to the interior of the forest, whereas the bunting occupies edge habitats; the cardinal is somewhat intermediate. After Whitcomb *et al.* (1981).

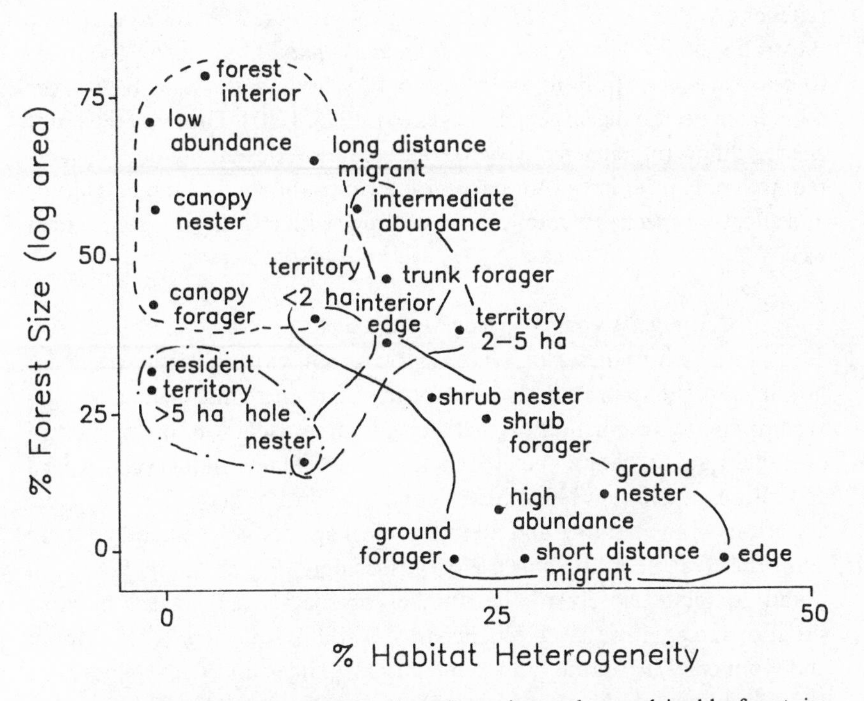


Fig. 5.6. The percentage of variance in species numbers explained by forest size and habitat heterogeneity for several ecological categories of birds occurring in Ontario forests. The groups included within the lines share at least 50% of their species and/or significantly more species than expected by chi-square tests. After Freemark and Merriam (1986).

based on patterns of distribution of several species that are brood parasites, predators, or potential competitors of the interior species over the range of woodlot sizes. Although the densities of several prime contenders for such roles (e.g. Brown-headed Cowbirds, Blue Jays, Gray Catbirds (*Dumetella carolinensis*)) did not vary significantly with forest-patch area, those of some other species (Starlings, Common Grackles (*Quiscalus quiscula*), and Red-winged Blackbirds (*Agelaius phoeniceus*)) did. Ambuel and Temple noted that the grackles may sometimes prey on the eggs and nestlings of other species and that 'all three species may compete with forest species for food and habitat' (1983: 1066). Further, although the cowbirds were no more abundant in the smaller woodlots, they may be more effective as brood parasites there (Brittingham and Temple 1983). The evidence supporting Ambuel and Temple's argument is largely inferential, but other studies involving experimental nest manipulations have indicated that predation rates may increase with fragmentation or be greater in smaller forest patches (Andrén *et al.* 1985, Wilcove 1985; see Fig. 3.3). In southeastern Australia, Noisy Miners (*Manorina melanocephala*) may aggressively exclude other species from forest fragments of < 10 ha but are unable to keep them from occupying larger patches (Loyn 1985, 1987). Thus, not only may species differ in their sensitivity to fragmentation depending on their area requirements or affinities for edge or interior habitats, but the magnitude and effects of species interactions may change with changes in patch area as well.

### *Community effects of fragmentation*

In view of these differences in the responses of species to fragmentation, it might seem futile to attempt to determine which of the changes accompanying fragmentation – changes in area, isolation, proportion of edge habitat, or habitat heterogeneity – have major influences on bird community attributes. Nonetheless, attempts have been made to correlate variations in the number and abundances of species in fragments to these and other fragment attributes. Not surprisingly, fragment area has been found to relate closely to the number of species ($S$) present in most situations (e.g. Howe 1984, Kitchener *et al.* 1982, Rafe *et al.* 1985, Moore and Hooper 1975, Ambuel and Temple 1983, Galli *et al.* 1976, Opdam *et al.* 1985, Dobkin and Wilcox 1986, van Dorp and Opdam 1987). This is simply an expression of the more general relationship between species number and area that holds whether or not the area represents a habitat fragment.

In some studies, however, a direct effect of fragment area on $S$ has not been found. In the California coniferous forests studied by Rosenberg and Raphael (1986), for example, $S$ increased in more fragmented stands, and Lovejoy *et al.* (1983, 1986) reported that $S$ was greater in small fragments of tropical rainforest immediately after their creation by clearing of surrounding areas than it was before the clearing. In both situations, however, the fragments were quite recent, and the increase in $S$ may have been produced as species from the surrounding disturbed areas were concentrated into the smaller area of remaining forest habitat. Haila *et al.* (1987) found that $S$ was greater (after rarefaction) in intermediate-sized fragments of old-growth coniferous forest in Finland than in small or large habitat blocks. Some of the changes in $S$ with decreases in area may also reflect changes in the relative amounts of edge and 'core' habitat available. Rosenberg and Raphael (1986) attributed the increases in $S$ with fragmentation that they observed to the colonization of the stands by species preferring edge habitat, and Temple (1986) and Gotfryd and Hansell (1986) found that the amount of core area (or, conversely, of edge habitat) in forest fragments is a better predictor of *S* than is fragment area alone.

The effects of isolation on the species richness of fragments are varied. In their studies of small forest fragments in agricultural areas of The Netherlands, Opdam *et al.* (1985; Opdam and Schotman 1987) found that isolation had no significant effect, paralleling the results of studies in forest reserves in the Western Australia wheatbelt (Kitchener *et al.* 1982). Lynch and Whigham (1984), on the other hand, found correlations between measures of patch isolation and the abundances of 16 of 31 common bird species in forest fragments in coastal Maryland. In an upland region in the same area, the degree of isolation of fragments was an order of magnitude greater than in the coastal plain, and isolation had even greater effects on bird distributions and abundances (Robbins 1980). In Connecticut forests, fragment area is the best predictor of total density and number of forest-interior species in small patches, but isolation is the best predictor for large fragments (Askins *et al.* 1987). Howe (1984) found that isolation contributed significantly to variations in species' abundances in forest patches in southern Wisconsin but not in eastern Australia, where the habitat is less severely fragmented on a regional scale.

Fragments are set apart from the surrounding matrix by virtue of major habitat differences. Differences in habitat features between the habitat fragments may also contribute to community patterns, although once again the relationships are not consistent among studies. In Australia, Kitchener *et al.* (1982) found that floristic diversity and plant life forms were more important than area in explaining the abundances of some passerine species, and Rafe *et al.* (1985) and Whitcomb *et al.* (1981) also found habitat variables to be important in their analyses. On the other hand, Ambuel and Temple (1983) detected no area-dependent changes in vegetation composition or structure that seemed likely to affect the bird community, and Dobkin and Wilcox (1986) found that area affected the distributions of many species in natural fragments of riparian habitat independent of the habitat diversity of the patches. In both of these situations, however, the range of habitats contained in the patches was fairly restricted.

Some of the differences among these studies of fragmentation effects reflect differences in procedures and variables measured. Multiple regression procedures are often used to identify presumed determinants of the patterns, but the results of such analyses are sensitive to which variables are included in the analysis. If some important variables are omitted, others may emerge as 'important' more or less by default. For example, Robbins

Table 5.1. *Patterns of correlations of bird species' abundances in forest fragments in coastal Maryland with variables describing several attributes of fragment structure or composition.*

The values are the percentages of significant partial correlations in stepwise linear regressions for each group of species.

| | Species group | | | |
|---|---|---|---|---|
| Fragment attribute | Neotropical migrants ($n=16$) | Short-distance migrants ($n=6$) | Residents ($n=8$) | Total ($n=30$) |
| Patch area | 38 | 17 | 13 | 27 |
| Patch isolation | 63 | 50 | 38 | 53 |
| Tree physiognomy | 50 | 67 | 50 | 53 |
| Understorey physiognomy | 19 | 0 | 13 | 13 |
| Pine abundance | 25 | 33 | 13 | 23 |
| Plant diversity | 63 | 83 | 13 | 53 |
| Median number significant per species | 2.9 | 2.7 | 1.6 | |
| Median value of $R$ (multiple correlation coefficient) | 0.40 | 0.40 | 0.25 | |

*Source:* From Lynch & Whigham (1984).

(1980) reported that area was the most frequent correlate of avian abundances in 80 census plots in northeastern USA, but isolation was not included in this analysis. Galli *et al.* (1976) also found that area had a strong effect in their studies in New Jersey and that habitat seemed unimportant. The only habitat variable measured, however, was foliage-height diversity. Moreover, sampling was more intense in the larger areas, possibly confounding the density comparisons among fragments. In some studies (e.g. Opdam *et al.* 1985, Howe 1984), patches were selected on the basis of similarity in habitat, and the possible influences of variations in habitat features were thus excluded from the analyses by design.

In most natural situations, of course, the effects of changes in area, the proportions of edge and core habitats, isolation, and habitat features that accompany fragmentation are interrelated. To emphasize the effects of one aspect of fragmentation (typically area) to the exclusion of others is therefore unrealistic (Haila 1986b). Lynch and Whigham (1984) surveyed 270 upland forest patches in coastal Maryland and subjected a large number of variables to stepwise regression analysis. Within the range of patch sizes (5–>1000 ha) and degrees of isolation (0.1–1 km) they considered, measures of

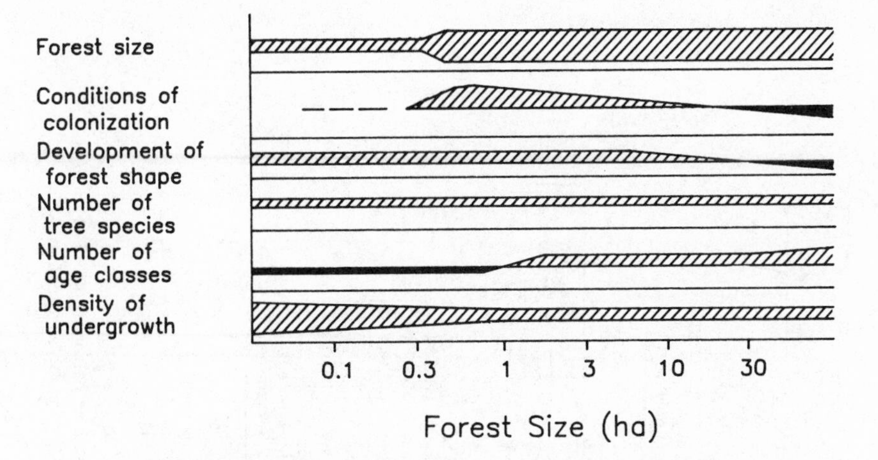


Fig. 5.7. Diagrammatic representation of the relative magnitudes of the effects of various features of forest fragments on the abundances of breeding birds, as a function of fragment area. After Cieślak (1985).

patch isolation, forest structure, and floristic diversity together accounted for 44 of 66 significant partial correlations with the abundances of 30 common bird species (Table 5.1). Area alone was the best descriptor of the abundance variations of only eight species. In pine forest fragments in Poland, Cieślak (1985) likewise found that a variety of factors influenced the relationship between the numbers and abundances of bird species present and fragment area (Fig. 5.7). This, of course, is what one might expect from the variation among species in their responses to various aspects of fragmentation. If area *per se* is important to some species, edge or interior habitat to others, isolation to others, and a combination of these factors to still others, it is unlikely that any single factor will emerge as the primary determinant of changes in entire bird communities with fragmentation. If one does, it is probably because it is a factor, such as area, that is closely correlated with changes in other factors, such as edge.

***Fragmentation and the extinction of local populations***

The consequence of fragmentation that is particularly disquieting to ecologists and conservationists is the local extinction of populations. As fragmentation occurs, some local populations are destroyed outright as blocks of their habitat are destroyed. The reduction in size and the increase in isolation of the remaining patches increase the influence of stochastic events on the availability of resources or the demography or genetic structure of populations within the patch, enhancing the likelihood that they will

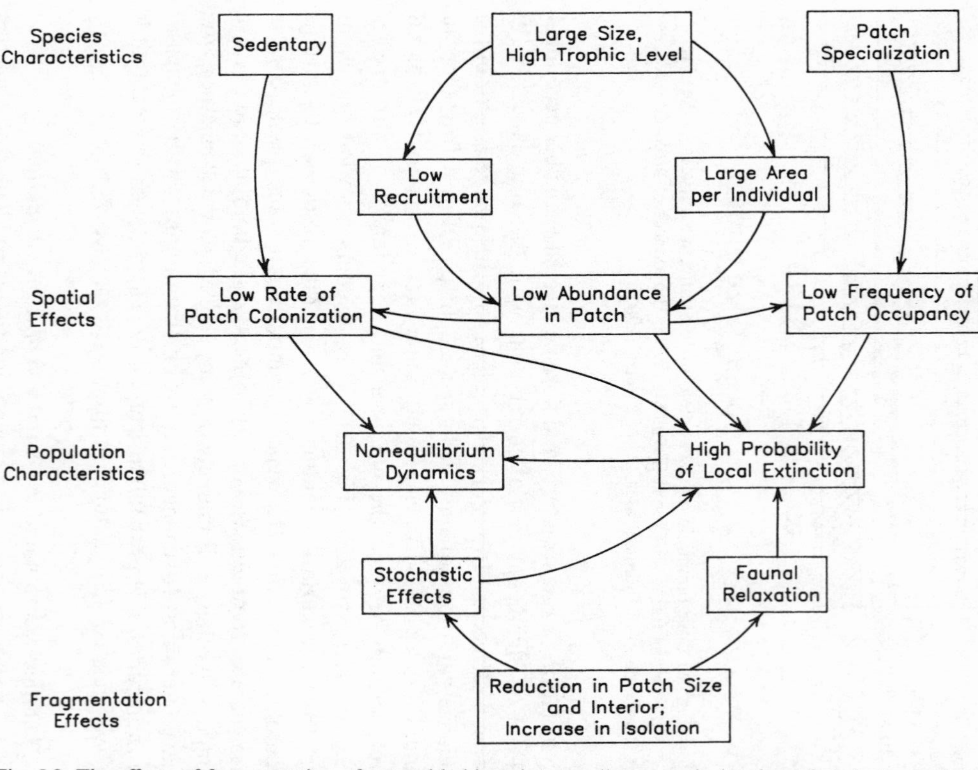


Fig. 5.8. The effects of fragmentation of natural habitats into smaller, more isolated patches (bottom of figure) on the likelihood of local extinction of populations of a species (middle of figure), as a function of various ecological characteristics of the species (top). From Wiens (1985a).

disappear from the patch (Fig. 5.8). The loss of area and habitat during fragmentation produces 'supersaturated' communities in the remaining patches, which through time will undergo faunal 'relaxation' as they lose populations. In a sense, this relaxation process is the opposite of community assembly, although the rate of loss of species is usually much more rapid than the rate of addition of species in community assembly.

The loss of species during relaxation might be essentially random or might follow a predictable sequence, according to the attributes of different species and their susceptibility to extinction. In the latter case, the impoverished community that results might have a similar species composition in similar-sized patches, in accordance with the minimal area requirements of the species. Several studies (Moore and Hooper 1975, Whitcomb *et al.* 1981, Forman *et al.* 1976, Galli *et al.* 1976) suggest that fragments of similar size and habitat do indeed tend to contain communities of similar composition, although the evidence is far from conclusive. On the other hand, the communities in similar-sized patches might differ substantially in species composition, depending on which species were contained in the fragments at the time of their isolation (a twist on the lottery hypothesis; Chapter 3).

There is little question, however, that the loss of species that accompanies fragmentation is usually nonrandom. As noted above, forest interior species are especially susceptible, and species that are large, occupy a high trophic position, or are habitat specialists may also be especially prone to local extinction because of their low frequency of occurrence among patches and their low abundance within patches (Fig. 5.8). In fragments of Amazonian forest created by clearing of adjacent areas, for example, two obligate army-ant-following birds, *Pithys albifrons* and *Gymnopithys rufigula*, disappeared from 1-ha patches within 2–4 months of isolation, and understorey frugivores were substantially rarer in both 1- and 10-ha patches in the second year following fragmentation (Lovejoy *et al.* 1983, 1986). In an 86-ha Javan woodland that was fragmented and isolated when the surrounding forest was cleared, roughly one-third of the 62 breeding species suffered local extinction in the following 50 years. These were primarily species with small initial population sizes or that were rare or absent in the surrounding region (Diamond *et al.* 1987).

Nonrandomness in community changes may also result if the species that are initially lost following fragmentation are important predators or prey in the community. Evolved predator–prey linkages may then be disrupted, leading to additional, 'secondary' extinctions as well (Terborgh and Winter 1980). This two-staged pattern of extinctions would lead to changes in the rate of local extinctions of species following fragmentation – first slow, as the 'primary' species disappear, and then more rapid, as the 'secondary' species follow. McLellan *et al.* (1986) used a simulation model to show that the same pattern of change in extinction rates may be produced as a consequence of the territory-size requirements and dispersal abilities of species. In their model, extinction rates are low as a previously continuous habitat suffers initial fragmentation. As fragmentation continues, a threshold is reached at which the remaining habitat no longer meets the area requirements of a number of species and interpatch dispersal can no longer prevent local extinction. The extinction rate may then increase rather abruptly. So far, there is no empirical evidence available to evaluate these ideas.

Although there is little doubt that fragmentation increases the rate of loss of populations from individual patches, its effects at a broader spatial scale are not so clear-cut. A population that is distributed over a number of patches, for example, may be less likely to suffer regional extinction than one occurring in a single, large, homogeneous area – this is the point of den Boer's (1981) spreading-of-risk argument (see also Andrewartha and Birch 1984, Quinn and Hastings 1987). Moreover, because individual patches may contain somewhat different communities, the overall diversity in a patchy, fragmented landscape may be greater than in a more homogeneous region of similar area (Noss 1983). In forests managed for Ruffed Grouse (*Bonasa umbellus*) by a prescribed cutting rotation in 1-ha patches, for example, the numbers of breeding species and total community densities were greater than in equivalent areas of unmanaged forest in the same region (Yahner 1984).

The key factor determining these regional-scale effects of fragmentation is dispersal. For a species to avoid extinction at a regional scale in the face of extirpation from local patches, movement of individuals among patches must be sufficiently great to repopulate patches following the disappearance of the species (Fig. 5.8). As patches become increasingly small and/or isolated, the likelihood of this 'rescue effect' (Brown and Kodric-Brown 1977) operating is lessened. If a population suffers extinction from all of the fragments in a region, it is effectively removed from the source pool of colonists. This alters the patterns of community response to fragmentation. In studies in New Jersey, for example, more than a third of the breeding bird species recorded in forest fragments showed no clear response to variations in fragment area (Forman *et al.* 1976, Galli *et al.* 1976). Many of these species were edge species, however, and several forest-interior species found elsewhere in this region were absent from *all* of the patches. This suggests that the extreme fragmentation of the New Jersey landscape may have led to the extirpation of these species prior to these studies, and the analyses may therefore have failed to reveal some of the most dramatic consequences of fragmentation.

These examples illustrate one of the major shortcomings of virtually all studies of fragmentation effects: the lack of suitable controls or of realistic *a priori* specifications of the expected patterns. The effects of fragmentation are usually assessed by comparisons among a set of samples of different sizes and degrees of isolation that are presumed to have once been part of a larger, more homogeneous area of habitat and to have now reached a new equilibrium. Neither assumption is fully justified, and some of the differences in the findings of fragmentation studies may reflect the fact that these assumptions have been violated to differing degrees. The most appropriate way to study fragmentation effects is through carefully planned investigations in which replicated areas of different sizes are surveyed before and after fragmentation occurs (Verner 1986). Aside from the project being conducted in Amazonian tropical forests by Lovejoy and his colleagues (Lovejoy and Oren 1981, Lovejoy *et al.* 1983, 1986), this has rarely been attempted. As a result, most statements regarding fragmentation effects on bird populations and communities are based on indirect, correlative analyses. Studying fragmentation before, during, and after the disturbances is difficult, especially at a broad scale. This is an area in which modeling studies (e.g. Urban and Shugart 1986, Seagle 1986, McLellan *et al.* 1986) may be especially useful.

**The landscape context of habitat fragments**

A focus exclusively on fragmentation of habitats misses the point that it is often the structure of an entire landscape mosaic rather than the size or shape of individual patches that is important to birds. The likelihood that dispersal can occur between fragments and forestall the extinction of sensitive species on a regional scale is influenced by the configuration of the fragments and the landscape mosaic in which they are embedded. If fragments are linked by narrow corridors of similar habitat, such as shelter belts, hedgerows, or streambeds for patches of woodlands or power-line or railway rights-of-way for fragments of grassland, dispersal among patches may be facilitated (MacClintock *et al.* 1977, Forman and Godron 1986, Simberloff and Cox 1987, Noss 1987). Wegner and Merriam (1979) found that birds travelled along fencerows between woodlots much more than they crossed open fields, and Johnson and Adkisson (1985) linked the role of jays in dispersing beech nuts between wooded patches in a fragmented landscape to their use of wooded fencerows connecting the patches. Small patches that are interconnected to larger fragments by corridors may thus contain species that would be absent were the patches totally isolated. In a sense, the effective area of a fragment is increased by corridors, and it may be more appropriate to consider the entire "archipelago" of linked patches as the fragment, rather than the individual patches (Haila and Hanski 1984). As in a hydraulic system, the effectiveness of corridors in promoting the flow of individuals among patches and the maintenance of populations in local areas is dependent on such factors as the width of the corridor, its length and continuity, and the degree to which the habitat in the corridor matches those in the connected patches (the habitat-induced 'resistance' to flow).

Unlike oceanic islands, fragments of habitat are not surrounded by a totally inhospitable environment. Thus, even if patches are completely isolated from areas of similar habitat, the dynamics of the populations they contain may be influenced by features of the surrounding habitats or distances to other patches of the same habitat. In the fragmented landscape of the central Netherlands studied by Opdam and Schotman (1987), for example, Nuthatches occupied all of the woodlots larger than 6 ha. Smaller woods were also occupied if they were close to areas of extensive forest, but those farther away from the forests were occupied infrequently (Fig. 5.9).

The spatial relationships among 'source' and 'sink' patches (Chapter 4) may also be important. If populations in some fragments persist because of immigration from other source areas, the dynamics of the local populations will be dependent on events occurring over the broader area that includes the source habitats (Haila *et al.* 1987, Diamond *et al.* 1987, Askins *et al.*, 1987). Thus, local breeding populations of Red-shouldered Hawks (*Buteo lineatus*) in some upland areas of Maryland are maintained by the larger populations occurring in woodlands along rivers (Snyder and Snyder 1975, Whitcomb *et al.* 1981), and a local population of Yellow-breasted Chats (*Icteria virens*) studied in Indiana by Thompson and Nolan (1973) was dependent on subsidization by recruits from other more productive areas.

In other situations, the interrelationships of habitat patches in a mosaic are less clearly related to population recruitment, but the effects on community attributes are nonetheless important. The composition of bird communities in wooded shelterbelts in Minnesota, for example, is influenced by various features of the nearby habitats (Yahner 1983). In Arizona, Szaro and Jakle (1985) found that the bird species characteristic of a riparian woodland fragment constituted as much as a third of the individuals found in adjacent desert washes and perhaps a sixth of the total density in bordering desert uplands. The species characteristic of the desert habitats,

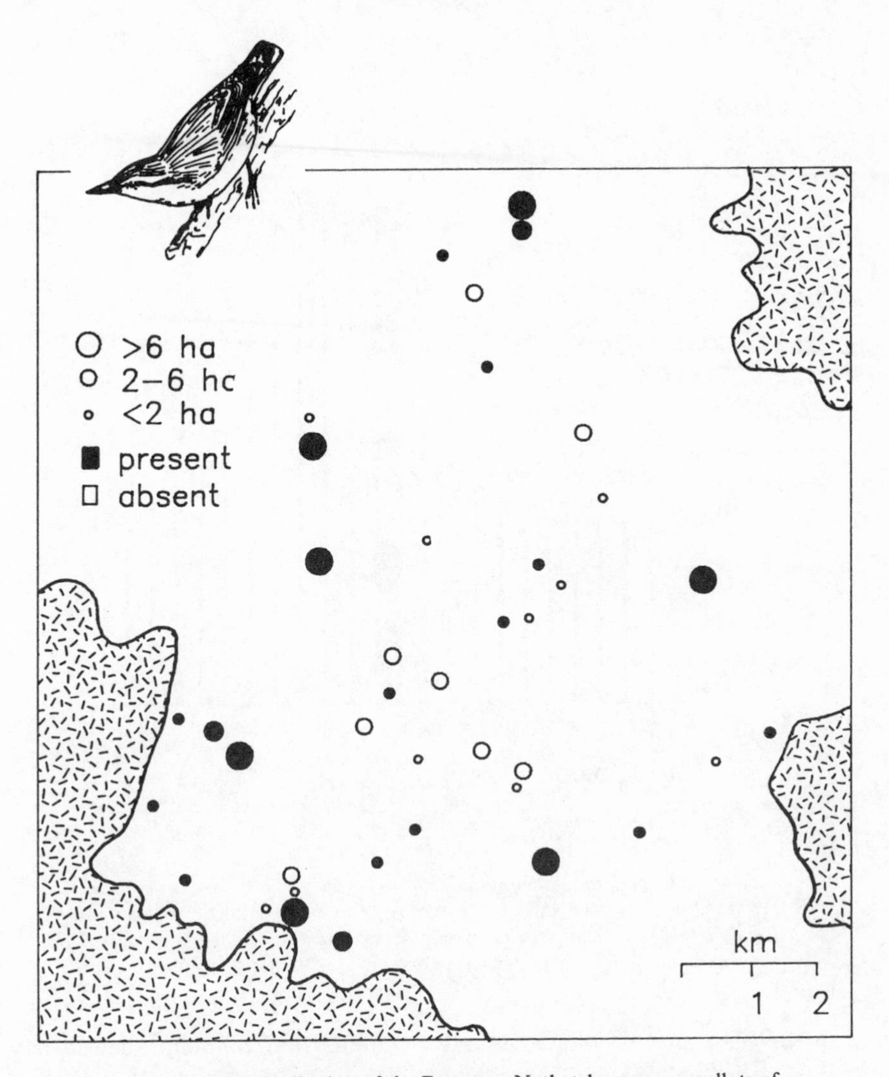


Fig. 5.9. The distribution of the European Nuthatch among woodlots of various sizes in relation to the distance to areas of extensive forest (shaded areas) in the central region of The Netherlands. All woodlots >6 ha in area contain at least one nuthatch territory, but many smaller woodlots contain nuthatches only if they are relatively close to extensive forests. After Opdam and Schotman (1987).

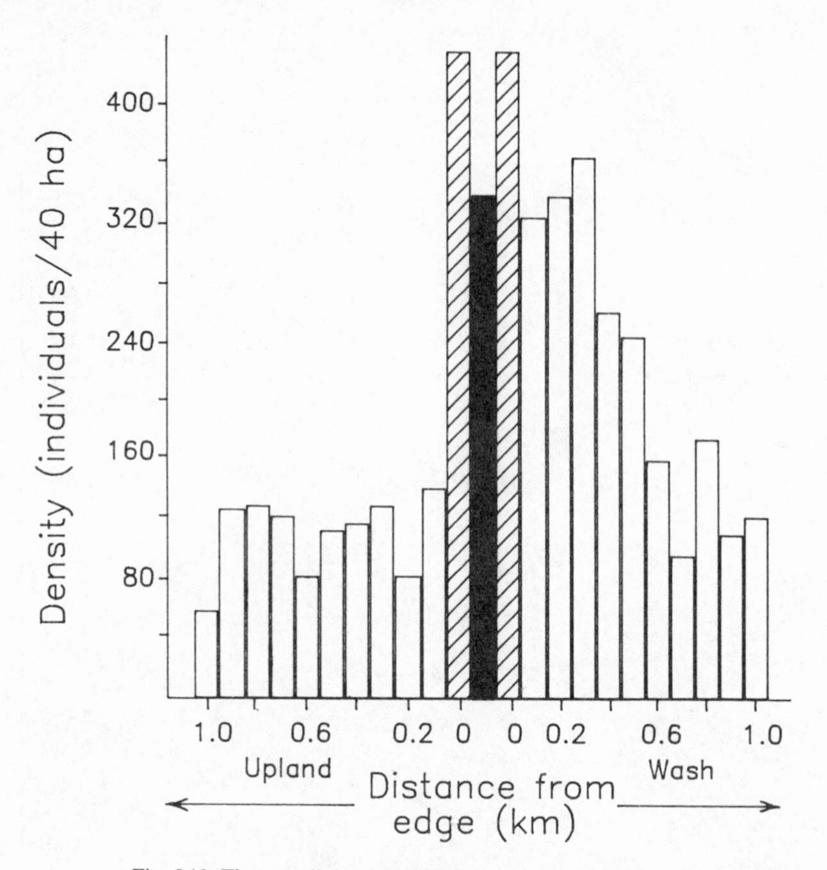


Fig. 5.10. The overall density of birds in a riparian woodland in Arizona (solid), in the woodland edge (hatched), and in adjacent desert upland and desert wash habitats as a function of increasing distance from the edge of the woodland. After Szaro and Jakle (1985).

on the other hand, accounted for only 1% of the total community density in the riparian habitat. Bird densities in the desert washes decreased dramatically with increasing distance from the riparian habitat (Fig. 5.10), reflecting limitations in the movement distances of riparian birds from their primary habitat.

The dynamics of sink populations are affected by the sizes of source areas, the proximity of these areas to the sink patches, and the nature of the surrounding habitat matrix. The willingness of individuals to cross boundaries between adjacent patches (boundary 'permeability') may be a function of the boundary sharpness or contrast between the patches (Cieślak 1983, Wiens *et al.* 1985). Thus, if source and sink patches are intermixed with a variety of habitat types that are totally different (woodlots interspersed with wheatfields or parking lots, for example), the movement of individuals between the patches may be much less frequent than if the habitat in the adjacent patches is more similar. In the American tropics, the composition of bird communities and abundances of species in forest fragments are increasingly sensitive to influences from adjacent habitats as their size decreases, but the influences are greater if the adjacent habitats are secondary forest than if they are cleared croplands, which do not support potentially invading species (Janzen 1983). Of course, if the source habitats suffer fragmentation and isolation, the populations in both the source and the nearby sink patches are affected. The consequences of disturbance of source patches in a landscape mosaic may thus be especially dramatic and wide-reaching.

Habitat patches are linked to the landscape mosaic in another way as well. Although we often think of species as being associated with particular patch types ('forest-interior birds', for example), many species range more widely and may be dependent on several distinctly different habitats during their life history. Hansson (1979) drew attention to the importance of such landscape heterogeneity for small mammals in Scandinavia, where forested habitats may provide a flush of food in summer but secondary successional stages or human cultivation provide richer sources of the seeds that are eaten in winter. The survival of individuals is dependent on their occupying an area that contains both patch types. Among birds, several species (perhaps many?) make similar use of habitat mosaics. In eastern North America, icterids of several species assemble in large roosts or breeding aggregations, from which they radiate in daily movements to feed in adjacent grasslands or agricultural crops (Wiens and Dyer 1975a). Starlings exhibit similar patterns of mosaic use in many areas of Britain and Europe (Feare 1984), as do Eared Doves (*Zenaida auriculata*) in Argentina (Bucher 1974, 1982, Murton *et al.* 1974). In Australia, Galahs (*Cacatua roseicapilla*) and Little Corellas (*C. sanguinea*) nest and roost in gum groves but forage extensively in shrub-desert habitats, grasslands, or grain fields, often several kilometres from the breeding sites (personal observation). Within the foraging areas, the use of particular habitat patches by the birds is related to the availability of native grass seeds, which in turn varies with rainfall (Fig. 5.11). Six of the hummingbird species that occur on Trinidad wander widely among patches of different habitat types during their foraging; these species are absent from the smaller island of Tobago, where the heterogeneity of the landscape mosaic is more restricted (Feinsinger *et al.* 1985). In general, such patterns of use of patches of different habitat in a landscape mosaic appear to be

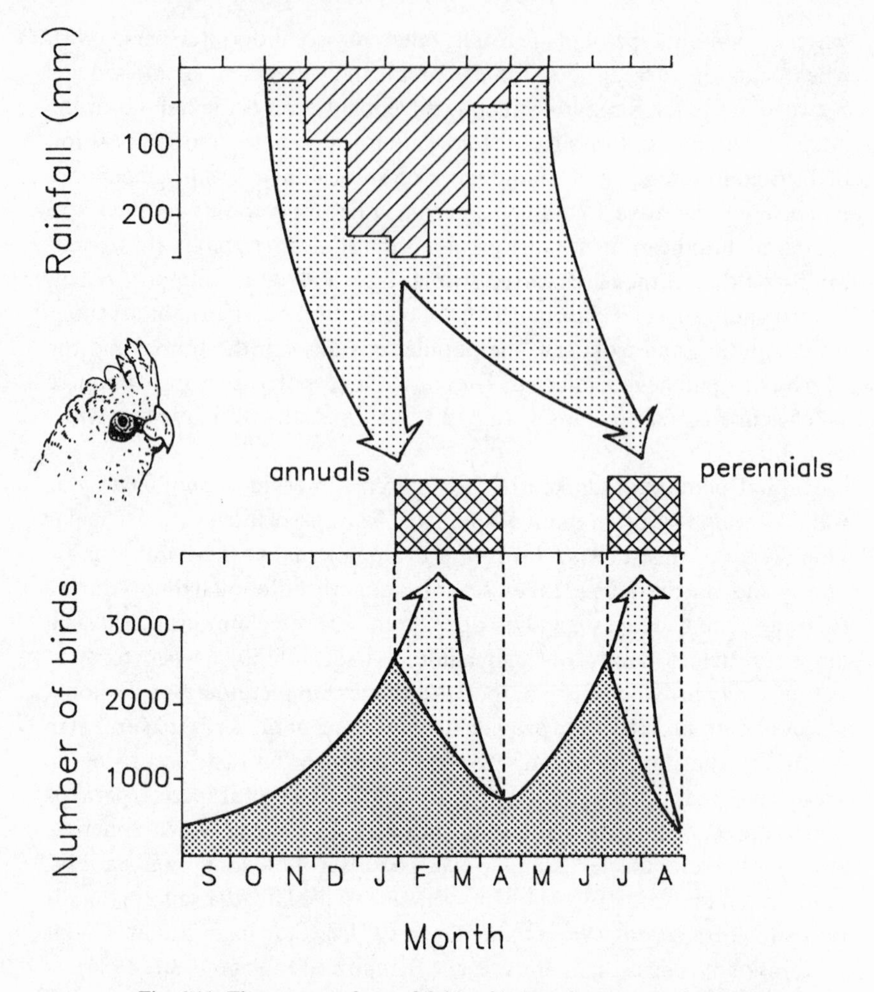


Fig. 5.11. The pattern of use of fields of irrigated sorgum and of native grasses by Little Corellas in northwestern Australia. Native seed production is high in February to mid-April (annuals) and in July and August (perennials) in response to seasonal rainfall (top and middle). At these times, the corellas move from the agricultural crops to feed on the preferred native grasses, but when the grasses are not available the birds concentrate in sorgum fields. Corella densities in sorgum fields are shown by dense stippling in the lower graph. After Recher *et al.* (1986) and Beeton (1977).

especially well-developed in granivorous species (Wiens and Johnston 1977), although this behavior may only be more conspicuous in these species than in others because of the pest status of many granivores.

These comments emphasize the point that the spatial arrangement of patches of habitat – their size, degree of isolation or linkage, and their context in a larger mosaic of habitats – may have substantial influences on community structure. Such influences have been considered in few studies, and these have focused largely on the effects of the reduction in area and increase in isolation accompanying fragmentation. The importance of the landscape context in which the fragments occur has only recently been appreciated (e.g. Harris 1984, Forman and Godron 1986, Wiens *et al.* 1985), but it should become a major focus of community investigations.

## Fragmentation and the design of natural reserves

Knowing how species respond to the size and configuration of patches of suitable habitat takes on particular importance when one considers how areas of habitat should be preserved as natural reserves. There will always be sociological, economic, and political factors that constrain the selection of areas to be set aside, but within those constraints ecologists should be able to offer suggestions as to which of several alternative designs might be best. Given the variable and equivocal nature of population and community responses to habitat fragmentation just discussed, it is not surprising that there is no clear consensus as to which approach might be 'best'. Because of the urgent need to establish such reserves, especially in the tropics, this uncertainty has produced considerable controversy (e.g. Whitcomb *et al.* 1976, Diamond 1976, 1984, Cole 1981, Higgs 1981, Frankel and Soulé 1981, Simberloff and Abele 1976, 1982, 1984, Margules *et al.* 1982, Wilcove *et al.* 1986, Murphy and Wilcox 1986, Zimmerman and Bierregaard 1986).

Much of the controversy relates to attempts to use the MacArthur–Wilson (1967) theory of equilibrium island biogeography to generate predictions about optimal reserve design (Wilson and Willis 1975, Diamond 1975b, Diamond and May 1976, Whitcomb *et al.* 1976). Several 'principles' have been suggested (Blouin and Connor 1985, Williams 1984). First, reserves should be as large as possible (to minimize extinctions). More specifically, a single large reserve is preferable to several small ones of the same total area. Second, reserves should be as close together as possible (to maximize patch colonizations). Third, reserves should be as circular as possible. This reduces the 'peninsula effect', in which species number is reduced in elongate areas in comparison with circular ones of the same size, and enhances the likelihood that the entire patch will be saturated with individuals.

There seems to be little disagreement about the second statement – if several reserves are scattered over a region, the probability that declining local populations in any one patch can be 'rescued' by immigration from the other patches is greater if the reserves are nearby than if they are widely separated. Corridors enhance this probability dramatically (but see Simberloff and Cox 1987). The third argument, that reserves should be as circular as possible, is somewhat at odds with the notion that elongate areas (or patches with elongate extensions) are likely to enhance dispersal through the corridor effect (Williams 1984). In one test of this principle involving data for various taxa from 33 island archipelagos around the world, island shape had no significant effect on variations in species richness after the influences of island area were removed by multiple regression (Blouin and Connor 1985).

There is also little dispute over the notion that larger areas will preserve more species than will smaller areas, other things being equal. This follows from well-documented species-area relationships. The relationship of area to species richness is often not very precise, however, and species-area regressions may therefore not provide a good prediction of the loss in species that would be expected to accompany a given loss of area (Boecklen and Gotelli 1984, Kangas 1987). Part of the problem, as Boecklen and Gotelli observe, is that MacArthur–Wilson island biogeography theory has led ecologists to interpret the $S/A$ regression line as the *equilibrium* number of species for a given area; points above the line are therefore 'supersaturated' and should be expected to undergo faunal relaxation to the equilibrium. But the $S/A$ line is empirically obtained by regression, and the more appropriate statistical interpretation is that it represents the *average* value of $S$ for a given $A$; a point above the line has a positive residual, perhaps due to local conditions of habitat heterogeneity, resource abundance, or community composition. To use $S/A$ regressions for precise predictions of the consequences of changes in area is fallacious.

Much of the debate has revolved about whether a single large area is necessarily better as a reserve than several smaller ones of the same total area. Simberloff and Abele (1976, 1982) proposed that several small reserves might maintain more species than a single large area provided that dispersal is sufficiently great to overcome the increased probability of local extinction of populations in the smaller patches. Their argument is essentially a restatement of den Boer's spreading-of-risk notion or, more colloquially, the idea that 'it may be wiser to avoid committing all available eggs to one basket, especially if by using several baskets the result is likely to be a greater choice of eggs' (Williams 1984: 9). The argument has intuitive appeal, especially if the patches are reasonably widely scattered (Forman and Godron 1986). Simberloff and Abele (1982) considered several data sets bearing on the issue and concluded that none of them unquivocally supported the superiority of a single large area and several showed greater overall species richness for several small ones (see also McLellan *et al.* 1986). Data from New Zealand birds, on the other hand, support both sides of the argument (Dawson 1984), although East and Williams (1984) and Diamond (1984) advocated large reserves for New Zealand species.

Simberloff and Abele's arguments have been challenged on several grounds. In their model, for example, only demographic stochasticity is considered as a cause of the enhanced extinction probability of populations in small areas. These areas may also be subject to greater environmental variation than are large areas and the populations they contain may be more susceptible to chance genetic changes (Wilcox and Murphy 1985, Wright and Hubbell 1983, Barrowclough and Shields 1984; but see Boecklen and Bell 1987). Wright and Hubbell's model analyses suggested that several small reserves may be equally effective in preserving a rare species as a single large one if the system is open to immigration of individuals from outside, but the large reserve is superior if there are no external source patches. Lovejoy and Oren (1981) suggested that smaller reserves might be superior if patches differ in species composition at the time of fragmentation but that a large reserve may conserve more species if the species composition of similar-sized patches is similar. McLellan *et al.* (1986) reached much the same conclusion from a model analysis, and they suggested in addition that a single large reserve is preferable if the slope of the species-area curve ($z$) is large. In Kobayashi's (1985) model analysis, the balance shifts toward several small reserves only if rare species have not yet suffered local extinction and/or if the dispersion patterns of species are highly clumped. In general, smaller patches are also less likely to contain species that prefer patch interior conditions, and, if all of the patches in a region are small, this set of species may be missing altogether. In the Australian wheatbelt, for example, species found in disturbed areas outside of forest reserves as well as in the reserves may be favored by smaller reserves, but species restricted to the reserves are lost disproportionately from small areas and are favored by larger reserves (Humphreys and Kitchener 1982).

The criticisms of Simberloff and Abele's advocacy of several small reserves as a possible management strategy are focused largely on factors ignored in their arguments that affect its generality. Simberloff and Abele's proposition, however, was made in response to the generalization that single large reserves are invariably best, and they explicitly disavowed any such generality for the alternative view. Species differ in their minimal area requirements, their responses to edge or interior conditions, their perception of patchiness, their dispersal capabilities, their migratory strategy, and so on. A set of medium-sized forest patches that are not too isolated from one another might constitute a reasonable reserve for populations of small migratory passerines but would be inappropriate for large, sedentary frugivores or raptors, for which a single very large reserve might be necessary (Lynch and Whigham 1984, East and Williams 1984). Because communities are composed of different complexes of species that respond to patch area, arrangement, and shape in different ways, it seems unlikely that any single set of general, theoretically derived principles would successfully predict the 'best' management strategy for a particular set of circumstances (McCoy 1982, Haila 1985, 1986a, Zimmerman and Bierregaard 1986).

Different conservation goals may also dictate different reserve designs. Is the objective the preservation or maintenance of an entire regional avifauna, of a stable number of species in an area, of particular webs of interactions among species, of a certain level of habitat diversity, of minimum viable populations of particular rare or endangered species, or something else? Whether a single large round reserve, several small scattered elongate or amoeboid areas, or some other configuration is most suitable depends on the goals of the reserve program. The idiosyncracies of particular situations and species also bear heavily on the decision, and these must be derived empirically, not theoretically. Diamond (1984) and Murphy and Wilcox (1986) have countered this emphasis on autecology by noting that conservation needs are immediate and pressing, that we have neither the time nor the money to conduct such detailed studies in order to determine the most appropriate reserve design in particular situations, and that the general principles derived from biogeography theory provide a valuable framework for reaching decisions. On the basis of their studies in the Brazilian Amazon, however, Zimmerman and Bierregaard (1986: 141) concluded that a knowledge of the natural history of the species involved is essential and that 'the calculation of reserve sizes based solely on species-area data can never be more than uninspired guessing'.

No matter what their size or shape, patches of habitat set aside as reserves do not exist in isolation from their surroundings but are part of a larger landscape mosaic – in Janzen's (1983) apt phrasing, 'no park is an island'. An area of tropical forest may function quite differently to maintain populations and species diversity if it is surrounded by secondary forest rather than slash-and-burn croplands. Species that are not extreme patch specialists may depend more on a particular mosaic of habitats than on specific habitat patches alone. This suggests that the emphasis on characteristics of habitat blocks in the above arguments may be misdirected – clusters of habitat patches rather than areas of relatively homogeneous habitat may be the appropriate units for many natural reserves. Harris (1984) has developed this 'landscape mosaic' approach to management in some detail, emphasizing the need to consider the nature of habitats in the matrix containing a given patch as well as patch size and isolation in assessing the potential value of the patch as a reserve.

All of these considerations lead to the conclusion that the MacArthur–Wilson theory and the design principles derived from it are of quite limited value in planning natural reserves (Margules *et al.* 1982, Reed 1983, Williams 1984, Dawson 1984, Simberloff and Abele 1984, Boecklen and Gotelli 1984, Lahti and Ranta 1985, Haila 1985, Seagle 1986). Moreover, the debate over whether a single large reserve is superior to several small reserves is probably largely irrelevant, for at least two reasons. First, the argument is often framed in the context of habitat fragmentation (e.g. Simberloff and Abele 1982). Fragmentation of habitats, however, involves a loss of total habitat area, and it is unlikely that several small fragments will contain more species than a single reserve of greater overall area. Second, the importance of the landscape context of habitat patches is ignored in most debates over the optimal size and number of reserves. It is not enough simply to propose that a larger reserve is more likely to contain greater habitat heterogeneity and thus is better. Aside from the fact that size and heterogeneity are not invariably correlated, it is not patchiness *per se* that is important in landscapes, but the particular spatial juxtapositioning of habitat units. To establish reserves according to ecological insights requires both a consideration of broad-scale landscape configurations and knowledge of the ecological requirements of the species that are important in particular situations.

## The importance of scale

Ecological patterns are not independent of the spatial or temporal scale on which they are viewed. Differences in scale are critically important, for they influence the questions that can be addressed, the procedures followed, the observations obtained, and how the results are interpreted (Wiens 1981a, Dayton and Tegner 1984, Wiens *et al.* 1986). In their studies of native nectarivores in Hawaii, for example, Pimm and Pimm (1982) observed aggressive interactions among species that they interpreted as evidence of resource partitioning and competition. Their studies were conducted on two 0.5-ha plots and were focused on the behavior of a few individuals. Mountainspring and Scott (1985) studied the same species in the same area but found no evidence of interactions among the species. Their conclusion, however, was based on correlations of densities over a broad habitat gradient. The behavioral interactions over resources that occurred at the very local scale apparently did not translate into negative associations between the species at a broader scale.

The distributional patchiness or dispersion of species is also affected by scale. On a sufficiently broad scale, animals of all sorts tend to be aggregated (Taylor *et al.* 1978, Taylor and Woiwod 1982, Anderson *et al.* 1982); within local breeding populations, however, territoriality may lead to a more uniform dispersion pattern. Even at this level, however, the patterns may be sensitive to changes of scale. Sherry and Holmes (1985) noted that the dispersion patterns of several breeding species in hardwood forests changed as the size of the area analyzed was changed. These changes reflected the influences of both intraspecific and interspecific social interactions and of fine-scale habitat patchiness within the forest. Schneider and Duffy (1985) found that seabirds in the upwelling region of the Benguela Current of the South Atlantic were aggregated at scales ranging from 0.3 to 23 km, although the extent of aggregation varied with scale within this range and differed substantially among species. A similar scale-dependency of aggregation has been reported for murres and puffins off the coast of Newfoundland; here, the magnitude of correlation between the aggregation of the seabirds and that of the schooling fish (capelin) on which they preyed was also scale-dependent (Schneider and Piatt 1986). The 'tracking scale' (the scale at which the degree of correlation between bird and fish dispersions was greatest) ranged from 2 to 6 km. Correlations conducted using very short or very long transects as units would fail to record what Schneider and Piatt believed to be the true pattern of spatial association between the birds and the fish.

Rotenberry and I have also examined the effects of changing the scale of study on the patterns of habitat association among shrubsteppe birds (Wiens and Rotenberry 1981a, Wiens 1985b, 1986, Rotenberry 1986, Wiens, Rotenberry, and Van Horne 1987). At a broad, biogeographic scale (Fig. 5.12A), variations in the densities of several species that commonly breed in shrubsteppe environments were significantly correlated with variations in several features of habitat structure (Table 5.2). When we changed the scale to focus on 14 sites located within the shrubsteppe region (Fig. 5.12B), these associations disappeared and the birds exhibited significant covariation with other attributes of habitat structure or floristic composi-

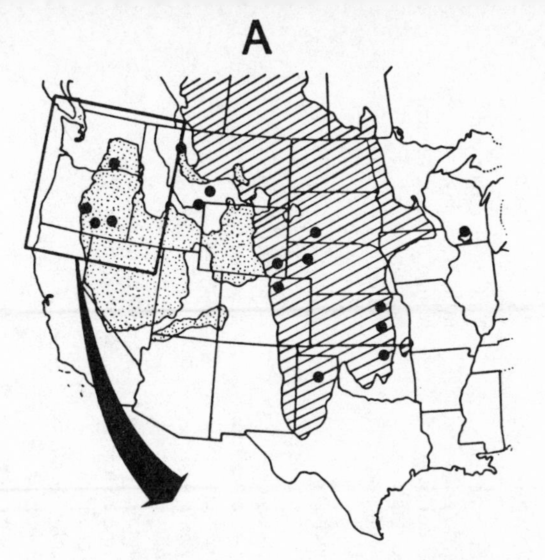


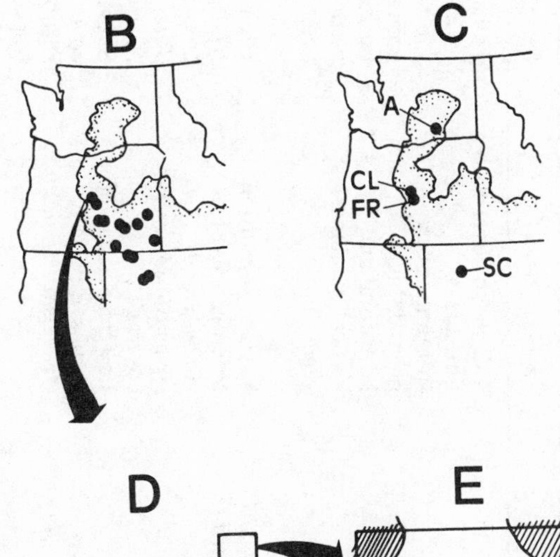


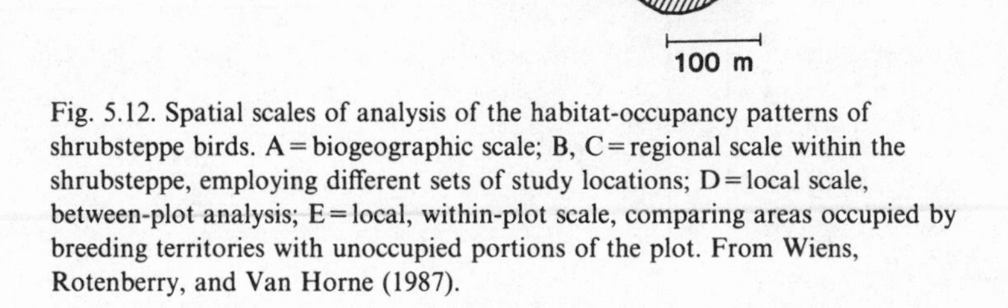


Fig. 5.12. Spatial scales of analysis of the habitat-occupancy patterns of shrubsteppe birds. A = biogeographic scale; B, C = regional scale within the shrubsteppe, employing different sets of study locations; D = local scale, between-plot analysis; E = local, within-plot scale, comparing areas occupied by breeding territories with unoccupied portions of the plot. From Wiens, Rotenberry, and Van Horne (1987).

Table 5.2. *Patterns of association between the distribution and abundance of four breeding birds of shrubsteppe environments, as revealed by analyses conducted at several spatial scales.*

| Scale | Horned Lark | Sage Thrasher | Sage Sparrow | Brewer's Sparrow |
|---|---|---|---|---|
| *Biogeographical* | Low stature, homogeneous vegetation | Tall shrub cover with little grass, substantial bare ground; high horizontal and vertical patchiness of vegetation | Tall shrub cover with little grass and substantial bare ground; high horizontal and vertical heterogeneity | Tall shrub cover with little grass; high horizontal and vertical patchiness |
| *Regional* | | | | |
| A. 14-site | High *Chrysothamnus* coverage, low *Atriplex* coverage | Vertical vegetation heterogeneity; low *Atriplex* and *Artemisia spinescens* coverage | High *Artemisia tridentata* coverage, low coverage of *Tetrydymia* and *Sarcobatus* | Low *Atriplex* and *Artemisia spinescens* coverage |
| B. 4-site | No significant correlations | Lower coverage of shrubs, especially *Artemisia tridentata* | Substantial grass cover and reduced shrub cover (especially *Atriplex* and *Artemisia spinescens*); low horizontal heterogeneity | No significant correlations |

| | | | | |
|---|---|---|---|---|
| *Local* | | | | |
| A. Between-plot | Absent from plot with greater shrub coverage (especially *Chrysothamnus*); present where standing dead vegetation more abundant | Densities somewhat greater where shrub coverage (especially *Chrysothamnus*) greater | No differences in abundance between plots | No differences in abundance between plots |
| B. Within-plot | No data | No consistent patterns | Occupied areas of greater shrub (especially *Chrysothamnus* and *Artemisia tridentata*) coverage and reduced grass coverage | No consistent patterns |

tion. In another regional-scale analysis that was based on eight plots at four locations (Fig. 5.12C), two of the species failed to evidence any significant habitat association patterns at all; the other two varied with habitat features in different ways from those in the other regional analysis (Table 5.2). Finally, we examined patterns at a local scale, restricting our attention to a single site (Fig. 5.12D). Patterns were evident in comparisons between the two study plots at this site for two of the species, but these patterns differed from those expressed at the broader scales. Only one of the species varied in its habitat-occupancy patterns at an even finer, within-plot scale (Table 5.2); these patterns were not evident for this species at the between-plot scale and contrasted with the patterns expressed at the 4-site regional scale but paralleled at least partially those found in the 14-site regional analysis. These changes are summarized diagramatically for one of the species, Sage Sparrows, in Fig. 5.13. The patterns of habitat associations we detected for these species were clearly dependent on the scale of investigation.

These and other examples (e.g. Gutzwiller and Anderson 1987, Morris 1987, Hengeveld 1987) illustrate some specific consequences of changing the scale on which a system is viewed. But ecological patterns and processes are more generally scale-dependent. Thus:

1. Populations appear to be much more stable and the species composition of communities more consistent if the patterns are derived on a regional rather than a local or plot scale, where the patterns seem much more stochastic (Chapter 4; Hamel *et al.* 1986). Urban and Shugart (1986) used a simulation model to show that the dynamics of habitat patches and species characteristics at a local scale produced considerable variation in population patterns as a result of local extinctions and recolonizations, but these local variations were not evident when the system was considered at the scale of an entire landscape mosaic. As chaos theory suggests, local disorder translates into regional or global order (Gleick 1987).
2. Temporal tracking of variations in food abundance, on the other hand, may sometimes be apparent at a local scale but absent at a broader geographical scale because individuals are unable to respond to regional differences in food supplies (Hutto 1985; but see Pulliam and Parker 1979, Dunning and Brown 1982).
3. The dynamics of a community that is closed to immigration at a particular scale (e.g. the Galápagos finches) may be quite different from those of an open community viewed at the same scale (e.g. shrubsteppe birds). In open systems, local patterns may be produced not only by local processes or events but also by the

dynamics of regional populations or events elsewhere (Wiens 1981a, Väisänen *et al.* 1986, Haila *et al.* 1987, Ricklefs 1987).

4. Patterns that are most evident at fairly broad regional or biogeographical scales may be a consequence of events at a more local scale, and an appreciation of the local scale is therefore necessary to interpret the broad-scale patterns correctly. The model analyses of Urban and Smith (unpublished), for example, demonstrate that realistic patterns of changes in species-area curves, species-abundance distributions, or species turnover during succession may be obtained in a bird community composed of random niches (ellipses in PCA space) from the dynamics of gap disturbance of small patches in a landscape mosaic.
5. The effects of disturbance on populations or communities may be quite different depending on whether the changes occur on a fine scale, such as gap formation, lead to fragmentation of habitats on a somewhat broader scale, or alter environments over an entire region.
6. If individuals of some species depend on several patch types in a habitat mosaic (e.g. Fig. 5.11), studies confined to a single habitat will produce misleading impressions of their ecology.
7. Because the size of a population considered demographically is not necessarily the same as the effective population size in genetic analyses, a scale suitable for the investigation of demography may be inappropriate for determining the genetic structure of the population. Different attributes of populations or communities may be expressed on different scales. The concepts of spreading-of-risk or of source-sink patch relationships, for example, may provide some insights into the extinction and colonization dynamics of populations at the scales of regions or landscape mosaics, but they are inappropriate at the level of biogeographic ranges (where Brown (1984) has mistakenly applied them) or at a very local, within-patch scale.
8. Processes such as competition or predation occur among individuals at a local scale, and their effects are therefore more likely to be evidenced at that scale than at a regional or biogeographical scale (Wiens 1981b, Haila and Hanski 1984, Connor and Bowers 1987).

This list could easily be extended. It is sufficient, however, to indicate that virtually all community patterns and processes are likely to exhibit some degree of scale-dependency.

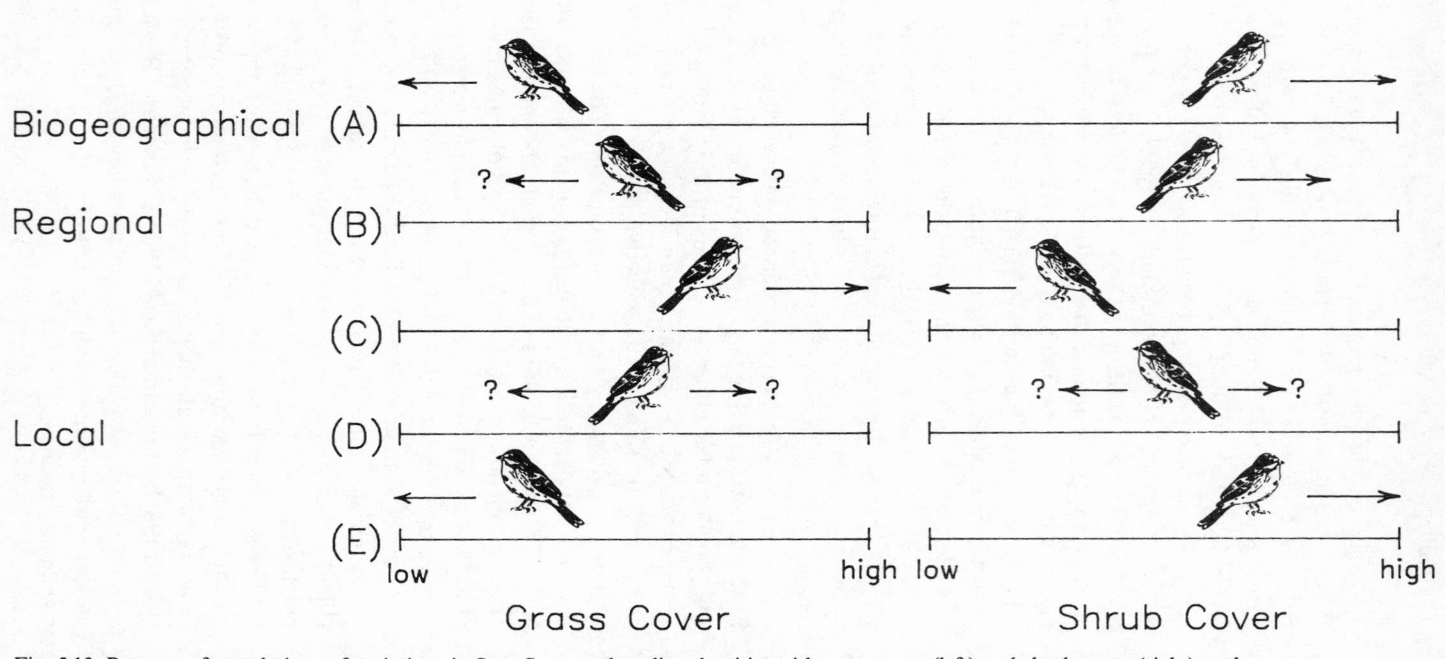


Fig. 5.13. Patterns of correlations of variations in Sage Sparrow breeding densities with grass cover (left) and shrub cover (right) at the spatial scales shown in Fig. 5.12. The arrows indicate the direction of correlation, and the positions of the birds and arrows show the relative strengths of the correlations. The patterns are described in greater detail in Table 5.2.

## Is there a right scale for studying bird communities?

### *Conceptual issues*

If the patterns of communities are so sensitive to the spatial scale on which they are viewed, how can we determine which scale (or scales) may provide the most accurate information? MacArthur (1972) offered conflicting advice on this question. He noted that the effects of competition might be most evident on a biogeographic scale, but he also observed that patterns derived at a biogeographic scale would be influenced by the complicating effects of speciation and history. An area that was too small, however, would contain an inadequate sample of species and individuals to determine community patterns. He concluded that an area of relatively homogeneous habitat just large enough to hold an adequate sample of species (i.e. a local habitat patch) would be most appropriate. Smith (1975) suggested that ecological systems characterized by high turnover or variance are 'improperly bounded'; he proposed expanding the scale of investigation until turnover is reduced to some acceptably low level.

The possible spatial scales at which communities can be investigated form a continuum. For convenience, however, we can partition this continuum of scales into several sections (Wiens *et al.* 1986): (1) the space occupied by a single individual over some relevant time period, (2) a local patch occupied by many individuals of several species, (3) a region that contains many patches or local populations that may or may not be linked by dispersal, and (4) a biogeographic scale that is large enough to encompass different climates, vegetation formations, and assemblages of species. Another scale, that of a closed system, cuts across these other scales: some systems may be closed at a local scale, and others remain open even at a biogeographic scale. As the degree of openness of a system increases, an increasingly larger area is required to study its dynamics.

The ecological patterns that are of interest or that can be investigated differ among these four scales, as does the likelihood that various ecological processes will contribute substantially to these patterns (Table 5.3). Patterns of resource partitioning may be most apparent at the scale of the local population, habitat separation among species at a regional scale, faunal turnover at a biogeographic scale, and so on. Competition and predation may have their greatest impact on patterns at the local and regional scales, whereas behavioral or physiological responses dominate patterns at the individual scale. Other aspects of communities, such as the form of disturbances, the features of the environment that are important, or the relative impact of stochastic effects on the patterns, may also change as one moves

Table 5.3. *Some attributes of avian communities that may vary with changes in the spatial scale on which an investigation is conducted. The listing is not intended to be complete, but it indicates how the focus of an investigation and the scale are interrelated.*

| | Scale of investigation | | | |
|---|---|---|---|---|
| Attribute | Individual | Local population | Regional | Biogeographical |
| Patterns of interest | Foraging<br>Behavioral interactions<br>Microhabitat use<br>Physiological adjustments<br>Time-energy budgeting | Resource partitioning<br>Resource limitation<br>Population density<br>Demography<br>Territorial relationships<br>Habitat selection<br>Social organization<br>Genetic structuring<br>Guild composition | Density compensation<br>Niche shifts<br>Habitat partitioning<br>Species-area relationships<br>Guild differences<br>Altitudinal distributions<br>Species-abundances<br>Diversity<br>Spreading of risk<br>Patch dynamics | Community convergence<br>Diversity gradients<br>Character displacement<br>Species-area relationships<br>Faunal turnover<br>Range dynamics/limits<br>Species-abundances<br>Species replacements<br>Island species richness |
| Important processes | Behavioral response<br>Physiological response | Competition<br>Predation<br>Parasitism<br>Extinction<br>Dispersal | Competition<br>Predation<br>Parasitism<br>Extinction<br>Dispersal | Evolution<br>Speciation<br>Extinction<br>Colonization |
| Disturbance | Gap formation | Habitat fragmentation | Habitat fragmentation | Tectonics<br>Vicariance<br>Glaciation |

| | | | | |
|---|---|---|---|---|
| Important environmental features | Microtopography<br>Microhabitat structure<br>Floristics<br>Microclimate | Habitat structure<br>Floristics | Landscape mosaic<br>Vegetation types<br>Physiognomy<br>Topography<br>Climate | Climatic zones<br>Geographical barriers<br>Vegetation formations |
| Scale of time-lags | Inconsequential (minutes, days) | Months, years | Years, decades | Historical (centuries, millennia) |
| Stochastic influences on patterns | Large | Moderate | Slight | Inconsequential |
| Methodology used | Observations<br>Experiments<br>Comparisons | Observations<br>Experiments<br>Comparisons | Comparisons | Comparisons |
| Testability of causal hypotheses | High | Moderate | Low | Unlikely |

from one scale to another (Table 5.3). The listing of factors in Table 5.3 could easily be expanded; the important point is that the attributes of communities that can be studied and the way an investigation can be designed vary as a function of scale. Put another way, certain aspects of communities may be studied effectively at certain scales but not at others.

These subdivisions of the scale continuum are defined by characteristics of the biota and do not correspond with specified areal dimensions. Because species differ in body size and in the magnitude of daily, seasonal, or lifetime movements, a given scale will translate into areas of quite different sizes for different species or assemblages of species. At the individual scale, a resident parid in boreal forests may spend its entire life in an area of a few square kilometres, a resident raptor may move over an area of hundreds or thousands of square kilometres, and a nomadic teal of ephermeral desert ponds in Australia may range over much of the continent during its lifetime. These interspecific differences in individual spatial scaling affect the areas that must be considered at each of the other scales. Thus, the area that corresponds to the scale of a 'local population' or a 'region' will be radically different for each of the species just mentioned. Species that are common may operate on different scales than do rare species (Bock 1987). If one considers a community composed of a mixture of resident and migrant or common and rare species, defining the area that corresponds to any single scale for the entire assemblage becomes problematic, because an area sufficient to match the individual scale for some species may represent several local populations, a region, or even a biogeographic scale for others.

There are substantial disagreements among ecologists regarding which of these scales might be most appropriate for community investigations. Because some important processes that occur on broad scales may influence the dynamics of populations or communities at a local scale (e.g. El Niño; Dayton and Tegner 1984, Schreiber and Schreiber 1984), it has been argued that a broad-scale perspective is essential (May 1981a, James *et al.* 1984). This is especially true of open systems. A focus at an excessively restricted scale may pass so far inside the boundaries of an interacting unit that one sees the parts of the system in great detail but misses entirely the interactions that bind them together (Allen and Starr 1982, Allen *et al.* 1984). It has also been argued that investigations at the scale of individuals or local communities are subject to the effects of individual idiosyncracies, sampling error, chance effects, or other variations that produce 'noise' and obscure the general patterns of communities. To the extent that this 'noise' is biologically meaningless, ecologists who focus at this scale may be studying 'nonevents' of limited importance. These idiosyncratic effects disappear when the patterns are derived by averaging over a broader scale, and simple explanations that promise satisfying generalizations are more likely to emerge.

Brown (1984, Brown *et al.* 1986) has argued this position especially ardently, noting that, although a study at a local scale

> can reveal the complexity of patterns and processes that characterize that particular system, it can provide little insight into which of these are specific to that system and which can be generalized to other communities. Furthermore, if there are general rules that govern the organization of communities, it may be impossible or impractical to adduce all of these from microscopic studies of the interactions of individual species . . . the statistical distributions of population densities, body sizes, rates of energy use, and areas of geographical ranges among the many species that comprise local communities may elucidate patterns and processes that cannot be discovered from microscopic experimental studies . . .. Large-scale geographical studies, especially comparisons of communities in widely separated regions inhabited by different taxa of organisms, should continue to provide a valuable perspective
>
> (Brown *et al.* 1986: 60).

These arguments can be countered by noting that studies at the broad scale are likely to overlook important details that account for the dynamics of local populations and communities. Studies at the scale of individuals or local populations may reveal what happens to individuals at the level where competitive interactions, predation, or resource use and limitation actually occur (Wiens 1983a, 1984a, 1986). Far from being idiosyncratic 'noise', the variations within or among local populations may contain important mechanistic information needed to test causal hypotheses. That these variations disappear when samples are averaged over a broader scale is not surprising – this is an inevitable consequence of averaging. The finer-scale, more detailed investigations may provide the information necessary to differentiate among competing hypotheses, whereas the capacity of broad-scale investigations to do this is quite limited, and inference and assertion may become the procedures of preference. Patterns at the broad scale may also be particularly sensitive to which samples form the basis for the analysis, and the possibility that some broad-scale patterns are myths, artifacts of the averaging procedures, cannot be ignored (Wiens 1984a). To derive generalizations from such patterns simply because they are expressed at a broad scale would be premature. At the local scale, experiments may be designed to unravel the various causal forces behind the patterns that are observed. Such an approach is precluded at broader scales, where one must rely entirely on comparisons or 'natural experiments', the drawbacks of which I discussed in Volume 1. The *ceteris paribus* assumption becomes especially critical at this broader scale.

The above arguments are an expression of the long-standing debate over the relative merits of reductionism and holism in ecology (e.g. Levins and Lewontin 1982, Simberloff 1982a, b, Allen and Starr 1982, McIntosh 1985). There are elements of truth in both positions, of course. The selection of a scale of investigation depends heavily on the nature of the questions that are asked, and to suggest that a particular scale is necessarily better than any other thus implies that some questions are better than others. This is not so. If one wishes to investigate the responses of individuals or populations to resource variations in space or the ways in which foraging behavior is influenced by interspecific interactions, a focus on the individual or local population scale is most appropriate, although some perspective from the other scales is also helpful. This is demonstrated nicely by the studies of Sabo and Holmes (1983) and Sherry and Holmes (1985) in forests of northeastern United States and by Haila's work (1986b, Haila *et al.* 1983, 1987) in Finland. If one's questions deal with the distributional patterns of species among habitat patches over a landscape, the regional scale may be most appropriate, although information from the local and biogeographic scales can provide important additional insights. If one is interested in explaining local diversity, regional as well as local scales must be considered, as patterns of local diversity reflect the characteristics of the regional species pool from which the local community is assembled as well as more proximate local conditions (Ricklefs 1987). The greatest insights may be obtained if one considers several levels of a hierarchy of scales (Allen *et al.* 1984, O'Neill *et al.* 1986).

Regardless of the scale or scales selected for study, one cannot pose a question that relates to a particular scale and then attempt to answer it by using information from a different scale or from indiscriminate, across-scale comparisons. Experimental perturbations imposed at the scale of individual territories, for example, may provide clear indications of factors influencing foraging behavior or individual patch use, but they are inappropriate for gauging the importance of the manipulated variables to local population dynamics (Wiens, Rotenberry, and Van Horne 1986). Findings about patterns and processes at one scale cannot be extrapolated to other scales without careful evaluation of the scale-dependency of the patterns or processes. Studies must be designed so that the scales of the questions asked, the processes believed to be important, and the interpretations derived from the results all match (Wiens 1981a, Haila and Hanski 1984).

### *Operational approaches*

There are two operational approaches to defining the spatial scale of an ecological investigation. On the one hand, an area of a certain size may be selected because it is logistically easy to establish and maintain. Avian ecologists customarily survey densities on plots of 1–10 ha because these can be marked off and censused in a reasonable period of time, whereas plots of several square kilometres would require substantially more effort. Regional comparisons are made among sites spanning a particular range of environmental conditions because that amount of area can be surveyed conveniently in the time available. The scale of investigation is determined arbitrarily, with little or no explicit consideration of the scale(s) on which the ecological patterns or processes of interest might actually be expressed. One hopes that the observational scale at least approximates that of the slice of nature studied. If this is not the case, distortions in the patterns are likely to occur.

Alternatively, the scales of investigation might be selected so as to be as similar as possible to those of the natural phenomena being studied. Doing this requires some previous knowledge of the system, and it may frequently result in an inconvenient or difficult research design. Units of measurement may have to be rescaled so as to be relevant to the rate dynamics of the system itself. Although it is a more arduous undertaking (and therefore one rarely followed), the latter approach would seem to offer greater potential for discovering how organisms actually do relate to their environments and to one another.

If the latter course is to be followed, how can one operationally define the areas that might correspond to a particular scale selected for investigation? One approach is to use the allometric relationships established for a wide array of organismal functions (e.g. Peters 1983, Calder 1984a, Schmidt-Nielsen 1984) to predict the area required by individuals of different sizes. Body mass bears a general relationship to home-range size, although the relationship differs for organisms of differing trophic status. If one knows the size of individuals of a species, these relationships may be used to obtain an approximation of the area required per individual, which may specify the area required to conduct a study at the individual scale. A similar approach might be used to determine scaling functions for population phenomena or for temporal dimensions (Calder 1984b).

Another approach is to search for discontinuities in environmental or community measures as a function of changes in the area considered. Schneider and Duffy (1985), for example, used the area of analysis ('frame size') that produced the maximum values of indices of aggregation of seabirds to determine the scale on which the species appeared to respond to environmental patchiness. A variety of statistical procedures, such as multivariate ordination (Allen and Starr 1982), spatial autocorrelation (Cliff and Ord 1973), or fractal analysis (Mandelbrot 1983, Orbach 1986), may help one determine the spatial scale on which the numerical behavior of a system changes. These changes may delineate the boundaries of domains of scale – portions of the scale spectrum over which basic patterns and processes are relatively invariant or scale-independent.

A third, conceptual approach has been suggested by Addicott *et al.* (1987), who focus on explicit criteria for determining the scale of patchiness in systems. Because patchiness may occur on a variety of spatial scales, it should be defined functionally in terms appropriate to specific organisms and processes. This can be done by defining an ecological 'neighborhood' that is specific to the operation of an ecological process, a time scale, and an organism's mobility or activity. Because a neighborhood defines a spatial unit that is specific to a particular process-organism combination, the sizes of neighborhoods for different processes or organisms differ. The neighborhoods for short-term processes, such as foraging, are thus smaller than those for long-term processes, such as demographic dynamics. To measure the attributes of neighborhoods, one must focus on the movement patterns of individuals and populations within and between patches in a landscape (Wiens *et al.* 1985). If one's focus is on populations, it is important to adjust such individual neighborhood-size estimates by some appropriate index of the minimal population size relevant to a particular question. In considerations of the genetic structure of populations, for example, effective population size ($N_e$) may be the relevant measure; Barrowclough and Shields (1984) suggested that values of $N_e > 100$ are generally necessary for the long-term persistence of the genetic structure of populations. Similar reasoning might be used to determine the minimal effective population size in terms of social organization or other attributes.

Because patterns or processes at any given scale are often influenced by events at other scales, it is often unclear exactly which scale may be most appropriate to answer a given question. A multiscale approach may therefore be most sensible (Wiens *et al.* 1986). It is fine to advocate studying a particular situation from the perspective afforded by several scales of resolution, but a more explicit recipe for such studies would be helpful. Several ecologists (e.g. Allen and Starr 1982, Allen *et al.* 1984, Maurer 1985,

O'Neill *et al.* 1986, Urban *et al.* 1987) have championed hierarchy theory as such a recipe. In a hierarchical system, the higher levels constrain and control the lower levels to various degrees, depending on the time constants of the dynamics of particular levels (Allen and Starr 1982). In ecological communities, the degree of openness of the system may thus determine in large part the degree to which the systems are hierarchical. In the sorts of breeding communities composed of mixtures of residents, short-distance migrants, and long-distance migrants modelled by May (1981b), linkages between different absolute spatial scales are evident.

'Scale' in a hierarchical system relates to three components of the system, that of the environment; that of the responding organisms, populations, or assemblages; and that of the human observer who attempts to study these interactions. One reason for the difficulty of dealing with scale in ecological systems is that the operational scales of each of these components may not coincide: environmental patchiness may be expressed on one scale, organisms may respond to that patchiness on some other scale, and we may arbitrarily select to investigate the linkages between the two on some third, unrelated scale. Once the hierarchical structure of these systems and the differences in relevant scales are recognized, they may be dissected by focusing simultaneously on more than a single scale or level of the hierarchy.

Both Maurer (1985) and Urban (1986, Urban and Smith, unpublished) have attempted to analyze bird communities explicitly in the framework of hierarchy theory. Maurer focused on the hierarchical structure of communities in Arizona grasslands and shrublands by calculating community measures of differing scales of resolution. Community-level attributes were represented by total density and total biomass; because the properties of individual species are summed or averaged in these measures, Maurer felt that they may reveal constraints operating on all species in each habitat. At an intermediate level, he considered the relative properties of species (relative abundance, species richness, dominance concentration); the individual species level was characterized by measures of the densities of each species. The three levels that Maurer defined may well be subject to different constraints, but it is doubtful that they are elements of a nested hierarchy in any other than an analytical sense. It is also not clear that general community-level measures such as total density or biomass indicate anything about 'general constraints' that apply to all species in a community.

In any case, Maurer found that that these attributes changed differently between seasons and between years of differing rainfall. He interpreted these differences as responses by different levels of the community on different scales and concluded that, by using appropriate combinations of observational scales, one can elucidate community structure and clarify the relationships among the structural components of the community. Stripped of its references to hierarchy theory, this approach is really not very different from that followed by avian ecologists for some time, and the observation that different measures of community or population attributes respond differently to environmental changes is really not too surprising, as they are calculated by different procedures.

Urban's approach to analyzing community patterns within a hierarchy framework was somewhat different. He focused on how the abundances of species in patches of a habitat mosaic are influenced by processes or factors occurring at a lower, within-patch scale (e.g. gap formation, the natality, mortality, and dispersal of individuals) and are constrained by features at a larger scale (e.g. the structure of the landscape mosaic and the metapopulation). This compartmentalization of factors is especially well suited to the modeling approach that Urban and his colleagues (Urban and Shugart 1986, Urban and Smith, unpublished) have followed, and it has demonstrated how some higher-level patterns may be produced by processes occurring at a lower level in the hierarchy. There have been few explicit applications of hierarchy theory to community studies, however, so its potential to produce fresh insights is unknown.

## Conclusions

Because species respond to environmental conditions in different ways, changing the place one studies a community is likely to lead to differences in the composition or structure of the community one encounters, and changes in the environment of a particular place over time will also result in community changes. Spatial variation in environments and the dynamics of patches of habitat over time influence virtually every aspect of communities. Simply noting the importance of spatial variation is not sufficient, however, for the influences of patchiness on communities are more complex. They vary in different environments or regions, for example, depending on the frequency and intensity of disturbances over evolutionary and historical time and the ways in which species have adapted or responded to these disturbances. Moreover, to understand how and why communities are structured as they are and behave as they do, we must know not only how the species relate to environmental features and to one another, but how these relationships are influenced by the spatial *configuration* of entire mosaics of habitat patches as well. If a species is dependent on two habitat types for different aspects of its life history or daily activities (e.g. nesting versus foraging), for example, knowing its breeding requirements alone will not enable us to predict its overall habitat occupancy or its role in community dynamics. If its movements are limited, its occurrence in an area will be determined by whether or not both habitat types occur within its zone of activity. If it refuses to enter a third habitat type, its use of the first two types will depend on whether they border one another or are separated by a patch of the third type. A major challenge of avian ecology, at both the population and community levels, involves determining how birds are affected by the spatial configuration – patch area, context, corridors, edges, etc. – of the landscape they do or do not occupy.

Consideration of these spatial relationships of birds and their environments is complicated by scale. We may view ecological systems at a wide array of scales, and it seems that the patterns that emerge differ at each scale we choose. In general, avian ecologists have been inattentive to the effects of scale, although recognition of their importance is becoming more widespread. Unfortunately, there is no single 'best' scale for investigating avian communities; the right scale depends on the question asked, the species and habitats involved, and the processes believed to be important. Arguments over the relative merits of a local versus a biogeographical focus are irrelevant; each is appropriate for certain kinds of investigations and inappropriate for others. One errs not by advocating that a particular scale may be useful for examining community patterns and processes, but by forcing investigations of specific patterns and processes into a scale of analysis that is improper. A second major challenge of avian ecology is to develop procedures for dealing with scale effects in terms that are relevant to the organisms being studied and the questions being asked.

The studies considered in this chapter indicate that we have made some important advances in recognizing and evaluating the importance of spatial variation and scaling in studies of bird communities. If we are to progress farther, however, we must consider especially the following points:

1. The degree to which a community is open or closed to external influences has major effects on how research should be designed or results interpreted. In a closed system, one may design field experiments with a reasonable degree of control and the results of the experiments or of comparative observations can be interpreted causally with reasonable confidence. In an open system, one is always haunted by the 'ghost of events elsewhere' that may invalidate an experiment and reduce causal explanations to speculative inferences.
2. The degree of openness of a community is largely a function of the

magnitude of dispersal or movements of individuals. Dispersal also bears importantly on our considerations of the dynamics of populations and communities in patchy landscapes, as it influences the probabilities of local extinction in habitat patches or fragments, of their subsequent recolonization, or of their 'rescue' before extinction occurs. We know relatively little about the dispersal of birds on landscape or regional scales.

3. Habitat patchiness is most dramatic when it is created suddenly, especially as a result of human activities. Disturbance at a wide variety of scales is a natural as well as an artificial phenomenon, however. This means that many species have encountered habitat patchiness and disruption before the recent wave of human activities. Without a doubt, some species are terribly sensitive to these disruptions, and the magnitude and speed of human-produced habitat changes poses a special threat. Other species, however, may be adapted to natural habitat fragmentation. Breeding male Dickcissels, for example, engage in a 'distant flight' behavior, in which they leave breeding territories for several hours and fly several kilometres away. Dickcissels occupy early stages of old-field or grassland succession, and Zimmerman (1971) has interpreted their behavior as searching for other patches of habitat that may be suitable for breeding in the future. Similar behavior has been observed in Bobolinks (*Dolichonyx oryzivorus*) and Lark Buntings (*Calamospiza melanocorys*), both of which occupy patchy, frequently disturbed grasslands (Martin 1971; personal observations). Certainly, we should be concerned about rampant destruction and fragmentation of natural habitats, especially in the tropics, but it is wrong to assume that fragmentation of any sort represents a threat to all bird species.
4. Because disturbance and patch dynamics are natural phenomena, we cannot assume that conditions in an area before it was fragmented by recent human activities were necessarily undisturbed and the populations and communities equilibrial and the habitats fully saturated. As a consequence, it is difficult to interpret the results of studies based on comparisons among habitat patches of different sizes without some knowledge of their prior histories. Such comparisons lack reasonable controls, in that the larger patches are simply presumed to represent the state of affairs in the smaller patches before their fragmentation.
5. Generally, the landscape context of the patches is ignored; to the

degree that this differs among patches as a function of patch size (as is likely), it will also complicate the comparisons.

6. If the data on species numbers or abundances in fragments of different sizes are not adjusted for area effects by rarefaction or a similar procedure, the resulting patterns may be misinterpreted. They will probably show, for example, that the number of species increases with patch area; it may also be important to determine how the number of species in, say, a 10–ha area changes with overall patch size, but this is not obvious from an unadjusted comparison.
7. Aside from the predictions of island biogeography theory and species-area relationships, we have little basis on which to develop *a priori* expectations of the effects of fragmentation in a particular situation, of the responses of species or communities to various landscape mosaic configurations, or of the amount of a given habitat or habitat mosaic necessary to accomplish a particular conservation objective. The controversies over the design of natural reserves indicate that the existing theory does not take us much beyond the common-sense conclusion that bigger is better. At the same time, it is clearly not possible to base the resolution of every specific conservation issue on intensive ecological study of every species present in a community, or even of the primary species of interest. There is a role for theory here, but we need new theory to play it. This new theory must take into consideration the nuances of spatial patchiness and scaling. The development of such theory is a third major challenge facing avian ecologists.

# PART III

## Prospects

During the past three decades, community ecology in general and avian community ecology in particular have undergone dramatic changes. The qualitative descriptions of community composition that characterized the 1950s gave way to the heady euphoria of the 1960s and early 1970s, when increasingly quantitative descriptions were linked with attractive conceptual or mathematical models of community structure and the role of competition in producing that structure. But then, during the mid-1970s, mutterings of dissatisfaction were heard here and there, and these increased to a sometimes chaotic clamor during the early 1980s. Investigations lost the unitary focus that competition-based community theory had provided and doubt in the validity of past studies or the wisdom of future studies of communities became widespread. Some investigators echoed the skepticism expressed in 1954 by Andrewartha and Birch, who had concluded that community studies were unlikely to contribute any understanding to the central questions of ecology.

Of necessity, much of my emphasis in this book and its companion volume has been on the problems of past studies and the inadequacies that have fueled this recent skepticism. In order to make progress in the difficult task of understanding how assemblages of organisms are put together and what processes act upon them, it has been necessary to examine where we have been and what we know versus what we only think we know. It would be easy to conclude that doing community ecology properly is an impossibly difficult task, that dealing with the logical problems and methodological pitfalls discussed in Volume 1 or the complicating effects of temporal and spatial variation detailed in Chapter 4 and 5 is an unattainable ideal. If one clings to old beliefs and old approaches, this is likely to be the case. But assemblages of organisms do exist, and they are no less interesting because they are variable or complex or difficult to study. Understanding these communities and the factors that influence them will not be easy. It will require new approaches, fresh theory, and considerable care, but it can be done. In the following chapter, I consider some factors that future studies of avian communities must consider and indicate some possible directions such studies might take. Rather than being discouraged by the difficulty of doing good community studies, we should be excited by the challenge. That, after all, is what makes science fun!

# 6

## Concluding comments: future directions in avian community ecology

In the past, avian ecologists documented simple patterns in communities and sought their explanations in the neat formulations of niche theory, Doing community ecology was exciting, fashionable, and fun. Increasingly, however, the complexity, variability, and ambiguity of nature have made community ecology more difficult and lessened its allure. In this and the previous volume, I have detailed a litany of problems, pitfalls, and perils that have plagued studies of avian communities. With the benefit of hindsight enriched by our increased awareness of the complexity of communities and the logical and methodological demands for doing science properly, it is easy to see where we may have gone astray. Such retrospection is useless, however, if we do not use it to redirect future activities in the discipline. We are faced with the prospect that many ecologists may become discouraged in their attempts to understand communities and will turn their attention to entirely different questions, while those who continue to study communities will become polarized into descriptive pattern-seekers or armchair theoreticians. In the meantime, the actual dynamics of communities may be left unattended because they are too fuzzy, too difficult to study, or not amenable to generalization or theory. We need to redirect our approach to community investigations in a way that will retain the excitement and satisfaction of the past yet not do violence to nature by oversimplification.

I attempt to provide some elements of this redirection in this chapter. Because theory is so central to our approach to community studies, I comment first on what we should expect of ecological theory. I then suggest some elements of future approaches to the study of bird communities.

### The role of theory

'Theories are the key to the scientific understanding of empircal phenomena' (Hempel 1977: 244); 'to do science means to construct theories and to adopt theoretical concepts in order to explain facts of nature' (Haila and Järvinen 1982: 261). These statements express the view, widely held in community ecology since the mid-1960s, that theory is the framework on which observations should be arrayed, the means to ordering and understanding phenomena. Indeed, some have suggested that the formalization of theory in ecology has, in large part, contributed to the increased respectability of the discipline as a *bona fide* science (Roughgarden 1983).

Although theory is the underpinning of much of what I have discussed in this book, I have not dealt with it in an explicit, formal sense. Because the value of theory (or models) is a matter of some debate among ecologists, however, it is appropriate to consider the nature of the disagreement and what roles we should expect theory to play in our investigations. McIntosh (1985) and Kingsland (1985) provide valuable reviews of the historical development of theory in ecology, and Oster and Wilson (1978) offer some enlightening comments about optimization theory in ecology.

Theory is an attempt to isolate and simplify subsets of a complex reality so that we can achieve some understanding and make reliable predictions. Niche theory, for example, provides a simplifed representation of how competition acts to produce and maintain a pattern of resource partitioning among coexisting species. In constructing a theory, irrelevant detail is stripped away; the resulting conceptual framework provides a way of organizing an 'otherwise indigestible mass' of observations (May 1976: 1), of avoiding being 'washed out to sea in an immense tide of unrelated information' (Watt 1971: 569). To a theoretician, the refusal to simplify ultimately leads to 'a glorification of the particular, to a celebration of the personality of each species and their interactions in each habitat' (Roughgarden 1983: 17). The key to developing theory is simplification, and this is achieved by making assumptions, at the expense of a full consideration of the details and complexity of nature.

This neglect of detail by theoreticians has engendered skepticism toward theory among many field ecologists, to whom the details are real and important. They often feel that ecological theory is developed as an end in itself, with little relevance to the real world, and they equate the simplicity of theoretical models with sloppiness and a lack of perception of nature's complexity (Roughgarden 1983). To many empirical ecologists, theory is often burdened by a variety of problems (Kikkawa 1977, James and McCulloch 1985, Slobodkin 1975, Pielou 1981, Bartholomew 1982, 1986). It relies on too many assumptions, many of which are biologically unrealistic. It fails to take into consideration multiple causes or indirect effects or may deal with processes that do not occur or are unimportant in specific situations. It generates untestable hypotheses or predictions that are shared by competing theories. At its worst, it develops biological nonsense with mathematical certainty and may lead one to disregard empirical data if they conflict with the statements of theory.

To a large extent, the disagreements about the value of theory in community ecology stem from overstated claims by some theoreticians and from a misunderstanding of the roles of theory among many field ecologists. With the explosion of ecological theory in the 1960s and 1970s, many community ecologists were overwhelmed by the neatness and elegance (and the mathematics) of theory. They took the equations quite seriously, 'as if there were hidden in them some great, but subtle, truth about nature. What was lost . . . was that the subtlety was mostly mathematical, and the truth they contained mostly allegorical' (Oster 1981: 831). Ecologists expected a close correspondence between theory and reality, and when it did not emerge they felt disappointed and almost betrayed (Brown 1981, Murray 1986). Thus, Simberloff (1982c: 241) concluded that competition theory 'caused a generation of ecologists to waste a monumental amount of time'.

Ecological theory comes in different forms, however, and it plays a variety of roles. Increasingly, mathematical equations have come to be regarded as the only form of true theory in ecology, but statistical, empirical, or verbal models also constitute valid and useful forms of theory. Strictly mathematical formulations are often not constrained by observations nor do they involve falsification or hypothetico-deductive testing; rather, they apply mathematical theorems and proofs to ecological questions. At the extreme, this is theory for theory's sake, tinkering with equations that is of greater interest to mathematicians than to field ecologists. Even this sort of theory, however, may contribute to ecological investigations, by stipulating idealized states against which observations from nature can be compared, by providing a conjecture about how nature operates, or by leading one to think in ways that otherwise might not have occurred and thereby producing fresh insights or new questions. Such 'pure theory' (Grant and Price 1981) may also be useful in organizing one's observations or bringing them into line with current ideas. It does not establish or show anything about nature, but it does specify the possibilities, given certain assumptions. This sort of theory does not necessarily lead directly to testable propositions or hypotheses.

On the other hand, what Grant and Price have called 'operational theory' does do so. Such theory may take a variety of forms, from mathematical equations to verbal models. It can serve a heuristic role, by focusing research on certain ideas or explanations of phenomena. It can generate predictions that are testable or are useful in management. Because the assumptions of such theories are (or should be) explicitly stated, one can look to the assumptions for probable causes for falsifying tests of hypotheses or predictions generated from the theory.

Must theory or its predictions be realistic and testable to be useful? Some, such as Bronowski (1977), Stearns and Schmid-Hempel (1987), and Fagerström (1987), say no. In their view, theories are not developed for the purpose of passing tests: 'whatever it is that we want theories for, it is not to test them . . . the test by falsification will diagnose when a theory falls sick, but it does not reflect what we ask a healthy theory to be or to do' (Bronowski 1977: 95). The physicist Paul Dirac commented that 'it's most important to have a *beautiful* theory. And if the observations don't support it, don't be too distressed, but wait a bit and see if some error in the observations doesn't show up' (in conversation, cited by Judson 1980: 198). These are the voices of advocates of pure theory. Others, such as Andrewartha and Birch (1984), Murray (1986), and Pierce and Ollason (1987), have emphasized more operational theory and have insisted on its testability and predictiveness, arguing that 'we must not sacrifice facts for the sake of beauty' (Van Valen and Pitelka 1974: 925). In order to be testable, of course, a theory must be biologically realistic. J.B.S. Haldane perhaps put it best: 'No scientific theory is worth anything unless it enables us to predict something which is actually going on. Until that is done, theories are a mere game of words, and not such a good game as poetry' (1937: 7).

As in most controversies, there are elements of truth in both views. By generating novel insights and fresh concepts, pure theory may be quite useful even if it is untestable or founded on unrealistic assumptions. To condemn a theory as useless because it is untestable or does not fit observed facts is to miss the point of the theory. To understand natural phenomena, we must (1) find an explanation and then (2) determine whether or not it is correct. Both pure and operational theory may contribute to the first step, but the second requires testing of some sort. Salt (1983) has suggested that the first step is the role of the observer, the second that of the theoretician or modeler, but I would put it just the other way: theoreticians can offer a variety of possible explanations for a phenomenon, from among which empiricists must determine which are likely to be correct (see also May 1981b, Connell 1983). In any case, it is apparent that the pessimism toward theory expressed by Simberloff and others is justified only if one expects all theory to be immediately testable and its predictions always founded on realistic assumptions. This view fails to recognize that more abstract, pure theory plays different roles in science, but roles that are nonetheless valuable.

The skepticism of many field ecologists toward theory may be as much a reaction to the generality of theory as to its oversimplification of reality. MacArthur, after all, held that 'science should be general in its principles' (1972: 1), and much of the theory developed during the past few decades has aimed to achieve such generality. The cost of this generality has been a limited applicability of the theory to specific situations, and this has disturbed those ecologists who have expected (incorrectly) that such general theory should explain *all* observations (Murray 1986). Clearly, however, a general theory should explain most (or at least a good many) observations to be useful. The problem is an expression of Levins' (1966) dilemma: how does one balance the needs for generality, realism, and precision in a model or theory? Several ecologists (e.g. Brown 1984, Levins and Lewontin 1982, Levin 1981, unpublished) have continued to emphasize the importance of generality in ecological theory. Murray (1986: 146), for example, lamented that 'the search for general theory languishes while hypotheses of limited applicability proliferate', presumably because many ecologists believe that the complexity of nature precludes general theory and justifies more specific theory.

Murray's lament is accurate, if not entirely justified. Perhaps in response to the failure of existing general theories to account for a great many observations and the growing awareness of the complexity of natural communities, there has been a recent shift toward theories of more limited and specific domain (Colwell 1984, Oster 1981, Price 1984, Maurer 1985, P. Price 1986). In Bartholomew's words, 'natural history tells us unequivocally that we are foolish to look for general answers to specific questions about how organisms perform' (1986: 328). This shift is reflected in the development of a so-called 'mechanistic approach' to community ecology, in which concepts based on the details of individual morphology, behavior, and physiology form the theoretical basis for explaining community patterns (Price 1986, Schoener 1986a).

Lack (1976, following Williams 1969) distinguished between what he called the 'distant' and 'close' views of ecological problems. The 'distant' view depends on *a priori* ideas, mathematical models, and selected examples, whereas the 'close' view involves the intensive, inductive study of particular situations. Lack favored the more specific 'close' view over the more general 'distant' view that dominated theory, perhaps because he was, primarily, a naturalist. Theoretical efforts are now increasingly coincident with the 'close' view; generality has been sacrificed to achieve greater realism. The peril of such reductionism in theory, of course, is that it may continue unabated into a proliferation of situation-specific, idiosyncratic statements with no degree of generality at all.

Not all ecologists feel that investigations should be guided by theory, or at least guided in the sense of testing hypotheses derived from theory. James and McCulloch (1985) have advocated the value of exploratory data analysis as a complement to hypothesis testing, and Pielou (1981) suggested that 'investigating', searching directly for empirical answers to single, clearcut questions without preconceptions (i.e. theory), is likely to contribute more to our knowledge of nature than modeling or theory. Each of these approaches represents a break from what has become the traditional form of hypothetico-deductive investigation in community ecology, but it is doubtful that either can be done entirely in the absence of theory, at least for very long. Whether we accept the teachings of the MacArthurian paradigm or not, as scientists we operate in a framework that is rich in ideas and concepts, and they guide our investigations and influence our observations whether we realize it or not. Neither exploratory data analysis nor investigating is likely to be entirely aimless; if anything, they are likely to be influenced more by pure theory than by operational theory, which is more explicitly focused on testing hypotheses.

In view of these comments about theory, what can be said about the contributions of the 'MacArthurian paradigm' to community ecology? Has it all been a waste of time, as Simberloff suggested? Of course not! Part of the art of science is asking, like the elephant's child, 'a new fine question that has never been asked before' (Kipling 1902). Using theory, MacArthur and those who followed his interests have done this often and well. As a result, our attention has been drawn to some interesting and important facets of community ecology, and the investigations of these phenomena have, in turn, opened avenues toward more intensive and specific exploration of the causes of the diversity and complexity of communities. The paradigm also fostered a respect (perhaps at times an awe) for theory and mathematics, and it has contributed importantly to the acceptance of these as useful tools in ecological studies. That the 'normal science' conducted within the paradigm was often not entirely logical or attentive to methodology is unfortunate, but perhaps inevitable. Still, community ecology today would be a poorer, less exciting discipline without the ideas and stimulation fostered by the paradigm.

## Future directions for avian community ecology

For a long time, avian community ecology was practiced within the framework of what I have called the MacArthurian paradigm. Many conceptual and empirical advances were made, but there was also a tendency for alternative viewpoints to be discouraged or suppressed. The paradigm set limits to the kinds of questions that were asked and approaches that were followed. Many ecologists fell prey to the 'tyranny of bright ideas' (Macfadyen 1975) and looked at problems in terms of answers they hoped to find rather than new and different questions they hoped to answer. Recently, the structure of this paradigm has been challenged on several fronts. In contrast to a Kuhnian scenario, however, no new paradigm has emerged to take its place. Instead, the traditional views have become transformed so that they are scarcely recognizable and the call for a 'pluralism of approaches' (e.g. Schoener 1986b, Diamond and Case 1986) has become almost a cliché.

Where do we go now? How do we restructure the discipline of avian community ecology to make it exciting, scientifically rigorous, and more relevant to the complexity of nature? How do we redirect our research so that it will produce a real understanding of bird communities? No unified approach has yet emerged, and I doubt that one will. To be successful, however, any new approaches should consider the following points – desiderata for a more rigorous community ecology, if you will (Table 6.1). The points apply equally to charting new directions in theory and in empirical studies.

1. Regardless of how one defines a 'community', the community being investigated and the criteria used to determine its membership should be described explicitly. If one's focus is on species interactions, it may be appropriate to restrict the membership of the 'community' studied to species that actually or potentially interact, but other objectives may dictate other demarcations of 'communities' that are just as valid or useful. Ecologists should not become overly concerned with the semantics of communities or with whether or not communities are 'real' or possess holistic properties. Multispecies assemblages do occur, whether or not they are repeatable, integrated, natural units, and much is to be gained by studying them in their own right. Our objectives determine whether it is most appropriate to restrict attention to a taxonomically defined subset of species, a resource-defined guild, the entire biota of an area, or some other set of species. There should be no ambiguity, however, about what is being studied and why it is an appropriate focus for a particular question.
2. Community 'macroparameters' such as species richness, diversity, and niche overlap should be de-emphasized and attention focused instead on concepts and measures relating more directly to individuals. The insights that can be gained from such macroparameters

Table 6.1. *Desiderata for a more rigorous community ecology. Modified and expanded from Wiens (1983a), Pearson (1986), and Schoener (1986a).*

1. Be more explicit about defining the 'community' studied and justifying that definition.
2. De-emphasize community macroparameters and focus on individuals, especially aspects relating to energetics, density effects, and habitat selection.
3. Use resource-defined guilds as a framework for intensive comparative studies.
4. Consider both ecological and evolutionary constraints on community patterns.
5. Consider all life stages in community analyses, and evaluate the effects of community openness versus closure.
6. Conduct studies, interpret the results, and generalize from them within the appropriate domains of scales in space and time for the phenomena or biota investigated.
7. Avoid thinking of communities as either equilibrium or nonequilibrium, but examine the dynamics and variability of community measures as features of interest in their own right.
8. Conduct long-term observational and experimental studies.
9. View communities in a landscape context, considering the effects of habitat-mosaic patterns and abandoning notions based on assumptions of spatial homogeneity.
10. Focus on the factors influencing community assembly as a conceptual framework for community studies.
11. Deal with the effects of multiple causes on community patterns.
12. Emphasize the importance of defining and measuring resources and testing the assumption of resource limitation.
13. Develop specific, mechanistically based theory.
14. Frame hypotheses in precise, testable terms whenever possible.
15. Take into account the effects of feedback relationships, indirect interactions, time lags, and nonlinear responses.
16. Avoid extrapolating from particular taxa or habitats to other taxa or habitats, and avoid especially a 'north-temperate bias' in thinking about communities.
17. Recognize the importance of replication in both observational and experimental studies.
18. Do not shun or avoid criticism and controversy.

are limited, and the description of community membership at this level may best be used to draw attention to subsets of species that may be studied with greater profit (Colwell and Winker 1984). Greater attention should be given to expressing community patterns in terms of physiology, behavior, or life-history traits. We might focus on attributes of species that influence their utilization of energy rather than on the effects of interactions on their population dynamics (Brown 1981, Martin 1986, Murray 1986). The effects of local, habitat- or patch-specific densities on community patterns and dynamics deserve closer study. Habitat selection by individuals is a critical process in the formation of local and regional communities, yet it has usually been considered primarily in autecological terms (e.g. Fretwell and Lucas 1969, Klopfer and Ganzhorn 1985). The prospects for building a new sort of community theory on this basis, however, are exciting, and Rosenzweig (1981, 1985) has made an encouraging start in this direction.

3. In some cases, a focus on guilds may be especially appropriate and productive, particularly if the guild is founded on the use of a discrete, quantifiable resource (e.g. nectar). Whether one follows Jaksić (1981) and defines guilds by resource use without regard to taxonomy or instead adopts taxonomic restrictions depends on one's objectives and questions. If one is interested in the effects on individual fitness or population demography that result from mutual exploitation of a limiting resource, it may be important to include distantly related taxa in the same guild. There are formidable logistic and design constraints on studying such guilds, however, and it may be no less interesting to determine how, say, different members of the same genus or various small bird species go about using a particular resource, whether or not it is limiting and even if other taxa also use that resource.
4. Ecological and evolutionary influences on community patterns should be considered in an integrative manner. The proposition that species respond independently and individualistically to environments, for example, may be an alternative to a competition-based hypothesis in a proximate sense, but it masks the fact that the species have an evolutionary history that undoubtedly has included species interactions, along with a great many other influences. These evolutionary effects on species' adaptations leave a mark on contemporary community patterns – what Connell (1980) has termed the 'ghost of competition past'. Many of the interesting questions of community ecology center around why particular species with particular phenotypic traits coexist under particular ecological conditions, and the answer will involve both proximate (ecological) and ultimate (evolutionary) components. Sherry (1984) has pursued this approach in his work on neotropical flycatchers, but evolutionary constraints on proximate ecological patterns are not often explicitly recognized.

5. Include all phases of the life cycle of species in community analyses. It is unrealistic, for example, to restrict measurements to one sex or to focus entirely on adults and ignore juveniles or subadults. It is also critical that the effects of system openness versus closure be considered. Interpretations of patterns in a community composed mostly of migrants that winter (or breed) elsewhere must be more guarded than those derived from a community closed to immigration and emigration (e.g. some islands). Open systems are sensitive to the 'ghost of events elsewhere'. We should devote more attention to dispersal and its effects on communities and incorporate them more explicitly into our community models. It would be interesting to develop predictive concepts relating to the differences we might expect between closed systems and communities of varying degrees of openness.
6. Studies should be conducted and their results interpreted and generalized within the appropriate domains of scales in space and time for the phenomena or organisms investigated. Unless one is aware of the scale-dependency of patterns, it is easy to believe that whatever scale one chooses for an investigation will reveal the true patterns, but this is not the case. Patterns (and their interpretations) change with changes in scale. The challenge is not simply to recognize that ecological patterns are scale-dependent, but to match the scale of an investigation to the question being asked, and then match the observational or experimental design to the scale of the study. Because the appropriate scales differ for different questions and differ for the same question asked of different sets of organisms in different environments, it is difficult to know at the outset of a study what scales may be most suitable. At one time I claimed that community investigations would be most profitably pursued at a local scale (Wiens 1983a). I still feel that answers to some questions will emerge most readily from the intensive study of local communities, but investigations at quite different scales can also provide important perspectives on other sorts of questions.
7. The notion that communities are normally at or close to a resource-defined equilibrium should be critically evaluated rather than accepted as an article of faith. Equilibrium may be a convenient theoretical notion but it is an unrealistic expectation for most communities. On the other hand, nonequilibrium is not a very useful concept either. Resource conditions vary in nature, and

individuals, populations, guilds, and communities also vary in the degree to which their status is closely linked to those resource states. The result is that the patterns and processes of assemblages may approach either end of the equilibrium–nonequilibrium spectrum at some times and be some distance away at others. Different sorts of assemblages exhibit different dynamics on this spectrum. Surely it is more important to attempt to understand the nature of these dynamics of communities than to pigeonhole them into 'equilibrium' or 'nonequilibrium' categories. By rejecting the idea that communities are in equilibrium unless proven otherwise, our attention is drawn more explicitly to the importance of variance in community measures. Rather than viewing variance in our data with despair, we should regard it with satisfaction, as a source of additional information and new insights (Bartholomew 1986). The growing emphasis on mechanistic approaches to community studies (Schoener 1986a) will enhance our awareness of the importance of variation, but we also need more ecological theory that incorporates sources of temporal and spatial variation and that seeks something other than equilibrium solutions.

8. In view of the temporal variability of nature, observational and experimental studies should be conducted over a sufficiently long period of time to reveal the normal dynamics of the system. The necessary duration of studies will vary as a function of the environments and organisms studied and the questions asked, but in most cases it will probably be longer than has been customary in the past. Short-term studies are likely to give us only glimpses of communities as they exist at a particular time. Roth's (1976: 777) hope that 'such glimpses can tell us something of the way species fit together as communities – if we look at enough of them to permit generalizations' is quixotic.
9. Communities should be considered in a landscape context. Whether they are entirely natural or are disturbed by human activities, environments occur as mosaics of habitat patches, and the sizes, configurations, and spatial relationships of elements within these mosaics may influence community patterns and dynamics. Notions about communities based on assumptions of spatial homogeneity should be abandoned.
10. Community assembly is an ongoing process. Instead of considering contemporary communities to be the finished products of assembly, we should consider them to be dynamic and look to the

factors and processes influencing community assembly as an appropriate conceptual framework for community studies. By considering the full range of factors affecting community assembly, we may avoid emphasizing particular factors (e.g. species interactions, area effects) to the exclusion of others. The role of chance and 'lottery' effects in assembly should receive greater attention in studies of bird communities.

11. Rarely is it likely that a community pattern is the result of a single process. Instead, patterns are probably often influenced by several processes acting simultaneously or in sequence. The same pattern may also be caused by different processes acting in different ways at different times. Community studies and, especially, community theory must be designed to consider the effects of multiple causes of patterns.
12. Although 'resources' are involved in most explanations of community patterns, all too often they have been defined in *ad hoc* ways, rarely measured directly, or inferred to be limiting on the basis of faith rather than evidence. Determining what the resources really are depends both on the kinds of organisms involved and the sorts of questions that are being asked, but it also requires that one distinguish between resource abundance, availability, and use and be certain which of these one is actually measuring. If the questions being asked relate to resource limitation, one should recognize that limitation is a function of resource availability and resource demand and of how that relationship influences the well-being of individuals or the demography of populations. It is no longer sufficient simply to assert that resources are limited because the organisms behave in accordance with the predictions of resource-limitation theory, or that resources are superabundant because there seem to be plenty of them. If the hypotheses we are testing rely on an assumption of resource limitation, that assumption should be tested as rigorously as possible, not just uncritically accepted or disregarded.
13. We should develop a pluralistic, domain-specific body of theory and devote less effort to universal, general theory. Theory and models should be formulated on the basis of their mechanistic or biological foundations rather than their mathematical convenience (Schoener 1986a, Tilman 1987).
14. The predictions of theory should be framed as testable hypotheses whenever possible. We must recognize, however, that some inter-

esting ecological questions are not testable using formal hypothetico-deductive procedures. Such questions or hypotheses should not be ignored simply because they are not entirely testable; they may still be examined by exploratory data analysis or 'investigating'. In such studies, however, the untestability of the hypotheses must be explicitly stated and interpretations of the results moderated accordingly.

15. Part of the complexity of natural systems is due to feedback relationships, indirect interactions, time lags in responses, and nonlinear or threshold relationships. These complicate attempts both to develop realistic theory about communities and to conduct observational or experimental examinations of pattern-process relationships. Nonetheless, they do occur, and we must take them more fully into account in our theoretical and empirical studies.
16. Studies of particular taxa or habitat types should not be extrapolated to other taxa or habitats simply for the sake of generalization or on the basis of the plausibility of such extrapolations. This is especially relevant to the 'north-temperate bias' in our thinking about communities. Because most studies of bird communities have been conducted in north-temperate areas and most theory has been developed to explain these communities, there is a tendency to regard them as 'normal' and communities elsewhere as 'peculiar' by comparison. In studies of community convergence, the differences between, say, south-temperate and north-temperate areas are often explained by cataloging apparent idiosyncrasies of the southern system rather than by regarding the northern communities as unusual (Morton 1985). We tend to regard north-temperate breeding species that winter in tropical areas as invaders from the temperate zone rather than integral parts of tropical communities that emigrate to temperate locations to breed (Gochfeld 1985, Rabøl 1987). We must recognize that birds in many parts of the world are not like typical north-temperate species and that we must therefore study them in different ways. Explanations of studies conducted in north-temperate locations (or in any other region of the world) are of limited generality.
17. The importance of replication in studies of communities must be more fully acknowledged. The need to replicate samples or treatments or plots is widely recognized (if not always followed) among experimental ecologists, but it is no less critical to observational or comparative studies. To derive patterns from comparisons of

unreplicated samples is to assume that there is no temporal or spatial variation in the attributes or relationships measured, and we know that this is not so. Well-designed community studies must include replication whenever possible. Comparisons using data sets gathered in different places by different observers using different methods, however, in no way constitute replication.

18. Although they sometimes make one uncomfortable, criticism and controversy are important elements of scientific progress. We should retain a healthy skepticism about the work of others, recognizing that every study has its limitations (Wiens 1981c). There may be honest mistakes in the methods that are used, the observations that are made, or the analyses employed; biases in the interpretation of the data or the logic that is followed; or deception in the way in which findings are presented, and we should always be alert to these possibilities. At the same time, however, we should be equally demanding in our evaluation of our own work and welcome rather than avoid criticism. One's ego should not be so entangled with one's science that it prevents objective evaluation and rational criticism. Controversy is not a mark of a discipline gone sour, but of one that is healthy and dynamic.

These desiderata represent a formidable challenge, but that is what makes the prospects for avian community ecology so exciting. As Olli Järvinen once remarked, 'we should let natural complexity be our delight rather than a discouraging adversary. If no more than a groping start has been made in studying communities, many good field trips await us.'

# References

Aarssen, L.W. (1984) On the distinction between niche and competitive ability: implications for the coexistence theory. *Acta Biotheoretica*, **33**, 67–83.

Abbott, I. (1977) The role of competition in determining differences between Victorian and Tasmanian passerine birds. *Australian Journal of Ecology*, **25**, 429–47.

(1978) Factors determining the number of land bird species on islands around South-western Australia. *Oecologia (Berl.)*, **33**, 221–33.

(1980) Theories dealing with the ecology of landbirds on islands. *Advances in Ecological Research*, **11**, 329–71.

(1981) The composition of landbird faunas of islands round southwestern Australia: is there evidence for competitive exclusion? *Journal of Biogeography*, **8**, 135–44.

Abbott, I, & Black, R. (1980) Changes in species composition of floras on islets near Perth, Western Australia. *Journal of Biogeography*, **7**, 399–410.

Abbott, I. & Grant, P.R. (1976) Nonequilibrial bird faunas on islands. *The American Naturalist*, **110**, 507–28.

Abbott, I., Abbott, L.K. & Grant, P.R. (1977) Comparative ecology of Galápagos ground finches (*Geospiza* Gould): Evaluation of the importance of floristic diversity and interspecific competition. *Ecological Monographs*, **47**, 151–84.

Abrams, P. (1980) Some comments on measuring niche overlap. *Ecology*, **61**, 44–9.

Abramsky, Z., Bowers, M.A. & Rosenzweig, M.L. (1986) Detecting interspecific competition in the field: testing the regression method. *Oikos*, **47**, 199–204.

Addicott, J.F., Aho, J.M., Antolin, M.F., Padilla, D.K., Richardson, J.S. & Soluk, D.A. (1987) Ecological neighborhoods: scaling environmental patterns. *Oikos*, **49**, 340–46.

Alatalo, R.V. (1980) Seasonal dynamics of resource partitioning among foliage-gleaning passerines in Northern Finland. *Oecologia*, **45**, 190–6.

(1981) Habitat selection of forest birds in the seasonal environment of Finland. *Annales Zoologici Fennici*, **18**, 103–14.

(1982) Multidimensional foraging niche organization of foliage-gleaning birds in northern Finland. *Ornis Scandinavica*, **13**, 56–71.

Alatalo, R.V. & Lundberg, A. (1983) Laboratory experiments on habitat separation and foraging efficiency in marsh and willow tits. *Ornis Scandinavica*, **14**, 115–22.

Alatalo, R.V., Eriksson, D., Gustafsson, L. & Larsson, K. (1987) Exploitation competition influences the use of foraging sites by tits: experimental evidence. *Ecology*, **68**, 284–90.

Alatalo, R.V., Gustafsson, L. & Lundberg, A. (1986) Interspecific competition and niche changes in tits (*Parus* spp): evaluation of nonexperimental data. *The American Naturalist*, **127**, 819–34.

Alatalo, R.V., Gustafsson, L., Lundberg, A. & Ulfstrand S. (1985) Habitat shift of the Willow Tit *Paus montanus* in the absence of the Marsh Tit *Parus palustris*. *Ornis Scandinavica*, **16**, 121–8.

Alerstam, T. (1985) Fågelsamhället i Borgens lövskogsområde. *Anser*, **24**, 213–34.
Allen, T.F.H. & Starr, T.B. (1982) *Hierarchy: Perspectives for ecological complexity.* Chicago: University of Chicago Press.
Allen, T.F.H., O'Neill, R.V. & Hoeckstra, T.W. (1984) Interlevel relations in ecological research and management: some working principles from hierarchy theory. *US Department of Agriculture Forest Service General Technical Report RM-110.*
Alley, T.R. (1982) Competition theory, evolution, and the concept of ecological niche. *Acta Biotheoretica*, **31**, 165–79.
Ambuel, B. & Temple, S.A. (1983) Area-dependent changes in the bird communities and vegetation of southern Wisconsin forests. *Ecology*, **64**, 1057–68.
Anderson, D.C. & MacMahon, J.A. (1986) An assessment of ground-nest depredation in a catastrophically disturbed region, Mount St Helens, Washington. *The Auk*, **103**, 622–6.
Anderson, R.M., Gordon, D.M., Crawley, M.J. & Hassell, M.P. (1982) Variability in the abundance of animal and plant species. *Nature*, **296**, 245–8.
Andrén, H., Angelstam, P., Lindstrom, E. & Widén, P. (1985) Differences in predation pressure in relation to habitat fragmentation: an experiment. *Oikos*, **45**, 273–7.
Andrewartha, H.G. & Birch, L.C. (1954) *The distribution and abundance of animals.* Chicago: University of Chicago Press.
(1984) *The ecological web. More on the distribution and abundance of animals.* Chicago: University of Chicago Press.
Angelstam, P. (1986) Predation on ground-nesting birds' nests in relation to predator densities and habitat edge. *Oikos*, **47**, 365–73.
Angelstam, P., Lindstrom, E. & Widén, P. (1984) Role of predation in short-term population fluctuations of some birds and mammals in Fennoscandia. *Oecologia*, **62**, 199–208.
(1985) Synchronous short-term population fluctuations of some birds and mammals in Fennoscandia – occurrence and distribution. *Holarctic Ecology*, **8**, 285–98.
Ankney, C.D. & MacInnes, C.D. (1978) Nutrient reserves and reproductive performance of female Lesser Snow Geese. *The Auk*, **95**, 459–71.
Arthur, W. (1982) The evolutionary consequences of interspecific competition. *Advances in Ecological Research*, **12**, 127–87.
Askenmo, C., von Bromssen, A., Ekman, J. & Jansson, C. (1977) Impact of some wintering birds on spider abundance in spruce. *Oikos*, **28**, 90–4.
Askins, R.A. & Philbrick, M.J. (1987) Effect of changes in regional forest abundance on the decline and recovery of a forest bird community. *The Wilson Bulletin*, **99**, 7–21.
Askins, R.A., Philbrick, M.J. & Sugeno, D.S. (1987) Relationship between the regional abundance of forest and the composition of forest bird communities. *Biological Conservation*, **39**, 129–52.
Baeyens, G. (1981) Magpie breeding success and Carrion Crow interference. *Ardea*, **69**, 125–39.
Baker, M.C. & Baker, A.E.M. (1973) Niche relationships among six species of shorebirds on their wintering and breeding ranges. *Ecological Monographs*, **43**, 193–212.
Balen, J.H. van. (1980) Population fluctuations of the Great Tit and feeding conditions in winter. *Ardea*, **68**, 143–64.
Balen, J.H. van, Booy, C.J.H., Franeker, J.A. van & Osieck, E.R. (1982) Studies on hole-nesting birds in natural nest sites. I. Availability and occupation of natural nest sites. *Ardea*, **70**, 1–24.
Barber, R.T. & Chavez, F.P. (1983) Biological consequences of El Niño. *Science*, **222**, 1203–10.
(1986) Ocean variability in relation to living resources during the 1982–83 El Niño. *Nature*, **319**, 279–85.

Barrowclough, G. F. & Shields, G.F. (1984) Karyotypic evolution and long-term effective population sizes in birds. *The Auk,* **101**, 99–102.
Bartholomew, G.A. (1982) Scientific innovation and creativity: a zoologist's point of view. *American Zoologist,* **22**, 227–35.
(1986) The role of natural history in contemporary biology. *BioScience,* **36**, 324–9.
Batten, L.A. & Marchant, J.H. (1977) Bird population changes for the years 1965–75. *Bird Study,* **24**, 55–61.
Beeton, R.J.S. (1977) The impact and management of birds on the Ord River development in western Australia. M. Nat. Res. Armidale NSW: University of New England.
Bejer, B. & Rudemo, M. (1985) Fluctuations of tits (Paridae) in Denmark and their relations to winter food and climate. *Ornis Scandinavica,* **16**, 29–37.
Bell, H.L. (1983) 'Resource-Partitioning Between Three Syntopic Thornbills (Acanthizidae: *Acanthiza* Vigors and Horsfield).' Ph.D. dissertation. Armidale, NSW: University of New England.
Bell, H.L. & Ford, H.A. (MS) Does food scarcity lead to increased or decreased foraging overlap? A test in Australian warblers *Acanthiza.*
Bender, E.A., Case, T.J. & Gilpin, M.E. (1984) Perturbation experiments in community ecology: theory and practice. *Ecology,* **65**, 1–13.
Bengtson, S-A. & Bloch, D. (1983) Island land bird population densities in relation to island size and habitat quality on the Faroe Islands. *Oikos,* **41**, 507–22.
Bent, A.C. (1968) Life histories of North American cardinals, grosbeaks, buntings, towhees, finches, sparrows, and allies. *US National Museum Bulletin,* **237**, 1–1889.
Berger, A.J. (1981) *Hawaiian birdlife*. Honolulu: University of Hawaii Press.
Berndt, R. von & Henss, M. (1967) Die kohlmeise, *Parus major*, als invasionsvogel. *Die Vogelwart,* **24**, 17–37.
Berry, M.P.S. & Crowe, T.M. (1985) Effects of monthly and annual rainfall on game bird populations in the northern Cape Province, South Africa. *South African Journal of Wildlife Research,* **15**, 69–76.
Berthold, P. (1973) Über starken Rückgang der Dorngrasmücke *Sylvia communis* und anderer Singvogelarten im westlichen Europa. *Journal für Ornithologie,* **114**, 348–60.
Best, L.B. (1979) Effects of fire on a field sparrow population. *The American Midland Naturalist,* **101**, 434–42.
Bethel, W.M. & Holmes, J.C. (1973) Altered evasive behavior and responses to light in amphipods harboring acanthocephalan cystacanths. *Journal of Parasitology,* **59**, 945–56.
(1977) Increased vulnerability of amphipods to predation owing to altered behavior induced by larval acanthocephalans. *Canadian Journal of Zoology,* **55**, 110–15.
Beven, S., Connor, E.F. & Beven, K. (1984) Avian biogeography in the Amazon basin and the biological model of diversification. *Journal of Biogeography,* **11**, 383–99.
Birch, L.C. (1957) The meanings of competition. *American Naturalist,* **91**, 5–18.
(1979) The effect of species of animals which share common resources on one another's distribution and abundance. *Fortschritte der Zoologie,* **25**, 197–221.
Blake, J.G. & Hoppes, W.G. (1986) Influence of resource abundance on use of tree-fall gaps by birds in an isolated woodlot. *The Auk,* **103**, 328–40.
Bleakney, J.S. (1972) Ecological implications of annual variation in tidal extremes. *Ecology,* **53**, 933–8.
Blouin, M.S. & Connor, E.F. (1985) Is there a best shape for nature reserves? *Biological Conservation,* **32**, 277–88.
Boag, P.T. & Grant, P.R. (1978) Heritability of external morphology in Darwin's finches. *Nature*, **274**, 793–4.

(1981) Intense natural selection in a population of Darwin's finches (Geospizinae) in the Galápagos. *Science,* **214**, 82–5.
(1984) Darwin's finches (*Geospiza*) on Isla Daphne Major, Galápagos: breeding and feeding ecology in a climatically variable environment. *Ecological Monographs*, **54**, 463–89.
Bock, C.E. (1987) Distribution-abundance relationships of some Arizona landbirds: a matter of scale. *Ecology,* **68**, 124–9.
Boecklen, W.J. & Bell, G.W. (1987) Consequences of faunal collapse and genetic drift for the design of nature reserves. In *Nature Conservation: The Role of Remnants of Native Vegetation*, ed. Saunders, D.A., Arnold, G.W., Burbidge, A.A. & Hopkins, A.J.M., pp. 141–9. Chipping Norton, NSW: Surrey Beatty & Sons.
Boecklen, W.J. & Gotelli, N.J. (1984) Island biogeographic theory and conservation practice: species-area or specious-area relationships? *Biological Conservation,* **29**, 63–80.
Bossema, I.A., Roell, A., Baeyens, G., Zeevalking, H. & Leever, H. (1976) Interspecifieke aggressie en sociale organixatie bij onze inheemse corviden. *Levende Nat,* **79**, 149–66.
Botkin, D.B. & Sobel, M.J. (1975) Stability in time-varying ecosystems. *The American Naturalist,* **109**, 625–46.
Boyden, T.C. (1978) Territorial defense against hummingbirds and insects by·tropical hummingbirds. *The Condor,* **80**, 216–21.
Bradley, R.A. (1983) Complex food webs and manipulative experiments in ecology. *Oikos,* **41**, 150–2.
Brawn, J.D., Boecklen, W.J. & Balda, R.P. (1987) Investigations of density interactions among breeding birds in ponderosa pine forests: correlative and experimental evidence. *Oecologia,* **72**, 348–57.
Brew, J.S. (1982) Niche shift and the minimisation of competition. *Theoretical Population Biology,* **22**, 367–81.
Brewer, R. (1963) Stability in bird populations. *Occasional Papers of the C.C. Adams Center for Ecological Studies,* **7**, 1–12.
Brittingham, M.C. & Temple, S.A. (1983) Have cowbirds caused forest songbirds to decline? *BioScience,* **33**, 31–5.
Brokaw, N.V.L. (1985) Treefalls, regrowth, and community structure in tropical forests. In *The ecology of natural disturbance and patch dynamics*, ed. Pickett, S.T.A. & White, P.S., pp. 53–69. New York: Academic Press.
Brömssen, A.V. & Jansson, C. (1980) Effects of food addition to willow tit *Parus montanus* and crested tit *P. cristatus* at the time of breeding. *Ornis Scandinavica,* **11**, 173–8.
Bronowski, J. (1977) *A sense of the future*. Cambridge, Massachusetts: MIT Press.
Brooks, D.R. (1985) Historical ecology: a new approach to studying the evolution of ecological associations. *Annals of the Missouri Botanical Garden,* **72**, 660–80.
Brown, J.H. (1981) Two decades of homage to Santa Rosalia: toward a general theory of diversity. *American Zoologist,* **21**, 877–88.
(1984) On the relationship between abundance and distribution of species. *The American Naturalist,* **124**, 255–79.
Brown, J.H. & Bowers, M.A. (1985) Community organization in hummingbirds: relationships between morphology and ecology. *The Auk,* **102**, 251–69.
Brown, J.H. & Gibson, A.C. (1983) *Biogeography.* St Louis: Mosby.
Brown, J.H. & Kodric-Brown, A. (1977) Turnover rates in insular biogeography: effect of immigration on extinction. *Ecology,* **58**, 445–9.
(1979) Convergence, competition, and mimicry in a temperate community of hummingbird-pollinated flowers. *Ecology*, **60**, 1022–35.

Brown, J.H., Calder, W.A. III & Kodric-Brown, A. (1978) Correlates and consequences of body size in nectar-feeding birds. *The American Zoologist*, **68**, 687–700.

Brown, J.H., Davidson, D.W., Munger, J.C. & Inouye, R.S. (1986) Experimental community ecology: the desert granivore system. In *Community ecology*, ed. Diamond, J. & Case, T.J., pp. 41–61. New York: Harper & Row.

Brown, J.H., Kodric-Brown, A., Whitham, T.G. & Bond, H.W. (1981) Competition between hummingbirds and insects for the nectar of two species of shrubs. *The Southwestern Naturalist*, **26**, 133–45.

Bucher, E.H. (1974) Bases ecológicas para el control de la paloma torcaza. *Universidad Nacional Cordoba Faculidad Ciencias Exatas Fiscas Natural Publicacion No.4.*

(1982) Colonial breeding of the Eared Dove (*Zenaida auriculata*) in northeastern Brazil. *Biotropica*, **14**, 255–61.

Burgess, R.L. & Sharpe, D.M., eds. (1981) *Forest island dynamics in man-dominated landscapes*. New York: Springer-Verlag.

Butcher, G.S., Niering, W.A., Barry, W.J. & Goodwin, R.H. (1981) Equilibrium biogeography and the size of nature preserves: an avian case study. *Oecologia*, **49**, 29–37.

Calder, W.A. III (1984a) *Size, Function, and Life History*. Cambridge: Harvard University Press.

(1984b) How long is a long-term study? *The Auk*, **101**, 893–4.

Calder, W.A. III & Booser, J. (1973) Hypothermia of Broad-tailed Hummingbirds during incubation in nature with ecological correlations. *Science*, **180**, 751–3.

Cane, M.A. (1983) Oceanographic events during El Niño. *Science*, **222**, 1189–95.

Carothers, J.H. (1986) Behavioral and ecological correlates of interference competition among some Hawaiian Drepanidinae. *The Auk*, **103**, 564–74.

Carothers, J.H. & Jaksić, F.M. (1984) Time as a niche difference: the role of interference competition. *Oikos*, **42**, 403–6.

Carpenter, F.L. (1978) A spectrum of nectar-eater communities. *The American Zoologist*, **18**, 809–19.

(1979) Competition between hummingbirds and insects for nectar. *The American Zoologist*, **19**, 1105–14.

Carpenter, F.L. & MacMillen, R.E. (1976) Threshold model of feeding territoriality and test with a Hawaiian honeycreeper. *Science*, **194**, 639–42.

(1978) Resource limitation, foraging strategies, and community structure in Hawaiian honeycreepers. *Proceedings XVII International Ornithological Congress*, 1100–4.

Carpenter, F.L., Paton, D.C. & Hixon, M.A. (1983) Weight gain and adjustment of feeding territory size in migrant hummingbirds. *Proceedings of the National Academy of Sciences, USA*, **80**, 7259–63.

Carrascal, L.M., Potte, J., & Sanchez-Aguado, F.J. (1987) Spatio-temporal organization of the bird communities in two mediterranean montane forests. *Holarctic Ecology*, **10**, 185–92.

Case, T.J. & Gilpin, M.E. (1974) Interference competition and niche theory. *Proceedings of the National Academy of Sciences, USA*, **71**, 3073–7.

Case, T.J., Faaborg, J. & Sidell, R. (1983) The role of body size in the assembly of West Indian bird communities. *Evolution*, **37**, 1062–74.

Cavé, A.J. (1983) Purple heron survival and drought in tropical West-Africa. *Ardea*, **71**, 217–24.

Cederholm, G. & Ekman, J. (1976) A removal experiment on Crested Tit *Parus cristatus* and Willow Tit *P. montanus* in the breeding season. *Ornis Scanidinavica*, **7**, 207–13.

Chavez, F.P. & Barber, R.T. (1984) Propagated temperature changes during onset and recovery of the 1982–83 El Niño. *Nature*, **309**, 47–9.

Chesson, P.L. (1981) Models for spatially distributed populations: the effects of within-patch variability. *Theoretical Population Biology,* **19**, 288–325.
(1985) Coexistence of competitors in spatially and temporally varying environments: a look at the combined effects of different sorts of variability. *Theoretical Population Biology,* **28**, 263–87.
(1986) Environmental variation and the coexistence of species. In *Community Ecology*, ed. Diamond, J. & Case, T.J., pp. 240–56. New York: Harper & Row.
Chesson, P.L. & Case, T.J. (1986) Overview: nonequilibrium community theories: chance, variability, history, and coexistence. In *Community Ecology*, ed. Diamond, J. & Case, T.J., pp. 229–39. New York: Harper & Row.
Chesson, P.L. & Warner, R.R. (1981) Environmental variability promotes coexistence in lottery competitive systems. *The America Naturalist,* **117**, 923–43.
Cieślak, M. (1983) Wstepna ocena czynnikow kszaltujacych zgrupowania ptakow legowych brzegu lasu. *Cziowiek i Srodowisko,* **7**, 449–59.
(1985) Influence of forest size and other factors on breeding bird species number. *Ekologia Polska,* **33**, 103–21.
Cliff, A.D. & Ord, J.K. (1973) *Spatial Autocorrelation.* London: Methuen.
Cody, M.L. (1974) *Competition and the structure of bird communities.* Princeton: Princeton University Press.
(1975) Towards a theory of continental species diversities: bird distributions over mediterranean habitat gradients, In *Ecology and Evolution of Communities*, ed. Cody, M.L. & Diamond, J.M., pp. 214–57. Cambridge: Harvard University Press.
(1979) Resource allocation patterns in Palaerarctic warblers (Sylviidae). *Fortschritte der Zoologie,* **25**, 223–34.
(1981) Habitat selection in birds: the roles of vegetation structure, competitors, and productivity. *BioScience,* **31**, 107–11.
(1983) Bird diversity and density in South African forests. *Oecologia*, **59**, 201–15.
Cole, B.J. (1981) Colonizing abilities, island size, and the number of species on archipelagoes. *The American Naturalist,* **117**, 629–38.
Collias, N.E. & Collias, E.C. (1984) *Nest Building and Bird Behavior*. Princeton: Princeton University Press.
Collins, B.G. (1985) Energetics of foraging and resource selection by honeyeaters in forest and woodland habitats of Western Australia. *New Zealand Journal of Zoology,* **12**, 577–87.
Collins, B.G. & Briffa, P. (1982) Seasonal variation of abundance and foraging of three species of Australian honeyeaters. *Australian Wildlife Research,* **9**, 557–69.
Collins, B.G. & Newland, C. (1986) Honeyeater population changes in relation to food availability in the Jarrah forest of Western Australia. *Australian Journal of Ecology,* **11**, 63–76.
Collins, B.G., Briffa, P. & Newland, C. (1984) Temporal changes in abundance and resource utilization by honeyeaters at Wongamine nature reserve. *The Emu,* **84**, 159–66.
Collins, S.L., James, F.C. & Risser, P.G. (1982) Habitat relationship of wood warblers (Parulidae) in northern central Minnesota. *Oikos,* **39**, 50–8.
Colwell, R.K. (1984) What's new? Community ecology discovers biology. In *A New Ecology. Novel Approaches to Interactive Systems*, ed. Price, P.W., Slobodchikoff, C.N. & Gaud, W.S., pp. 387–96. New York: John Wiley & Sons.
Colwell, R.K. & Fuentes, E.R. (1975) Experimental studies of the niche. *Annual Review of Ecology and Systematics,* **6**, 281–310.
Colwell, R.K. & Winkler, D.W. (1984) A null model for null models in biogeography. In *Ecological Communities. Conceptual Issues and the Evidence*, ed. Strong, D.R. Jr, Simberloff, D., Abele, L.G. & Thistle, A.B., pp. 344–59. Princeton: Princeton University Press.

Colwell, R.K., Betts, B.J., Bunnell, P., Carpenter, F.L. & Feinsinger, P. (1974) Competition for the nectar of *Centropogon valerii* by the hummingbird *Colibri thalassinus* and the flower-piercer *Diglossa plumbea*, and its evolutionary implications. *The Condor*, **76**, 447–52.

Connell, J.H. (1975) Some mechanisms producing structure in natural communities: a model and evidence from field experiments. In *Ecology and Evolution of Communities*, ed. Cody, M.L. & Diamond, J.M., pp. 460–90. Cambridge: Harvard University Press.

(1980) Diversity and coevolution of competitors, or the ghost of competition past. *Oikos*, **35**, 131–8.

(1983) On the prevalence and relative importance of interspecific competition: evidence from field experiments. *The American Naturalist*, **122**, 661–96.

Connell, J.H. & Sousa, W.P. (1983) On the evidence needed to judge ecological stability or persistence. *The American Naturalist*, **121**, 789–824.

Connor, E.F. & Bowers, M.A. (1987) The spatial consequences of interspecific competition. *Annales Zoologici Fennici*, **24**, 213–26.

Connor, E.F. & Simberloff, D. (1978) Species number and compositional similarity of the Galápagos flora and avifauna. *Ecological Monographs*, **48**, 219–48.

(1983) Interspecific competition and species co-occurrence patterns on islands: null models and the evaluation of evidence. *Oikos*, **41**, 455–65.

Cracraft, J. (1983) Cladistic analysis and vicariance biogeography. *The American Scientist*, **71**, 273–81.

(1985a) Biological diversification and its causes. *Annals of the Missouri Botanical Garden*, **72**, 794–822.

(1985b) Historical biogeography and patterns of differentiation in the South American avifauna: areas of endemism. *Ornithological Monographs*, **36**, 49–84.

(1986) Origin and evolution of continental biotas: speciation and historical congruence within the Australian avifauna. *Evolution*, **40**, 977–96.

Craig, D.L. & MacMillen, R.E. (1985) Honeyeater ecology: an introduction. *New Zealand Journal of Zoology*, **12**, 565–8.

Craig, J.L. & Douglas, M.E. (1986) Resource distribution, aggressive asymmetries and variable access to resources in the nectar feeding bellbird. *Behavioral Ecology and Sociobiology*, **18**, 231–40.

Craig, R.J. (1984) Comparative foraging ecology of Louisiana and Northern waterthrushes. *The Wilson Bulletin*, **96**, 173–83.

Crowell, K.L. (1983) Islands – insight or artifact?: Population dynamics and habitat utilization in insular rodents. *Oikos*, **41**, 442–54.

Crowell, K.L. & Pimm, S.L. (1976) Competition and niche shifts of mice introduced onto small islands. *Oikos*, **27**, 251–8.

Cruz, A., Manolis, T. & Wiley, J.W. (1985) The shiny cowbird: a brood parasite expanding its range in the Caribbean region. *Ornithological Monographs*, **36**, 607–20.

Curtis, J.T. (1959) *The Vegetation of Wisconsin*. Madison: The University of Wisconsin Press.

Davidson, N.C. & Evans, P.R. (1982) Mortality of redshanks and oystercatchers from starvation during severe weather. *Bird Study*, **29**, 183–8.

Davis, J. (1973) Habitat preferences and competition of wintering juncos and golden-crowned sparrows. *Ecology*, **54**, 174–80.

Davis, M.B. (1986) Climatic instability, time lags, and community disequilibrium. In *Community ecology*, ed. Diamond, J. & Case, T.J., pp. 269–84. New York: Harper & Row.

Dawson, D.G. (1984) Principles of ecological biogeography and criteria for reserve design (extended abstract). *Journal of the Royal Society of New Zealand*, **14**, 11–15.

Dayton, P.K. & Tegner, M.J. (1984) The importance of scale in community ecology: a kelp forest example with terrestrial analogs. In *A new ecology. Novel approaches to interactive systems*, ed. Price, P.W., Slobodchikoff, C.N. & Gaud, W.S., pp. 457–81. New York: John Wiley & Sons.

DeAngelis, D.L. & Waterhouse, J.C. (1987) Equilibrium and nonequilibrium concepts in ecological models. *Ecological Monographs*, **57**, 1–21.

den Boer, P.J. (1981) On the survival of populations in a heterogeneous and variable environment. *Oecologia*, **50**, 39–53.

den Held, J.J. (1981) Population changes of the Purple Heron in relation to drought in the wintering area. *Ardea*, **69**, 185–91.

Denslow, J.S. (1985) Disturbance-mediated coexistence of species. In *The ecology of natural disturbance and patch dynamics*, ed. Pickett, S.T.A. & White, P.S., pp. 307–23. New York: Academic Press.

DesGranges, J.-L. (1979) Organization of a tropical nectar-feeding bird guild in a variable environment. *The Living Bird*, **17**, 199–236.

Dhondt, A.A. (1977) Interspecific competition between Great and Blue tit. *Nature*, **268**, 521–3.

(1983) Variations in the number of overwintering Stonechats possibly caused by natural selection. *Ringing and Migration*, **4**, 155–8.

Dhondt, A.A. & Eyckerman, R. (1980) Competition between the Great Tit and the Blue Tit outside the breeding season in field experiments. *Ecology*, **61**, 1291–6.

Diamond, J.M. (1969) Avifaunal equilibria and species turnover rates on the Channel Islands of California. *Proceedings of the National Academy of Sciences USA*, **64**, 57–63.

(1971) Comparison of faunal equilibrium turnover rates on a tropical island and a temperate island. *Proceedings of the National Academy of Sciences, USA*, **68**, 2742–5.

(1975a) Assembly of species communities. In *Ecology and evolution of communities*, ed. Cody, M.L. & Diamond, J.M., pp. 342–444. Cambridge, Massachusetts: Harvard University Press.

(1975b) The island dilemma: lessons of modern biogeographic studies for the design of natural reserves. *Biological Conservation*, **7**, 129–45.

(1976) Island biogeography and conservation: strategy and limitations. *Science*, **193**, 1027–9.

(1978) Niche shifts and the rediscovery of interspecific competition. *The American Scientist*, **66**, 322–31.

(1979) Population dynamics and interspecific competition in bird communities. *Fortschritte der Zoologie*, **25**, 389–402.

(1982) Effect of species pool size on species occurrence frequencies: musical chairs on islands. *Proceedings of the National Academy of Sciences, USA*, **79**, 2420–4.

(1984) Distributions of New Zealand birds on real and virtual islands. *New Zealand Journal of Ecology*, **7**, 37–55.

(1986) Overview: Laboratory experiments, field experiments, and natural experiments. In *Community ecology*, ed. Diamond, J. & Case, T.J., pp. 3–22. New York: Harper & Row.

Diamond, J.M. & Case, T.J., eds. (1986) *Community ecology*. New York: Harper & Row.

Diamond, J.M. & May, R.M. (1976) Island biogeography and the design of nature reserves. In *Theoretical Ecology: principles and applications*, ed. May, R.M., pp. 163–86. Philadelphia: W.B. Saunders.

(1977) Species turnover rates on islands: dependence on census intervals. *Science*, **197**, 266–70.

Diamond, J.M., Bishop, K.D. & van Balen, S. (1987) Bird survival in an isolated Javan woodland: island or mirror? *Conservation Biology*, **1**, 132–42.

Dobkin, D.S. & Wilcox, B.A. (1986) Analysis of natural forest fragments: riparian birds in the Toiyabe Mountains, Nevada. In *Wildlife 2000. Modeling Habitat Relationships of Terrestrial Vertebrates*, ed. Verner, J., Morrison, M.L. & Ralph, C.J., pp. 293–9. Madison: University of Wisconsin Press.

Dorp, D. van & Opdam, P.F.M. (1987) Effects of patch size, isolation and regional abundance on forest bird communities. *Landscape Ecology*, **1**, 59–73.

Dueser, R.D. & Hallett, J.G. (1980) Competition and habitat selection in a forest-floor small mammal fauna. *Oikos*, **35**, 293–7.

Duffy, D.C. (1983) Competition for nesting space among Peruvian guano birds. *The Auk*, **100**, 680–8.

Duffy, D.C. & Merlen, G. (1986) Seabird densities and aggregations during the 1983 El Niño in the Galápagos Islands. *The Wilson Bulletin*, **98**, 588–91.

Dunn, E.K. (1977) Predation by weasels (*Mustella nivalis*) on breeding tits (*Parus* spp.) in relation to the density of tits and rodents. *Journal of Animal Ecology*, **46**, 633–52.

Dunning, J.B. Jr & Brown, J.H. (1982) Summer rainfall and winter sparrow densities: a test of the food limitation hypothesis. *The Auk*, **99**, 123–9.

Eadie, J.M. & Keast, A. (1982) Do goldeneye and perch compete for food? *Oecologia*, **55**, 225–30.

East, R. & Williams, G.R. (1984) Island biogeography and the conservation of New Zealand's indigenous forest-dwelling avifauna. *New Zealand Journal of Ecology*, **7**, 27–35.

Ekman, J. (1986) Tree use and predator vulnerability of wintering passerines. *Ornis Scandinavica*, **17**, 261–7.

Elton, C. (1930) *Animal Ecology and Evolution*. New York: Oxford University Press.

Emlen, J.T. (1977) Land bird communities of Grand Bahama Island: the structure and dynamics of an avifauna. *Ornithological Monographs*, **24**, 1–129.

(1979) Land bird densities on Baja California islands. *The Auk*, **96**, 152–67.

(1980) Interactions of migrant and resident land birds in Florida and Bahama pinelands. In *Migrant Birds in the Neotropics: Ecology, Behavior, Distribution, and Conservation*, ed. Keast, A. & Morton, E.S., pp. 133–43. Washington, D.C.: Smithsonian Institution Press.

Emlen, J.T. & DeJong, M.J. (1981) Intrinsic factors in the selection of foraging substrates by Pine Warblers: a test of an hypothesis. *The Auk*, **98**, 294–8.

Endler, J.A. (1982a) Problems in distinguishing historical from ecological factors in biogeography. *The American Zoologist*, **22**, 441–52.

(1982b) Pleistocene forest refuges: fact or fancy? In *Biological Diversification in the Tropics*, ed. Prance, G., pp. 641–57. New York: Columbia University Press.

Enemar, A. (1966) A ten-year study on the size and composition of a breeding passerine bird community. *Var Fagelvarld, Supplement*, **4**, 47–94.

Enemar, A., Nilsson, L. & Sjostrand, B. (1984) The composition and dynamics of the passerine bird community in a subalpine birch forest, Swedish Lapland. A 20-year study. *Annales Zoologici Fennici*, **21**, 321–38.

Eriksson, M.O.G. (1979) Competition between freshwater fish and Goldeneyes *Bucephala clangula* (L.) for common prey. *Oecologia*, **41**, 99–107.

Erlinge, S., Göransson, G., Hansson, L., Högstedt, G., Liberg, O., Nilsson, I.N., Nilsson, T., von Schantz, T. & Sylvén, M. (1983) Predation as a regulating factor on small rodent populations in southern Sweden. *Oikos*, **40**, 36–52.

Erlinge, S., Göransson, G., Högstedt, G., Jansson, G., Liberg, O., Loman, J., Nilsson, I.N., von Schantz, T. & Sylvén, M. (1984) Can vertebrate predators regulate their prey? *The American Naturalist*, **123**, 125–33.

Evans, P.R. & Dugan, P.J. (1984) Coastal birds: numbers in relation to food resources. In *Coastal waders and wildfowl in winter*, ed. Evans, P.R., Goss-Custard, J.D. & Hale, W.G., pp. 8–28. Cambridge: Cambridge University Press.

Evans, P.R. & Smith, P.C. (1975) Studies of shorebirds at Lindisfarne, Northumberland. 2. Fat and pectoral muscles as indicators of body condition in the Bar-tailed Godwit. *Wildfowl*, **26**, 37–46.

Faaborg, J. Arendt, W.J. & Kaiser, M.S. (1984) Rainfall correlates of bird population fluctuations in a Puerto Rican dry forest: a nine year study. *The Wilson Bulletin*, **96**, 575–93.

Fagerström, T. (1987) On theory, data and mathematics in ecology. *Oikos*, **50**, 258–61.

Feare, C.J. (1972) The seasonal pattern of feeding in the Rook (*Corvus frugilegus*) in northeast Scotland. *Proceedings of the 15th International Ornithological Congress*, 643.

(1984) *The Starling*. Oxford: Oxford University Press.

Feare, C.J., Dunnett, G.M. & Patterson, I.J. (1974) Ecological studies of the Rook (*Corvus frugilegus* L.) in northeast Scotland: food intake and feeding behavior. *Journal of Applied Ecology*, **11**, 867–96.

Feinsinger, P. (1976) Organization of a tropical guild of nectarivorous birds. *Ecological Monographs*, **46**, 257–91.

(1978) Ecological interactions between plants and hummingbirds in a successional tropical community. *Ecological Monographs*, **48**, 269–87.

Feinsinger, P. & Colwell, R.K. (1978) Community organization among neotropical nectar-feeding birds. *The American Zoologist*, **18**, 779–95.

Feinsinger, P. & Swarm, L.A. (1982) "Ecological release," seasonal variation in food supply, and the hummingbird *Amazilia tobaci* on Trinidad and Tobago. *Ecology*, **63**, 1574–87.

Feinsinger, P., Colwell, R.K., Terborgh, J. & Chaplin, S.B. (1979) Elevation and the morphology, flight energetics and foraging ecology of tropical hummingbirds. *The American Naturalist*, **113**, 481–97.

Feinsinger, P., Swarm, L.A. & Wolfe, J.A. (1985) Nectar-feeding birds on Trinidad and Tobago: comparison of diverse and depauperate guilds. *Ecological Monographs*, **55**, 1–28.

Feinsinger, P., Wolfe, J.A. & Swarm, L.A. (1982) Island ecology: reduced hummingbird diversity and the pollination biology of plants, Trinidad and Tobago, West Indies. *Ecology*, **63**, 494–506.

Fenton, M.B. & Fleming, T.H. (1976) Ecological interactions between bats and nocturnal birds. *Biotropica*, **8**, 104–10.

Fischer, D.H. (1981) Wintering ecology of thrashers in southern Texas. *The Condor*, **83**, 340–6.

Fitzgerald, B.M. & Veitch, C.R. (1985) The cats of Herekopare Island, New Zealand; their history, ecology and affects on birdlife. *New Zealand Journal of Zoology*, **12**, 319–30.

Folse, L.J.Jr. (1982) An analysis of avifauna-resource relationships on the Serengeti Plains. *Ecological Monographs*, **52**, 111–27.

Fonstad, T. (1984) Reduced territorial overlap between the Willow Warbler *Phylloscopus trochilus* and the Brambling *Fringilla montifringilla* in heath forest: competition or different habitat preferences? *Oikos*, **42**, 314–22.

Ford, H.A. (1979) Interspecific competition in Australian honeyeaters – depletion of common resources. *Australian Journal of Ecology*, **4**, 145–64.

(1983) Relation between number of honeyeaters and intensity of flowering near Adelaide, South Australia. *Corella*, **7**, 25–31.

(1985) A synthesis of the foraging ecology and behavior of birds in eucalypt forests and woodlands. In *Birds of Eucalypt Forests and Woodlands: Ecology, Conservation, Management*, ed. Keast, A., Recher, H.F., Ford, H. & Saunders, D., pp. 249–54. Chipping Norton, NSW: Surrey Beatty & Sons.

Ford, H.A. & Paton, D.C. (1977) The comparative ecology of ten species of honeyeaters in South Australia. *Australian Journal of Ecology*, **2**, 399–407.
(1982) Partitioning of nectar sources in an Australian honeyeater community. *Australian Journal of Ecology*, **7**, 149–59.
(1985) Habitat selection in Australian honeyeaters, with special reference to nectar productivity. In *Habitat Selection in Birds*, ed. Cody, M.L., pp. 367–88. New York: Academic Press.
Forman, R.T.T. & Godron, M. (1986) *Landscape Ecology*. New York: John Wiley & Sons.
Forman, R.T.T., Galli, A.E. & Leck, C.F. (1976) Forest size and avian diversity in New Jersey woodlots with some land use implications. *Oecologia*, **26**, 1–8.
Foster, W.L. & Tate, J. Jr. (1966) The activities and coactions of animals at sapsucker trees. *The Living Bird*, **5**, 87–113.
Frankel, O.H. & Soulé, M.E. (1981) *Conservation and Evolution*. Cambridge: Cambridge University Press.
Franzblau, M.A. & Collins, J.P. (1980) Test of hypothesis of territory regulation in an insectivorous bird by experimentally increasing prey abundance. *Oecologia*, **46**, 164–70.
Freed, L.A. (1987) Rufons-and-White Wrens kill House Wren nestlings during a food shortage. *The Condor*, **89**, 195–7.
Freemark, K.E. & Merriam, H.G. (1986) Importance of area and habitat heterogeneity to bird assemblages in temperate forest fragments. *Biological Conservation*, **36**, 115–41.
Fretwell, S.D. (1972) *Populations in a Seasonal Environment*. Princeton: Princeton University Press.
Fretwell, S.D. & Lucas, H.L. (1969) On territorial behavior and other factors influencing habitat distribution in birds. I. Theoretical development. *Acta Biotheoretica*, **19**, 16–36.
Friedmann, H. (1929) *The Cowbirds*. Baltimore: Charles C. Thomas.
Fritz, R.S. (1980) Consequences of insular population structure: distribution and extinction of Spruce Grouse populations in the Adirondack Mountains. In *Acta XVII Congressus Internationalis Ornithologici*, ed. Nohring, R., pp. 757–63. Berlin: Verlag der Deutschen Ornithologen-Gesellschaft.
Gaines, D. (1974) A new look at the nesting riparian avifauna of the Sacramento Valley, California. *Western Birds*, **5**, 61–80.
Galli, A.E., Leck, C.F. & Forman, R.T.T. (1976) Avian distribution patterns in forest islands of different sizes in central New Jersey. *The Auk*, **93**, 356–64.
Garcia, E.F.J. (1983) An experimental test of competition for space between Blackcaps *Sylvia atricapilla* and Garden Warblers *Sylvia borin* in the breeding season. *Journal of Animal Ecology*, **52**, 795–805.
Gass, C.L. & Lertzman, K.P. (1980) Capricious mountain weather: a driving variable in hummingbird territorial dynamics. *Canadian Journal of Zoology*, **58**, 1964–8.
Gass, C.L. & Sutherland, G.D. (1985) Specialization by territorial hummingbirds on experimentally enriched patches of flowers: energetic profitability and learning. *Canadian Journal of Zoology*, **63**, 2125–33.
Gass, C.L., Angehr, G. & Centa, J. (1976) Regulation of food supply by feeding territoriality in the Rufous Hummingbird. *Canadian Journal of Zoology*, **54**, 2046–54.
Gaubert, H. (1985) Étude comparée de la croissance pondérale des jeunes de deux populations de mésanges bleues, *Parus caeruleus* L., en Corse et en Provence: augmentation expérimentale de la taille des nichées corses. *Acta Oecologica Oecologica Generale*, **6**, 305–16.
George, T.L. (1987) Greater land bird densities on island vs mainland: relation to nest predation level. *Ecology*, **68**, 1393–400.

Gibb, J. (1960) Populations of tits and goldcrests and their food supply in pine plantations. *Ibis*, **102**, 163–208.
Gibbs, H.L. & Grant, P.R. (1987a) Adult survivorship in Darwin's Ground Finch (Geospiza) populations in a variable environment. *Journal of Animal Ecology*, **56**, 797–813.
(1987b) Oscillating selection on Darwin's finches. *Nature*, **327**, 511–13.
(1987c) Ecological consequences of an exceptionally strong El Niño event on Darwin's finches. *Ecology*, **68**, 1735–46.
Gibbs, H.L., Latta, S.C. & Gibbs, J.P. (1987) Effects of the 1982–83 El Niño event on Blue-footed and Masked booby populations on Isla Daphne Major, Galápagos. *The Condor*, **89**, 440–2.
Gibo, D.I., Stephens, R., Culpeper, A. & Dew, H. (1976) Nest-site preferences and nesting success of the starling *Sturnus vulgaris* L. in marginal and favorable habitats in Mississauga, Ontario, Canada. *American Midland Naturalist*, **95**, 493–9.
Gill, F.B. (1978) Proximate costs of competition for nectar. *The American Zoologist*, **18**, 753–63.
Gill, F.B. & Wolf, L.L. (1977) Nonrandom foraging by sunbirds in a patchy environment. *Ecology*, **58**, 1284–96.
(1978) Comparative foraging efficiencies of some montane sunbirds in Kenya. *The Condor*, **80**, 391–400.
(1979) Nectar loss by golden-winged sunbirds to competitors. *The Auk*, **96**, 448–61.
Gill, F.B., Mack, A.L. & Ray, R.T. (1982) Competition between Hermit Hummingbirds Phaethorninae and insects for nectar in a Costa Rican rain forest. *Ibis*, **124**, 44–9.
Giller, P.S. (1984) *Community Structure and the Niche*. London: Chapman and Hall.
Gilpin, M.E. (1979) Spiral chaos in a predator-prey model. *The American Naturalist*, **113**, 306–8.
Glasser, J.W. (1979) The role of predation in shaping and maintaining the structure of communities. *The American Naturalist*, **113**, 631–41.
Gleason, H.A. (1926) The individualistic concept of the plant association. *Bulletin of the Torrey Botanical Club*, **53**, 1–20.
Gleick, J. (1987) *Chaos: making a new science*. New York: Viking.
Głowacinski, Z. & Järvinen, O. (1975) Rate of secondary succession in forest bird communities. *Ornis Scandinavica*, **6**, 33–40.
Gochfeld, M. (1985) Numerical relationships between migrant and resident bird species in Jamaican woodlands. *Ornithological Monographs*, **36**, 654–62.
Goldwasser, S., Gaines, D. & Wilbur, S.R. (1980) The Least Bell's Vireo in California: a de facto endangered race. *American Birds*, **34**, 742–5.
Gotfryd, A. & Hansell, R.I.C. (1986) Prediction of bird-community metrics in urban woodlots. In *Wildlife 2000. Modeling Habitat Relationships of Terrestrial Vertebrates*, ed. Verner, J., Morrison, M.L. & Ralph, C.J., pp. 321–6. Madison: University of Wisconsin Press.
Grant, K.A. & Grant, V. (1968) *Hummingbirds and their Flowers*. New York: Columbia University Press.
Grant, P.R. (1972) Convergent and divergent character displacement. *Biological Journal of the Linnean Society*, **4**, 39–68.
(1985) Climatic fluctuations on the Galápagos Islands and their influence on Darwin's finches. *Ornithological Monographs*, **36**, 471–83.
(1986a) *Ecology and Evolution of Darwin's Finches*. Princeton: Princeton University Press.
(1986b) Interspecific competition in fluctuating environments. In *Community Ecology*, ed. Diamond, J. & Case, T.J., pp. 173–91. New York: Harper & Row.

Grant, P.R. & Grant, B.R. (1980) Annual variation in finch numbers, foraging and food supply on Isla Daphne Major, Galápagos. *Oecologia*, **46**, 55–62.

(1987) The extraordinary El Niño event of 1982–83: effects on Darwin's finches on Isla Genovesa, Galápagos. *Oikos*, **49**, 55–66.

Grant, P.R. & Price, T.D. (1981) Population variation in continuously varying traits as an ecological genetics problem. *The American Zoologist*, **21**, 795–811.

Grant, P.R. & Schluter, D. (1984) Interspecific competition inferred from patterns of guild structure. In *Ecological Communities. Conceptual Issues and the Evidence*, ed. Strong, D.R. Jr, Simberloff, D., Abele, L.G. & Thistle, A.B., pp. 201–33. Princeton: Princeton University Press.

Graves, G.R. & Gotelli, N.J. (1983) Neotropical land-bridge avifaunas: new approaches to null hypotheses in biogeography. *Oikos*, **41**, 322–33.

Grossman, G.D., Moyle, P.B. & Whitaker, J.O. Jr. (1982) Stochasiticity in structural and functional characteristics of an Indiana stream fish assemblage: a test of community theory. *The American Naturalist*, **120**, 423–54.

Gunnarsson, B. (1983) Winter mortality of spruce–living spiders: effect of spider interactions and bird predation. *Oikos*, **40**, 226–33.

Gustafsson, L. (1987) Interspecific competition lowers fitness in Collared Flycatchers *Ficedula albicollis*: an experimental demonstration. *Ecology*, **68**, 291–6.

Gutzwiller, K.J. & Anderson, S.H. (1987) Multiscale associations between cavity-nesting birds and features of Wyoming streamside woodlands. *The Condor*, **89**, 534–48.

Haffer, J. (1969) Speciation in Amazonian forest birds. *Science*, **165**, 131–7.

(1982) General aspects of the refuge theory. In *Biological Diversification in the Tropics*, ed. Prance, G.T., pp. 6–24. New York: Columbia University Press.

(1985) Avian zoogeography of the neotropical lowlands. *Ornithological Monographs*, **36**, 113–46.

Haila, Y. (1985) Birds as a tool in reserve planning. *Ornis Fennica*, **62**, 96–100.

(1986a) On the semiotic dimension of ecological theory: the case of island biogeography. *Biology and Philosophy*, **1**, 377–87.

(1986b) North European land birds in forest fragments: evidence for area effects? In *Wildlife 2000. Modeling Habitat Relationships of Terrestrial Vertebrates*, ed. Verner, J., Morrison, M.L. & Ralph, C.J., pp. 315–9. Madison: University of Wisconsin Press.

Haila, Y. & Hanski, I.K. (1984) Methodology for studying the effect of habitat fragmentation on land birds. *Annales Zoologici Fennici*, **21**, 393–7.

Haila, Y. & Järvinen, O. (1981) The underexploited potential of bird censuses in insular ecology. *Studies in Avian Biology*, **6**, 559–65.

(1982) The role of theoretical concepts in understanding the ecological theater: a case study on island biogeography. In *Conceptual Issues in Ecology*, ed. Saarinen, E., pp. 261–78. Boston: D. Reidel Publishing Company.

Haila, Y., Hanski, I.K. & Raivio, S. (1987) Breeding bird distribution in fragmented coniferous taiga in southern Finland. *Ornis Fennica*, **64**, 90–106.

Haila, Y., Järvinen, O., & Kuusela, S. (1983) Colonization of islands by land birds: prevalence functions in a Finnish archipelago. *Journal of Biogeography*, **10**, 499–531.

Haila, Y., Järvinen, O. & Väisänen, R.A. (1979) Long-term changes in the bird community of farmland in Åland, SW Finland. *Annales Zoologici Fennici*, **16**, 23–7.

Hairston, N.G. (1985) The interpretation of experiments on interspecific competition. *The American Naturalist*, **125**, 321–5.

Haldane, J.B.S. (1937) *Adventures of a Biologist*. New York: Harper and Brothers.

Hall, G.A. (1984a) A long-term bird population study in an Appalachian spruce forest. *The Wilson Bulletin*, **96**, 228–40.

(1984b) Population decline of neotropical migrants in an Appalachian forest. *American Birds,* **38**, 14–8.

Hallett, J.G. (1982) Habitat selection and the community matrix of a desert small-mammal fauna. *Ecology,* **63**, 1400–10.

Hallett, J.G. & Pimm, S.L. (1979) Direct estimation of competition. *The American Naturalist,* **113**, 593–600.

Hallett, J.G. O'Connell, M.A. & Honeycutt, R.L. (1983) Competition and habitat selection: a test of a theory using small mammals. *Oikos*, **40**, 175–81.

Hamel, P.B., Cost, N.D. & Sheffield, R.M. (1986) The consistent characteristics of habitats: a question of scale. In *Wildlife 2000. Modeling Habitat Relationships of Terrestrial Vertebrates*, ed. Verner, J., Morrison, M.L. & Ralph, C.J., pp. 121–8. Madison: The University of Wisconsin Press.

Hansson, L. (1979) On the importance of landscape heterogeneity in northern regions for the breeding population densities of homeotherms: a general hypothesis. *Oikos*, **33**, 182–9.

Harner, E.J. & Whitmore, R.C. (1977) Multivariate measures of niche overlap using discriminant analysis. *Theoretical Population Biology,* **12**, 21–36.

Harris, L.D. (1984) *The Fragmented Forest. Island Biogeography Theory and the Preservation of Biotic Diversity*. Chicago: The University of Chicago Press.

Hastings, A. (1980) Population dynamics in patchy environments. In *Modeling and Differential Equations in Biology*, ed. Burton, T.A., pp. 217–23. New York: Marcel Dekker, Inc.

(1987) Can competition be detected using species co-occurrence data? *Ecology,* **68**, 117–23.

Heinemann, D. (1984) 'Interactions among Rufous Hummingbirds, Hymenopterans, and a Shared Resource: Exploitative Exclusion of a Vertebrate from a Nectar Source, *Scrophularia montana*, by Insects.' Ph.D. Dissertation. Albuquerque: University of New Mexico.

Helle, P. & Järvinen, O. (1986) Population trends of north Finnish land birds in relation to their habitat selection and changes in forest structure. *Oikos*, **46**, 107–15.

Hempel, C.G. (1977) Formulation and formalization of scientific theories: a summary-abstract. In *The Structure of Scientific Theories*, ed. Suppe, F., pp. 244–54. Urbana: University of Illinois Press.

Hengeveld, R. (1987) Scales of variation: their distinction and ecological importance. *Annales Zoologici Fennici,* **24**, 195–202.

Herrera, C.M. (1985) Determinants of plant–animal coevolution: the case of mutualistic dispersal of seeds by vertebrates. *Oikos,* **44**, 132–41.

Herrera, C.M. & Hiraldo, F. (1976) Food-niche and tropic relationships among European owls. *Ornis Scandinavica,* **7**, 29–41.

Higgs, A.J. (1981) Island biogeography theory and nature reserve design. *Journal of Biogeography,* **8**, 117–24.

Higuchi, H., Tsukamoto, Y., Hanawa, S., & Takeda, M. (1982) Relationship between forest areas and the number of bird species. *Strix,* **1**, 70–8.

Hildén, O., Järvinen, A., Lehtonen, L. & Soikkeli, M. (1982) Breeding success of Finnish birds in the bad summer of 1981. *Ornis Fennica,* **59**, 20–31.

Hixon, M.A., Carpenter, F.L. & Paton, D.C. (1983) Territory area, flower density, and time budgeting in hummingbirds: an experimental and theoretical analysis. *The American Naturalist,* **122**, 366–91.

Hogstad, O. (1971) Stratification in winter feeding of the Great Spotted Woodpecker *Dendrocopos major* and the Three-toed Woodpecker *Picoides tridactylus. Ornis Scandinavica,* **2**, 143–6.

(1975) Quantitative relations between hole-nesting and open-nesting species within a passerine breeding community. *Norwegian Journal of Zoology*, **23**, 261–7.
Högstedt, G. (1980a) Resource partitioning in Magpie *Pica pica* and Jackdaw *Corvus monedula* during the breeding season. *Ornis Scandinavica*, **11**, 110–5.
(1980b) Prediction and test of the effects of interspecific competition. *Nature*, **283**, 64–6.
Holling, C.S. (1973) Resilience and stability of ecological systems. *Annual Review of Ecology and Systematics*, **4**, 1–23.
Holmes, J.C. & Price, P.W. (1986) Communities of parasites. In *Community Ecology. Pattern and Process*, ed. Kikkawa, J. & Anderson, D.J., pp. 187–213. Oxford: Blackwell Scientific Publications.
Holmes, R.T. & Pitelka, F.A. (1968) Food overlap among coexisting sandpipers on northern Alaskan tundra. *Systematic Zoology*, **17**, 305–18.
Holmes, R.T. & Sturges, F.W. (1975) Bird community dynamics and energetics in a northern hardwoods ecosystem. *Journal of Animal Ecology*, **44**, 175–200.
Holmes, R.T., Schultz, J.C. & Nothnagle, P. (1979) Bird predation on forest insects: an exclosure experiment. *Science*, **206**, 462–3.
Holmes, R.T., Sherry, T.W. & Sturges, F.W. (1986) Bird community dynamics in a temperate deciduous forest: long-term trends at Hubbard Brook. *Ecological Monographs*, **56**, 201–20.
Holt, R.D. (1977) Predation, apparent competition, and the structure of prey communities. *Theoretical Population Biology*, **12**, 197–229.
(1984) Spatial heterogeneity, indirect interactions, and the coexistence of prey species. *The American Naturalist*, **124**, 377–406.
Howe, H.F. (1983) Annual variation in a neotropical seed-dispersal system. In *Tropical Rain Forest: Ecology and Management*, ed. Sutton, S.L., Whitmore, T.C. & Chadwick, A.C., pp. 211–27. Oxford: Blackwell Scientific Publications.
Howe, R.W. (1984) Local dynamics of bird assemblages in small forest habitat islands in Australia and North America. *Ecology*, **56**, 1585–601.
Humphreys, W.F. & Kitchener, D.J. (1982) The effect of habitat utilization on species-area curves: implications for optimal reserve area. *Journal of Biogeography*, **9**, 391–6.
Hunter, M.L., Jones, J.J., Gibbs, K.E. & Moring, J.R. (1986) Duckling responses to lake acidification: do Black Ducks and fish compete? *Oikos*, **47**, 26–32.
Hutchinson, G.E. (1957) Concluding Remarks. *Cold Spring Harbor Symposia on Quanitative Biology*, **22**, 415–27.
(1961) The paradox of the plankton. *American Naturalist*, **95**, 145–59.
Hutto, R.L. (1985) Habitat selection by nonbreeding, migratory land birds. In *Habitat Selection in Birds*, ed. Cody, M.L., pp. 455–76. New York: Academic Press.
Jaksić, F.M. (1981) Abuse and misuse of the term "guild" in ecological studies. *Oikos*, **37**, 397–400.
James, F.C. & Boecklen, W.J. (1984) Interspecific morphological relationships and the densities of birds. In *Ecological Communities. Conceptual Issue and the Evidence*, ed. Strong, D.R.Jr. Simberloff, D., Abele, L.G. & Thistle, A.B., pp. 458–77. Princeton: Princeton University Press.
James, F.C. & McCulloch, C.E. (1985) Data analysis and the design of experiments in ornithology. *Current Ornithology*, **2**, 1–63.
James, F.C., Johnston, R.F., Wamer, N.O., Niemi, G.J., & Boecklen, W.J. (1984) The Grinnellian niche of the Wood Thrush. *The American Naturalist*, **124**, 17–30.
Jansson, C., Ekman, J. & von Bromssen, A. (1981) Winter mortality and food supply in tits *Parus* spp. *Oikos*, **37**, 313–22.
Janzen, D.H. (1983) No park is an island: increase in interference from outside as park size decreases. *Oikos*, **41**, 402–10.

Järvinen, A. (1983) Breeding strategies of hole-nesting passerines in northern Lapland. *Annales Zoologici Fennici,* **20**, 129–49.
(1985) Predation causing extended low densities in microtine cycles: implications from predation on hole-nesting passerines. *Oikos,* **45**, 157–8.
Järvinen, A. & Väisänen, R.A. (1984) Reproduction of Pied Flycatchers (*Ficedula hypoleuca*) in good and bad breeding seasons in a northern marginal area. *The Auk,* **101**, 439–50.
Järvinen, O. (1978) Are northern bird communities saturated? *Anser,* **3**, 112–16.
(1979) Geographical gradients of stability in European land bird communities. *Oecologia,* **38**, 51–69.
(1980a) Dynamics of North European bird communities. In *Acta XVII Congressus Internationalis Ornithologici,* ed. Nohring, R., pp. 770–6. Berlin: Deutsche Ornithologen-Gesellschaft.
(1980b) 'Ecological Zoogeography of northern European bird communities' Dissertation, Helsinki: University of Helsinki.
Järvinen, O. & Haila, Y. (1984) Assembly of land bird communities on northern islands: a quantitative analysis of insular impoverishment. In *Ecological Communities. Conceptual Issues and the Evidence,* ed. Strong, D.R.Jr., Simberloff, D., Abele, L.G. & Thistle, A.B., pp. 138–47. Princeton: Princeton University Press.
Järvinen, O. & Ulfstrand, S. (1980) Species turnover of a continental bird fauna: Northern Europe, 1850–1970. *Oecologia,* **46**, 186–95.
Järvinen, O. & Väisänen, R.A. (1977a) Recent quantitative changes in the populations of Finnish land birds. *Polish Ecological Studies,* **3**, 177–88.
(1977b) Long-term changes of the North European land bird fauna. *Oikos,* **29**, 225–8.
(1978a) Recent changes in forest bird populations in northern Finland. *Annales Zoologici Fennici,* **15**, 279–89.
(1978b) Long-term population changes of the most abundant south Finnish forest birds during the past 50 years. *Journal für Ornithologie,* **119**, 441–9.
(1979) Climatic changes, habitat changes, and competition: dynamics of geographical overlap in two pairs of congeneric bird species in Finland. *Oikos,* **33**, 261–71.
Järvinen, O., Kuusela, K. & Väisänen, R.A. (1977) Metsien rakenteen muutoksen vaikutus pesimalinnustoomme viimeisten 30 vuoden aikana. *Silva Fennica,* **11**, 284–94.
Jedraszko-Dabrowska, D. (1979) Rotation of individuals in breeding populations of dominant species of birds in a pine forest. *Ecologia Polska,* **27**, 545–69.
Jehl, J.R. Jr & Parkes, K.C. (1983) 'Replacements' of landbird species on Socorro Island, Mexico. *The Auk,* **100**, 551–9.
Jenkins, D., Watson, A. & Miller, G.R. (1967) Population fluctuations in the Red Grouse *Lagopus lagopus scoticus. Journal of Animal Ecology,* **36**, 97–122.
Joern, A. (1986) Experimental study of avian predation on coexisting grasshopper populations (Orthoptera: Acrididae) in sandhills grassland. *Oikos,* **46**, 243–9.
Johnson, W.C. & Adkisson, C.S. (1985) Dispersal of beech nuts by Blue Jays in fragmented landscapes.*The American Midland Naturalist,* **113**, 319–24.
Jones, H.L. & Diamond, J.M. (1976) Short-time-base studies of turnover in breeding bird populations on the California Channel Islands. *The Condor,* **78**, 526–49.
Jones, P.J. & Ward, P. (1979) A physiological basis for colony desertion by Red-billed Queleas (*Quelea quelea*). *Journal of Zoology,* **189**, 1–19.
Judson, H.F. (1980) *The Search for Solutions.* New York: Holt, Rinehart and Winston.
Kalela, O. (1938) Über die regionale Verteilung der Brutvogelfauna im Flussegebiet des Kokemäenjoki. *Ann. Zool. Soc. Zool. Bot. Fenn. Vanamo,* **5(9)**, 1–15, 1–297.
(1949) Changes in geographic ranges in the avifauna of northern and central Europe in relation to recent changes in climate. *Bird-Banding,* **20**, 77–103.

Källander, H. (1981) The effects of provision of food in winter on a population of the Great Tit *Parus major* and the Blue Tit *P. caeruleus*. *Ornis Scandinavica*, **12**, 244–8.
Kangas, P. (1987) On the use of species area curves to predict extinctions. *Bulletin of the Ecological Society of America*, **68**, 158–62.
Kareiva, P. (1986) Patchiness, dispersal, and species interactions: consequences for communities of herbivorous insects. In *Community Ecology*, ed. Diamond, J. & Case, T.J., pp. 192–206. New York: Harper & Row.
Karr, J.R. (1982a) Avian extinction on Barro Colorado Island, Panama: a reassessment. *The American Naturalist*, **119**, 220–39.
(1982b) Population variability and extinction in the avifauna of tropical land bridge island. *Ecology*, **63**, 1975–8.
Karr, J.R. & Freemark, K.E. (1983) Habitat selection and environmental gradients: dynamics in the 'stable' tropics. *Ecology*, **64**, 1481–94.
(1985) Disturbance and vertebrates: an integrative perspective. In *The Ecology of Natural Disturbance and Patch Dynamics*, ed. Pickett, S.T.A. & White, P.S., pp. 153–68. New York: Academic Press.
Keast, A. (1980) Spatial relationships between migratory parulid warblers and their ecological counterparts in the neotropics. In *Migrant Birds in the Neotropics. Ecology, Behavior, Distribution, and Conservation*, ed. Keast, A. & Morton, E.S., pp. 109–30. Washington, D.C: Smithsonian Institition Press.
Kendeigh, S.C. (1979) Invertebrate populations of the deciduous forest: fluctuations and relations to weather. *Illinois Biological Monographs*, **50**, 1–107.
(1982) Bird populations in east central Illinois: fluctuations, variations, and development over a half-century. *Illinois Biological Monographs*, **52**, 1–136.
Kendeigh, S.C., Dol'nik, V.R. & Gavrilov, V.M. (1977) Avian energetics. In *Granivorous Birds in Ecosystems*, ed. Pinowski, J. & Kendeigh, S.C., pp. 127–204. Cambridge: Cambridge University Press.
Kerr, R.A. (1984) The moon influences western US drought. *Science*, **224**, 587.
Kikkawa, J. (1977) Ecological paradoxes. *Australian Journal of Ecology*, **2**, 121–36.
Kingsland, S.E. (1985) *Modeling Nature. Episodes in the History of Population Ecology*. Chicago: University of Chicago Press.
Kipling, R. (1902) *Just So Stories for Little Children*. London: Macmillan.
Kitchener, D.J., Dell, J. & Muir, B.G. (1982) Birds in western Australian wheatbelt reserves – implications for conservation. *Biological Conservation*, **22**, 127–63.
Klomp, H. (1980) Fluctuations and stability in Great Tit populations. *Ardea*, **68**, 205–24.
Klopfer, P.H. & Ganzhorn, J.U. (1985) Habitat selection: behavioral aspects. In *Habitat Selection in Birds*, ed. Cody, M.L., pp. 435–53. New York: Academic Press.
Knopf, F.L. & Sedgwick, J.A. (1987) Latent population responses of summer birds to a catastrophic, climatological event. *The Condor*, **89**, 869–73.
Kobayashi, S. (1985) Species diversity preserved in different numbers of nature reserves of the same total area. *Researches on Population Ecology*, **27**, 137–43.
Kodric-Brown, A. & Brown, J.H. (1978) Influence of economics, interspecific competition, and sexual dimorphism on territoriality of migrant Rufous Hummingbirds. *Ecology*, **59**, 285–96.
(1979) Competition between distantly related taxa in the coevolution of plants and pollinators. *The American Zoologist*, **19**, 1115–27.
Kodric-Brown, A., Brown, J.H., Byers, G.A. & Gori, D.F. (1984) Organization of a tropical island community of hummingbirds and flowers. *Ecology*, **65**, 1358–68.
Korpimäki, E. (1985) Rapid tracking of microtine populations by their avian predators: possible evidence for stabilizing predation. *Oikos*, **45**, 281–4.
Krebs, C., Keller, B., & Tamarin, R. (1969) *Microtus* population biology. *Ecology*, **50**, 587–607.

Krebs, J.R. (1971) Territory and breeding density in the Great Tit, *Parus major* L. *Ecology,* **52**, 2–22.

Krebs, J.R. & Perrins, C. (1978) Behavior and population regulation in the Great Tit (*Parus major*). In *Population control by social behaviour*, ed. Ebling, F. J. & Stoddart, D.M., pp. 23–47. London: Institute of Biology.

Kricher, J.C. (1975) Diversity in two wintering bird communities: possible weather effects. *The Auk,* **92**, 766–77.

Lack, D. (1945) The Galápagos finches (Geospizinae): a study in variation. *Occasional Papers of the California Academy of Sciences,* **21**, 1–159.

(1954) *The Natural Regulation of Animal Numbers.* Oxford: Oxford University Press.

(1966) *Population Studies of Birds.* Oxford: Oxford University Press.

(1971) *Ecological Isolation in Birds.* Cambridge: Harvard University Press.

(1976) *Island Biology Illustrated by the Land Birds of Jamaica.* Oxford: Blackwell Scientific Publications.

Lahti, T. & Ranta, E. (1985) The SLOSS principle and conservation practice: an example. *Oikos,* **44**, 369–70.

Laird, M. & van Riper, C. III. (1981) Questionable reports of *Plasmodium* from birds in Hawaii, with recognition of *P. relictum* spp. *capistranoae* (Russell, 1932) as the avian malaria parasite there. *Society of Parasitology Special Publication,* **1**, 159–65.

Lance, A.N. (1978) Territories and the food plant of individual Red Grouse. *Journal of Animal Ecology,* **47**, 307–13.

Landres, P.B. & MacMahon, J.A. (1983) Community organization of arboreal birds in some oak woodlands of western North America. *Ecological Monographs,* **53**, 183–208.

Laurance, W.F. & Yensen, E. (1985) Rainfall and winter sparrow densites: a view from the northern Great Basin. *The Auk,* **102**, 152–8.

Laurent, J.L. (1986) Winter foraging behavior and resource availability for a guild of insectivorous gleaning birds in a southern alpine larch forest. *Ornis Scandinavia,* **17**, 347–55.

Laverty, T.M. & Plowright, R.C. (1985) Competition between hummingbirds and bumble bees for nectar in flowers of *Impatiens biflora. Oecologia,* **66**, 25–32.

Lawton, J.H. & Hassell, M.P. (1981) Asymmetrical competition in insects. *Nature,* **289**, 793–5.

Leisler, B. & Thaler, E. (1982) Differences in morphology and foraging behaviour in the goldcrest *Regulus regulus* and firecrest *R. ignicapillus. Annales Zoologici Fennici,* **19**, 277–84.

Levenson, J.B. (1981) Woodlots as biogeographic islands in southeastern Wisconsin. In *Forest Island Dynamics in Man-dominated Landscapes*, ed. Burgess, R.L. & Sharpe, D.M., pp. 13–39. New York: Springer-Verlag.

Levin, S.A. (1974) Dispersion and population interactions. *The American Naturalist,* **108**, 207–28.

(1981) The role of theoretical ecology in the description and understanding of populations in heterogeneous environments. *The America Zoologist,* **21**, 865–75.

Levin, S.A., Cohen, D. & Hastings, A. (1984) Dispersal strategies in patchy environments. *Theoretical Population Biology,* **26**, 165–91.

Levins, R. (1966) The strategy of model building in population biology. *The American Scientist*, **54**, 421–31.

Levins, R. & Lewontin, R. (1982) Dialectics and reductionism in ecology. In *Conceptual Issues in Ecology*, ed. Saarinen, E., pp. 107–38. Boston: D. Reidel Publishing Company.

Lidicker, W.Z. Jr. (1975) The role of dispersal in the demography of small mammals. In *Small Mammals: Their Productivity and Population Dynamics*, ed. Golley, F.B.,

Petrusewicz, K. & Ryszkowski, L., pp. 103–28. Cambridge: Cambridge University Press.

Llewellyn, J.B. & Jenkins, S.H. (1987) Patterns of niche shift in mice: seasonal changes in microhabitat breadth and overlap. *The American Naturalist*, **129**, 365–81.

Loiselle, B.A. & Hoppes, W.G. (1983) Nest predation in insular and mainland lowland rainforest in Panama. *The Condor*, **85**, 93–5.

Loman, J. (1980) Habitat distribution and feeding strategies of four south Swedish corvid species during winter. *Ekologia Polska*, **28**, 95–109.

Lovejoy, T.E. & Oren, D. (1981) The minimum critical size of ecosystems. In *Forest Island Dynamics in Man-dominated Landscapes*, ed. Burgess, R.L. & Sharpe, D.M., pp. 7–13. New York: Springer–Verlag.

Lovejoy, T.E., Bierregaard, R.O., Rankin, J.M. & Schubart, H.O.R. (1983) Ecological dynamics of tropical forest fragments. In *Tropical Rainforest: Ecology and Management*, ed. Sutton, S.L., Whitmore, T.C. & Chadwick, A.C., pp. 377–84. Oxford: Blackwell Scientific Publishers.

Lovejoy, T.E., Bierregaard, R.O., Rylands, A.B., Malcolm, J.R., Quintela, C.E., Harper, L.H., Brown, K.S. Jr., Powell, A.H., Powell, G.V.N., Schubart, H.O.R. & Hays, M.B. (1986) Edge and other effects of isolation on Amazon forest fragments. In *Conservation Biology: the Science of Scarcity and Diversity*, ed. Soulé, M.E., pp. 257–85. Sunderland, Massachusetts: Sinauer Associates.

Loyn, R.H. (1985) Birds in fragmented forests in Gippsland, Victoria. In *Birds of Eucalypt Forests and Woodlands*, ed. Keast, A., Recher, H.F., Ford, H. & Saunders, D., pp. 323–31. Chipping Norton, NSW: Surrey Beatty & Sons.

(1987) Effects of patch area and habitat on bird abundances, species numbers and tree health in fragmented Victorian forests. In *Nature Conservation: the Role of Remnants of Native Vegetation*, ed. Saunders, D.A., Arnold, G.W., Burbidge, A.A. & Hopkins, A.J.M., pp. 65–77. Chipping Norton, NSW: Surrey Beatty & Sons.

Loyn, R.H., Runnalls, R.G. & Forward, G.Y. (1983) Territorial Bell Miners and other birds affecting populations of insect prey. *Science*, **221**, 1411–13.

Lynch, J.F. (1987) Responses of breeding bird communities to forest fragmentation. In *Nature Conservation: the Role of Remnants of Native Vegetation*, ed. Saunders, D.A., Arnold, G.W., Burbidge, A.A. & Hopkins, A.J.M., pp. 123–40. Chipping Norton, NSW: Surrey Beatty & Sons.

Lynch, J.F. & Johnson, N.K. (1974) Turnover and equilibria in insular avifaunas, with special reference to the California Channel Islands. *The Condor*, **76**, 370–84.

Lynch, J.F. & Whigham, D.F. (1984) Effects of forest fragmentation on breeding bird communities in Maryland, USA *Biological Conservation*, **28**, 287–324.

Lynch, J.F. & Whitcomb, R.F. (1978) Effects of the insularization of the eastern deciduous forest on avifaunal diversity and turnover. In *Classification, inventory, and analysis of fish and wildlife habitat. FWS/OBS-78/76*, ed. Marmelstein, A., pp. 461–89. Washington, DC: US Dept. Interior, Fish and Wildlife Service, Office of Biological Services.

Lyon, D.L. & Chadek, C. (1971) Exploitation of nectar resources by hummingbirds, bees (*Bombus*) and *Diglossa baritula* and its role in the evolution of *Penstamon kunthii*. *The Condor*, **73**, 246–8.

Lyon, D.L., Crandall, J. & McKone, M. (1977) A test of the adaptiveness of interspecific territoriality in the Blue-throated Hummingbird. *The Auk*, **94**, 448–54.

MacArthur, J.W. (1975) Environmental fluctuations and species diversity. In *Ecology and Evolution of Communities*, ed. Cody, M.L. & Diamond, J.M., pp. 74–80. Cambridge: Harvard University Press.

MacArthur, R.H. (1958) Population ecology of some warblers in northeastern coniferous forests. *Ecology*, **39**, 599–619.

(1972) *Geographical Ecology*. New York: Harper & Row.
MacArthur, R.H. & Wilson, E.O. (1967) *The Theory of Island Biogeography*. Princeton: Princeton University Press.
MacClintock, L., Whitcomb, R.F. & Whitcomb, B.L. (1977) Island biogeography and 'habitat islands' of eastern forest. Part II. Evidence for the value of corridors and minimization of isolation in preservation of biotic diversity. *American Birds*, **31**, 6–16.
Macfadyen, A. (1975) Some thoughts on the behaviour of ecologists. *Journal of Animal Ecology*, **44**, 351–63.
MacMillen, R.E. & Carpenter, F.L. (1980) Evening roosting flights of the honeycreepers *Himatione sanguinea* and *Vestiaria coccinea* on Hawaii. *The Auk*, **97**, 28–37.
Mac Nally, R.C. (1983) On assessing the significance of interspecific competition to guild structure. *Ecology*, **64**, 1646–52.
Maher, W.J. (1979) Nestling diets of prairie passerine birds at Matador, Saskatchewan, Canada. *Ibis*, **121**, 437–52.
Mandelbrot, B. (1983) *The Fractal Geometry of Nature*. San Francisco: W.H. Freeman and Company.
Mares, M.A. & Rosenzweig, M.L. (1978) Granivory in North and South American deserts: rodents, birds, and ants. *Ecology*, **59**, 235–41.
Margules, C., Higgs, A.J. & Rafe, R.W. (1982) Modern biogeographic theory: Are there any lessons for nature reserve design? *Biological Conservation*, **24**, 115–28.
Marsh, C.P. (1986a) Impact of avian predators on high intertidal limpet populations. *Journal of Experimental Marine Biology and Ecology*, **104**, 185–201.
(1986b) Rocky intertidal community organization: the impact of avian predators on mussel recruitment. *Ecology*, **67**, 771–86.
Martin, S.G. (1971) 'Polygyny in the Bobolink: Habitat Quality and the Adaptive Complex.' Ph.D. Dissertation. Corvallis: Oregon State University.
Martin, T.E. (1981) Limitation in small habitat islands: chance or competition? *The Auk*, **98**, 715–34.
(1986) Competition in breeding birds: on the importance of considering processes at the level of the individual. *Current Ornithology*, **4**, 181–210.
(1987) Artificial nest experiments: effects of nest appearance and type of predator. *The Condor*, **89**, 925–8.
(1988a) Habitat and area effects on forest bird assemblages: is nest predation an influence? *Ecology*, **69**, 74–84.
(1988b) Processes organizing open-nesting bird assemblages: competition or nest predation? *Evolutionary Ecology*, **2**, 37–50.
Martin, T.E. & Karr, J.R. (1986) Patch utilization by migrating birds: resource oriented? *Ornis Scandinavica*, **17**, 165–74.
Maurer, B.A. (1984) Interference and exploitation in bird communities. *The Wilson Bulletin*, **96**, 380–95.
(1985) Avian community dynamics in desert grasslands: observational scale and hierarchical structure. *Ecological Monographs*, **55**, 295–312.
May, R.M. (1973) *Stability and Complexity of Model Ecosystems*. Princeton: Princeton University Press.
ed. (1976) *Theoretical Ecology. Principles and Applications*. Philadelphia: W.B. Saunders Company.
(1981a) Modeling recolonization by neotropical migrants in habitats with changing patch structure, with notes on the age structure of populations. In *Forest Island Dynamics in Man-Dominated Landscapes*, ed. Burgess, R.L. & Sharpe, D.M., pp. 207–13. New York: Springer-Verlag.
(1981b) The role of theory in ecology. *The American Zoologist*, **21**, 903–10.

May, R.M. & Robinson, S.K. (1985) Population dynamics of avian brood parasitism. *The American Naturalist*, **126**, 475–94.
Mayfield, H.F. (1978) Brood parasitism: reducing interactions between Kirtland's Warblers and Brown-headed Cowbirds. In *Endangered Birds. Management Techniques for Preserving Threatened Species*, ed. Temple, S.A., pp. 85–91. Madison: University of Wisconsin Press.
Maynard Smith, J. (1974) *Models in Ecology*. Cambridge: Cambridge University Press.
Mayr, E. (1963) *Animal Species and Evolution*. Cambridge: Harvard University Press.
Mayr, E. & O'Hara, R.J. (1986) The biogeographic evidence supporting the Pleistocene forest refuge hypothesis. *Evolution*, **40**, 55–67.
McCoy, E.D. (1982) The application of island-biogeographic theory to forest tracts: problems in the determination of turnover rates. *Biological Conservation*, **22**, 217–27.
McFarland, D.C. (1985a) Flowering biology and phenology of *Banksia integrifolia* and *B. spinulosa* (Proteaceae) in New England National Park, NSW. *Australian Journal of Botany*, **33**, 705–14.
(1985b) 'Community Organisation and Territorial Behaviour of Honeyeaters in an Unpredictable Environment.' Ph.D. thesis. Armidale, NSW: University of New England.
(1986a) The organization of honeyeater community in an unpredictable environment. *Australian Journal of Ecology*, **11**, 107–20.
(1986b) Seasonal changes in the abundance and body condition of honeyeaters (Meliphagidae) in response to infloresence and nectar availability in the New England National Park, New South Wales. *Australian Journal of Ecology*, **11**, 331–40.
(1986c) Determinants of feeding territory size in the New Holland Honeyeater *Phylidonyris novaehollandiae*. *The Emu*, **86**, 180–5.
McIntosh, R.P. (1985) *The Background of Ecology. Concept and Theory*. Cambridge: Cambridge University Press.
McLellan, C.H., Dobson, A.P., Wilcove, D.S. & Lynch, J.F. (1986) Effects of forest fragmentation on New-and Old-World bird communities: empirical observations and theoretical implications. In *Wildlife 2000. Modeling Habitat Relationships of Terrestrial Vertebrates*, ed. Verner, J., Morrison, M.L. & Ralph, C.J., pp. 305–13. Madison: University of Wisconsin Press.
McNaughton, S.J. & Wolf, L.L. (1979) *General Ecology*, 2nd edn. New York: Holt, Rinehart and Winston.
Merikallio, E. (1929) *Äyräpäänjärvi*. Helsinki: Otava.
Merton, D.V. (1975) Success in reestablishing a threatened species: the Saddleback – its status and conservation. *Bulletin of the International Council for Bird Preservation*, **12**, 150–8.
Milinski, M. (1984) Parasites determine a predator's optimal feeding strategy. *Behavioural Ecology and Sociobiology*, **15**, 35–7.
(1985) Risk of predation of parasitized sticklebacks (*Gasterosteus aculeatus* L.) under competition for food. *Behaviour*, **93**, 203–16.
Miller, G.R., Jenkins, D. & Watson, A. (1966) Heather performance and Red Grouse populations. I. Visual estimates of heather performance. *Journal of Applied Ecology*, **3**, 313–26.
Miller, G.R., Watson, A. & Jenkins, D. (1970) Responses of Red Grouse populations to experimental improvement of their food. In *Animal Populations in Relation to their Food Resources*, ed. Watson, A., pp. 323–35. Oxford: Blackwell Scientific Publishers.
Miller, R.S. (1967) Pattern and process in competition. *Advances in Ecological Research*, **4**, 1–74.

(1968) Conditions of competition between redwings and yellowheaded blackbirds. *Journal of Animal Ecology*, **37**, 43–61.
(1969) Competition and species diversity. *Brookhaven Symposia in Biology*, **22**, 63–70.
Miller, R.S. & Nero, R.W. (1983) Hummingbird-sapsucker associations in northern climates. *Canadian Journal of Zoology*, **61**, 1540–6.
Milne, A. (1961) Definition of competition among animals. In *Mechanisms in Biological Competition. Symposia of the Society for Experimental Biology Number 15*, pp. 40–61. Cambridge: Cambridge University Press.
Minot, E.O. (1981) Effects of interspecific competition for food in breeding Blue and Great tits. *Journal of Animal Ecology*, **50**, 375–85.
Minot, E.O. & Perrins, C.M. (1986) Interspecific interference competition – nest sites for Blue and Great tits. *Journal of Animal Ecology*, **55**, 331–50.
Møller, A.P. (1987) Egg predation as a selective factor for nest design: an experiment. *Oikos*, **50**, 91–4.
Moore, J. (1984) Parasites and altered host behavior. *Scientific American*, **250**, 108–15.
Moore, N.W. & Hooper, M.D. (1975) On the number of bird species in British woods. *Biological Conservation*, **8**, 239–50.
Moreau, R.E. (1966) *The Bird Faunas of Africa and its Islands*. New York: Academic Press.
Morris, D.W. (1987) Ecological scale and habitat use. *Ecology*, **68**, 362–9.
Morse, D.H. (1975) Ecological aspects of adaptive radiation in birds. *Biological Review*, **50**, 167–214.
(1977) The occupation of small islands by passerine birds. *The Condor*, **79**, 399–412.
(1980) Foraging and coexistence of spruce-woods warblers. *The Living Bird*, **18**, 7–25.
Morton, S.R. (1982) Dasyurid marsupials of the Australian arid zone: an ecological review. In *Carnivorous marsupials*, ed. Archer, M., pp. 117–30. Sydney: Royal Zoological Society of New South Wales.
(1985) Granivory in arid regions: comparison of Australia with North and South America. *Ecology*, **66**, 1859–66.
Moss, R. (1969) A comparison of Red Grouse (*Lagopus lagopus scoticus*) stocks with the production and nutritive value of heather (*Calluna vulgaris*). *Journal of Animal Ecology*, **38**, 103–12.
Moulton, M.P. (1985) Morphological similarity and coexistence of congeners: an experimental test with introduced Hawaiian birds. *Oikos*, **44**, 301–5.
Moulton, M.P. & Pimm, S.L. (1983) The introduced Hawaiian avifauna: biogeographic evidence for competition. *The American Naturalist*, **121**, 669–90.
(1986) The extent of competition in shaping an introduced avifauna, In *Community Ecology*, ed. Diamond, J. & Case, T.J., pp. 80–97. New York: Harper & Row.
Mountainspring, S. (1986) An ecological model of the effects of exotic factors in limiting Hawaiian honeycreeper populations. *Ohio Journal of Science*, **86**, 95–100.
Mountainspring, S. & Scott, J.M. (1985) Interspecific competition among Hawaiian forest birds. *Ecological Monographs*, **55**, 219–39.
Mueller–Dombois, D., Bridges, K.W. & Carson, H.L., eds. (1981) *Island Ecosystems: Biological Organization in Selected Hawaiian Communities*. Stroudsburg, Pennsylvania: Hutchinson Ross.
Murdoch, W.W. (1969) Switching in general predators: experiments on predator specificity and stability of prey populations. *Ecological Monographs*, **39**, 335–354.
Murphy, D.D. & Wilcox, B.A. (1986) On island biogeography and conservations. *Oikos*, **47**, 385–7.
Murray, B.G. Jr. (1971) The ecological consequences of interspecific territorial behavior in birds. *Ecology*, **52**, 414–23.

(1986) The structure of theory, and the role of competition in community dynamics. *Oikos*, **46**, 145–58.

Murray, K.G., Feinsinger, P., Busby, W.H., Linhart, Y.B., Beach, J.H. & Kinsman, S. (1987) Evaluation of character displacement among plants in two tropical pollination guilds. *Ecology*, **68**, 1283–93.

Murton, R.K., Bucher, E.H., Nores, M., Gomes, E. & Reartes, J. (1974) The ecology of the Eared Dove (*Zenaida auriculata*) in Argentina. *The Condor*, **76**, 80–8.

Murton, R.K., Isaacson, A.J. & Westwood, N.J. (1966) The relationship between Woodpigeons and their clover food supply and the mechanism of population control. *Journal of Applied Ecology*, **3**, 55–96.

Murton, R.K., Westwood, N.J. & Isaacson, A.J. (1964) A preliminary investigation of the factors regulating population size in the Woodpigeon. *Ibis*, **106**, 482–507.

Nelson, G. & Platnick, N. (1981) *Systematics and Biogeography. Cladistics and Vicariance*. New York: Columbia University Press.

Newton, I. (1967) The adaptive radiation and feeding ecology of some British finches. *The Ibis*, **109**, 33–98.

(1972) *Finches*. London: Collins.

(1980) The role of food in limiting bird numbers. *Ardea*, **68**, 11–30.

Nilsson, S.G. (1977) Density compensation and competition among birds breeding on small islands in a South Swedish lake. *Oikos*, **28**, 170–6.

(1984) The evolution of nest-site selection among hole-nesting birds: the importance of nest predation and competition. *Ornis Scandinavica*, **15**, 167–75.

(1986) Are bird communities in small biotope patches random samples from communities in large patches? *Biological Conservation*, **38**, 179–204.

Nilsson, S.G. & Nilsson, I.N. (1983) Are estimated species turnover rates on islands largely sampling errors? *The American Naturalist*, **121**, 595–7.

Nilsson, S.G., Bjorkman, C., Forslund, P. & Hoglund, J. (1985) Egg predation in forest bird communities on islands and mainland. *Oecologia*, **66**, 511–5.

Nisbet, R.M. & Gurney, W.S.C. (1982) *Modelling Fluctuating Populations*. New York: John Wiley & Sons.

Noon, B.R. (1981) The distribution of an avian guild along a temperate elevational gradient: the importance and expression of competition. *Ecological Monographs*, **51**, 105–24.

Noon, B.R., Dawson, D.K. & Kelly, J.P. (1985) A search for stability gradients in North American breeding bird communities. *The Auk*, **102**, 64–81.

Norberg, R.Å. (1978) Energy content of some spiders and insects on branches of spruce (*Picea abies*) in winter; prey of certain passerine birds. *Oikos*, **31**, 222–9.

Norberg, U.M. (1979) Morphology of the wings, legs and tail of three coniferous forest tits, the goldcrest, and the treecreeper in relation to locomotor pattern and feeding station selection. *Philosophical Transactions of the Royal Society of London (B)*, **287**, 131–65.

(1981) Flight, morphology and the ecological niche in some birds and bats. *Symposia of the Zoological Society of London*, **48**, 173–97.

Noss, R.F. (1983) A regional landscape approach to maintain diversity. *BioScience*, **33**, 700–6.

(1987) Corridors in real landscapes: a reply to Simberloff and Cox. *Conservation Biology*, **1**, 159–64.

Nudds, T.D. (1983) Niche dynamics and organization of waterfowl guilds in variable environments. *Ecology*, **64**, 319–30.

Nunney, L. (1983) Resource recovery time: just how destabilizing is it? In *Population Biology. Lecture Notes in Biomathematics 52*, ed. Freedman, H.I. & Strobeck, C., pp. 407–13.

(1985) Short-time delays in population models: a role in enhancing stability. *Ecology*, **66**, 1849–58.
Oberholser, H.C. (1974) *The Bird Life of Texas*. Austin: University of Texas Press.
O'Connor, R.J. (1985) Behavioural regulation of bird populations: a review of habitat use in relation to migration and residency. In *Behavioural Ecology: Ecological Consequences of Adaptive Behaviour*, ed. Sibley, R.M. & Smith, R.H., pp. 105–42. Oxford: Blackwell Scientific Publications.
(1986) Dynamical aspects of avian habitat use. In *Wildlife 2000. Modeling Habitat Relationships of Terrestrial Vertebrates*, ed. Verner, J., Morrison, M.L. & Ralph, C.J., pp. 235–40. Madison: University of Wisconsin Press.
Odum, E.P. (1971) *Fundamentals of Ecology*, 3rd edn. Philadelphia: W.B. Saunders Company.
Okubo, A. (1980) *Diffusion and ecological problems: mathematical models*. New York: Springer-Verlag.
Olson, S.L. & James, H.F. (1982) Fossil birds from the Hawaiian Islands: evidence for wholesale extinction by man before western contact. *Science*, **217**, 633–5.
O'Neill, R.V., DeAngelis, D.L., Waide, J.B. & Allen, T.F.H. (1986) *A Hierarchical Concept of Ecosystems*. Princeton: Princeton University Press.
Opdam, P. (1975) Inter- and intraspecific differentiation with respect to feeding ecology in two sympatric species of the genus *Accipiter*. *Ardea*, **63**, 30–54.
Opdam, P. & Schotman, A. (1987) Small woods in rural landscapes as habitat islands for woodland birds. *Acta Oecologia/Oecologia Generalis*, **8**, 269–74.
Opdam, P., Rijsdijk, G., & Hustings, F. (1985) Bird communities in small woods in an agricultural landscape: effects of area and isolation. *Biological Conservation*, **34**, 333–52.
Orbach, R. (1986) Dynamics of fractal networks. *Science*, **231**, 814–9.
Orians, G.H. & Collier, G. (1963) Competition and blackbird social systems. *Evolution*, **17**, 449–59.
Oster, G. (1981) Predicting populations. *The American Zoologist*, **21**, 831–44.
Oster, G.F. & Wilson, E.O. (1978) *Caste and Ecology in the Social Insects*. Princeton: Princeton University Press.
Paine, R.T. (1966) Food web complexity and species diversity. *The American Naturalist*, **100**, 65–75.
(1969) The *Pisaster-Tegula* interaction: prey patches, predator food preference, and intertidal community structure. *Ecology*, **50**, 950–61.
Paine, R.T. & Levin, S.A. (1981) Intertidal landscapes: disturbance and the dynamics of pattern. *Ecological Monographs*, **51**, 145–78.
Palmgren, P. (1930) Quantitative Untersuchungen über die Vogelfauna in den Wäldern Südfinnlands, mit besonderer Berücksichtingung Ålands. *Acta Zoologica Fennica*, **7**, 1–218.
(1936) Über die Vogelfauna der Binnengewasser Ålands. *Acta Zoologica Fennica*, **17**, 1–59.
Partridge, L. (1976a) Field and laboratory observations on the foraging and feeding techniques of Blue Tits (*Parus caeruleus*) and Coal Tits (*P. ater*) in relation to their habitats. *Animal Behaviour*, **24**, 534–44.
(1976b) Some aspects of the morphology of Blue Tits (*Parus caeruleus*) and Coal Tits (*Parus ater*) in relation to their behaviour. *Journal of Zoology, London*, **179**, 121–33.
Paton, D.C. (1979) 'The behaviour and feeding ecology of the New Holland Honeyeater, *Phylidonyrus novaehollandiae* in Victoria.' Ph.D. thesis. Melbourne: Monash University.
(1980) The importance of manna, honeydew and lerp in the diet of honeyeaters. *The Emu*, **80**, 213–26.

(1985) Food supply, population structure, and behaviour of New Holland Honeyeaters *Phylodonyris novaehollandiae* in woodland near Horsham, Victoria. In *Birds of Eucalypt Forests and Woodlands: Ecology, Conservation, Management*, ed. Keast, A., Recher, H.F., Ford, H. & Saunders, D., pp. 219–30. Chipping Norton, NSW: Surrey Beatty & Sons.

Paton, D.C. & Carpenter, F.L. (1984) Peripheral foraging by territorial Rufous Hummingbirds: defense by exploitation. *Ecology*, **65**, 1808–19.

Payne, R.B. & Groschupf, K.D. (1984) Sexual selection and interspecific competition: a field experiment on territorial behavior of nonparental finches (*Vidua* spp.). *The Auk*, **101**, 140–5.

Pearson, D.L. (1975) The relation of foliage complexity to ecological diversity of three Amazonian bird communities. *The Condor*, **77**, 453–66.

(1977) A pantropical comparison of bird community structure on six lowland forest sites. *The Condor*, **79**, 232–44.

(1986) Community structure and species co-occurrence: a basis for developing broader generalizations. *Oikos*, **46**, 419–23.

Perrins, C.M. (1979) *British Tits*. London: Collins.

Perrins, C.M. & Geer, T.A. (1980) The effect of Sparrowhawks on tit populations. *Ardea*, **68**, 133–42.

Peters, R.H. (1983) *The Ecological Implications of Body Size*. Cambridge: Cambridge University Press.

Pianka, E.R. (1966) Latitudinal gradients in species diversity: a review of concepts. *The American Naturalist*, **100**, 33–46.

Pickett, S.T.A. & White, P.S. (1985) Patch dynamics: a synthesis. In *The Ecology of Natural Disturbance and Patch Dynamics*, ed. Pickett, S.T.A. & White, P.S., pp. 371–84. New York: Academic Press.

Picman, J. (1977) Destruction of eggs by the Long-billed Marsh Wren (*Telmatodytes palustris palustris*). *Canadian Journal of Zoology*, **55**, 1914–20.

(1980) Impact of marsh wrens on reproductive strategy of Red-winged Blackbirds. *Canadian Journal of Zoology*, **58**, 337–50.

(1984) Experimental study on the role of intra- and inter-specific competition in the evolution of nest-destroying behavior in marsh wrens. *Canadian Journal of Zoology*, **62**, 2352–6.

Pielou, E.C. (1972) 2k contingency tables in ecology. *Journal of Theoretical Biology*, **34**, 337–52.

(1974) *Population and Community Ecology: Principles and Methods*. New York: Gordon and Breach Science Publishers.

(1977) *Mathematical Ecology*, New York: John Wiley & Sons.

(1981) The usefulness of ecological models: a stock-taking. *The Quarterly Review of Biology*, **56**, 17–31.

Pierce, G.J. & Ollason, J.G. (1987) Eight reasons why optimal foraging theory is a complete waste of time. *Oikos*, **49**, 111–8.

Piersma, T. (1987) Production by intertidal benthic animals and limits to their predation by shorebirds: a heuristic model. *Marine Ecology Progress Series*, **38**, 187–96.

Pimm, S.L. (1978) An experimental approach to the effects of predictability on community structure. *The American Zoologist*, **18**, 797–808.

(1984) Food chains and return times. In *Ecological Communities. Conceptual Issues and the Evidence*, ed. Strong, D.R.Jr, Simberloff, D., Abele, L.G. & Thistle, A.B., pp. 397–412. Princeton: Princeton University Press.

(1986) Community stability and structure. In *Conservation Biology. The Science of Scarcity and Diversity*, ed. Soulé, M.E., pp. 309–29. Sunderland, Massachusetts: Sinauer Associates.

Pimm, S.L. & Pimm, J.W. (1982) Resource use, competition, and resource availability in Hawaiian honeycreepers. *Ecology*, **63**, 1468–80.
Pimm, S.L. & Rosenzweig, M.L. (1981) Competitors and habitat use. *Oikos*, **37**, 1–6.
Pimm, S.L., Rosenzweig, M.L., & Mitchell, W. (1985) Competition and food selection: field tests of a theory. *Ecology*, **66**, 798–807.
Platnick, N.I. & Nelson, G. (1978). A method of analysis for historical biogeography. *Systematic Zoology*, **27**, 1–16.
Pontin, A.J. (1982) *Competition and Coexistence of Species*. London: Pitman Books.
Post, W. (1981) The influence of rice rats *Oryzomys palustris* on the habitat use of the seaside sparrow *Ammospiza maritima*. *Behavioural Ecology and Sociobiology*, **9**, 35–40.
Post, W. & Wiley, J.E. (1976) The Yellow-shouldered Blackbird – present and future. *American Birds*, **30**, 13–20.
(1977) Reproductive interactions of the Shiny Cowbird and the Yellow-shouldered Blackbird. *The Condor*, **79**, 176–184.
Prance, G.T. (1982) Forest refuges: evidence from woody angiosperms. In *Biological Diversification in the Tropics*, ed. Prance, G.T., pp. 137–57. New York: Columbia University Press.
Pregill, G.K. & Olson, S.L. (1981) Zoogeograpy of West Indian vertebrates in relation to Pleistocene climatic cycles. *Annual Review of Ecology and Systematics*, **12**, 75–98.
Price, M.V. (1986) Introduction to the symposium: mechanistic approaches to the study of natural communities. *The American Zoologist*, **26**, 3–4.
Price, P.W. (1984) Alternative paradigms in community ecology. In *A New Ecology. Novel Approaches to Interactive Systems*, ed. Price, P.W., Slobodchikoff, C.N., & Gaud, W.S., pp. 353–83. New York: John Wiley & Sons.
Price, P.W., Westoby, M., Rice, B., Atsatt, P.R., Fritz, R.S., Thompson, J.N. & Mobley, K. (1986) Parasite mediation in ecological interactions. *Annual Review of Ecology and Systematics*, **17**, 487–505.
Price, T.D., Grant, P.R., Gibbs, H.L. & Boag, P.T. (1984) Recurrent patterns of natural selection in a population of Darwin's finches. *Nature*, **309**, 787–9.
Primack, R.B. & Howe, H.F. (1975) Interference competition between a hummingbird (*Amazilia tzacatl*) and skipper butterflies (Hesperiidae). *Biotropica*, **7**, 55–8.
Pulich, W.M. (1976) *The Golden-cheeked Warbler, a bioecological study*. Austin: Texas Parks and Wildlife Department.
Pulliam, H.R. (1985) Foraging efficiency, resource partitioning, and the coexistence of sparrow species. *Ecology*, **66**, 1829–36.
(1986) Niche expansion and contraction in a variable environment. *The American Zoologist*, **26**, 71–9.
Pulliam, H.R. & Dunning, J.B. (1987) The influence of food supply on local density and diversity of sparrows. *Ecology*, **68**, 1009–14.
Pulliam, H.R. & Enders, F. (1971) The feeding ecology of five sympatric finch species. *Ecology*, **52**, 557–66.
Pulliam, H.R. & Parker, T.A. III. (1979) Population regulation of sparrows. *Fortschritte der Zoologie*, **25**, 137–47.
Puttick, G.M. (1981) Sex-related differences in foraging behaviour of Curlew Sandpipers. *Ornis Scandinavica*, **12**, 13–7.
Pyke, G.H. (1978) Optimal foraging in hummingbirds: testing the marginal value theorem. *The American Zoologist*, **18**, 739–52.
(1980) The foraging behaviour of honeyeaters: a review and some comparisons with hummingbirds. *Australian Journal of Ecology*, **5**, 343–69.

(1983) Seasonal pattern of abundances of honeyeaters and their resources in heathland areas near Sydney. *Australian Journal of Ecology*, **8**, 217–33.

(1985) The relationships between abundances of honeyeaters and their food resources in open forest areas near Sydney. In *Birds of Eucalypt Forests and Woodlands: Ecology, Conservation, Management*, ed. Keast, A., Recher, H.F., Ford, H. & Saunders, D., pp. 65–77. Chipping Norton, N.S.W.: Surrey Beatty & Sons.

Pyke, G.H. & Recher, H.F. (1986) Relationship between nectar production and seasonal patterns of density and nesting of resident honeyeaters in heathland near Sydney. *Australian Journal of Ecology*, **11**, 195–200.

Quammen, M.L. (1982) Influence of subtle differences on feeding by shorebirds on intertidal mudflats. *Marine Biology*, **71**, 339–43.

Quinn, J.F. & Hastings, A. (1987) Extinction in subdivided habitats. *Conservation Biology*, **1**, 198–208.

Quinn, W.H. & Neal, V.T. (1982) Long-term variations in the southern oscillation, El Niño, and the Chilean subtropical rainfall. *Fishery Bulletin*, **81**, 363–74.

Rabøl, J. (1987) Coexistence and competition between overwintering Willow Warblers *Phylloscopus trochilus* and local warblers at Lake Naivasha, Kenya, *Ornis Scandinavica*, **18**, 101–21.

Rafe, R.W., Usher, M.B. & Jefferson, R.G. (1985) Birds on reserves: the influence of area and habitat on species richness. *Journal of Applied Ecology*, **22**, 327–35.

Raffaele, H.A. (1977) Comments on the extinction of *Loxigilla portoricensis grandis* in St Kitts, Lesser Antilles. *The Condor*, **79**, 389–90.

Raphael, M.G. Morrison, M.L. & Yoder-Williams, M.P. (1987) Breeding bird populations during twenty-five years of postfire succession in the Sierra Nevada. *The Condor*, **89**, 614–26.

Rasmusson, E.M. & Wallace, J.M. (1983) Meteorological aspects of the El Niño/southern oscillaton. *Science*, **222**, 1195–1202.

Real, L.A. (1975) A general analysis of resource allocation by competing individuals. *Theoretical Population Biology*, **8**, 1–11.

Recher, H.F., Lunney, D., & Dunn, I. (1986) *A Natural Legacy. Ecology in Australia*, 2nd edn. Sydney: Pergamon Press.

Reed, T.M. (1980) Turnover frequency in island birds. *Journal of Biogeography*, **7**, 329–35.

(1982) Interspecific territoriality in the Chaffinch and Great Tit on islands and the mainland of Scotland: playback and removal experiments. *Animal Behaviour*, **30**, 171–81.

(1983) The role of species-area relationships in reserve choice: a British example. *Biological Conservation*, **25**, 263–71.

Reichman, O.J. (1979) Concluding remarks. *The American Zoologist*, **19**, 1173–5.

Rey, J.R. (1984) Experimental tests of island biogeographic theory. In *Ecological Communities. Conceptual Issues and the Evidence*, ed. Strong, D.R.Jr, Simberloff, D., Abele, L.G. & Thistle, A.B., pp. 101–12. Princeton: Princeton University Press.

Reynoldson, T.B. & Bellamy, L.S. (1971) The establishment of interspecific competition in field populations, with an example of competition in action between *Polycelis nigra* (Mull.) and *P. tenuis* (Ijima) (Turbellaria, Tricladida). In *Dynamics of Populations. Proceedings of the Advanced Study Institute on 'Dynamics of Numbers in Populations', Oosterbeek, The Netherlands, 7–18 September 1970*, ed. den Boer, P.J. & Gradwell, G.R., pp. 282–97. Wageningen: Centre for Agricultural Publishing and Documentation.

Rice, J., Ohmart, R.D. & Anderson, B.W. (1983a) Habitat selection attributes of an avian community: a discriminant analysis investigation. *Ecological Monographs*, **53**, 263–90.

(1983b) Turnovers in species composition of avian communities in contiguous riparian habitats. *Ecology*, **64**, 1444–55.

Ricklefs, R.E. (1969) An analysis of nesting mortality in birds. *Smithsonian Contributions in Zoology*, **9**, 1–48.

(1975) Competition and the structure of bird communities [review]. *Evolution*, **29**, 581–5.

(1977) A discriminant function analysis of assemblages of fruit-eating birds in Central America. *The Condor*, **79**, 228–31.

(1979) *Ecology*, 2nd edn. New York: Chiron Press.

(1987) Community diversity: relative roles of local and regional processes. *Science*, **235**, 167–71.

Ricklefs, R.E. & Cox, G.W. (1977) Morphological similarity and ecological overlap among passerine birds on St Kitts, British West Indies. *Oikos*, **29**, 60–6.

Risser, P.G., Birney, E.C., Blocker, H.D., May, S.W., Parton, W.J. & Wiens, J.A. (1981) *The True Prairie Ecosystem*. Stroudsburg, Pennsylvania: Hutchinson Ross Publishing Company.

Robbins, C.S. (1980) Effect of forest fragmentation on breeding bird populations in the Piedmont of the Mid-Atlantic region. *Atlantic Naturalist*, **33**, 31–6.

Robinson, S.K. & Holmes, R.T. (1984) Effects of plant species and foliage structure on the foraging behavior of forest birds. *The Auk*, **101**, 672–84.

Roff, D.A. (1974) The analysis of a population model demonstrating the importance of dispersal in a heterogeneous environment. *Oecologia*, **15**, 259–75.

Rosenberg, K.V. & Raphael, M.G. (1986) Effects of forest fragmentation on vertebrates in Douglas-fir forests. In *Wildlife 2000. Modeling Habitat Relationships of Terrestrial Vertebrates*, ed. Verner, J., Morrison, M.L. & Ralph, C.J., pp. 263–72. Madison: University of Wisconsin Press.

Rosenzweig, M.L. (1979) Optimal habitat selection in two-species competitive systems. *Fortschritte fur Zoologie*, **25**, 283–93.

(1981) A theory of habitat selection. *Ecology*, **62**, 327–35.

(1985) Some theoretical aspects of habitat selection. In *Habitat Selection in Birds*, ed. Cody, M.L., pp. 517–40. New York: Academic Press.

Rosenzweig, M.L., Abramsky, Z. & Brand, S. (1984) Estimating species interactions in heterogeneous environments. *Oikos*, **42**, 329–40.

Rotenberry, J.T. (1978) Components of avian diversity along a multifactorial climatic gradient. *Ecology*, **59**, 693–9.

(1980) Dietary relationships among shrubsteppe passerine birds: competition or opportunism in a variable environment? *Ecological Monographs*, **50**, 93–110.

(1986) Habitat relationships of shrubsteppe birds: even 'good' models cannot predict the future. In *Wildlife 2000: Modeling Habitat Relationships of Terrestrial Vertebrates*, ed. Verner, J., Morrison, M.L. & Ralph, C.J., pp. 217–21. Madison: University of Wisconsin Press.

Rotenberry, J.T. & Wiens, J.A. (1978) Nongame bird communities in northwestern rangelands. In *Proceedings of the Workshop on Nongame Bird Habitat Management in Coniferous Forests of the Western United States. USDA Forest Service General Technical Report PNW-64*, ed. DeGraaf, R.M., pp. 32–46. Portland, Oregon: Pacific Northwest Forest and Range Experiment Station.

(1980a) Habitat structure, patchiness, and avian communities in North American steppe vegetation: a multivariate analysis. *Ecology*, **61**, 1228–50.

(1980b) Temporal variation in habitat structure and shrubsteppe bird dynamics. *Oecologia*, **47**, 1–9.

Rotenberry, J.T., Fitzner, R.E. & Rickard, W.H. (1979) Seasonal variation in avian community structure: differences in mechanisms regulating diversity. *The Auk*, **96**, 499–505.

Roth, R.R. (1976) Spatial heterogeneity and bird species diversity. *Ecology*, **57**, 773–82.

Rothstein, S.I., Verner, J. & Stevens, E. (1980) Range expansion and diurnal changes in dispersion of the Brown-headed Cowbird in the Sierra Nevada. *The Auk*, **97**, 253–67.

Roughgarden, J. (1979) *Theory of Population Genetics and Evolutionary Ecology: an Introduction*. New York: Macmillan.

(1983) Competition and theory in community ecology. *The American Naturalist*, **122**, 583–601.

(1986) A comparison of food-limited and space-limited competition communities. In *Community Ecology*, ed. Diamond, J. & Case, T.J., pp. 492–516. New York: Harper & Row.

Roughgarden, J. & Feldman, M. (1975) Species packing and predation pressure. *Ecology*, **56**, 489–92.

Royama, T. (1970) Factors governing the hunting behaviour and selection of food by the Great Tit (*Parus major* L.). *Journal of Animal Ecology*, **39**, 619–68.

Sabo, S.R. & Holmes, R.T. (1983) Foraging niches and the structure of forest bird communities in contrasting montane habitats. *The Condor*, **85**, 121–38.

Safina, C. & Burger, J. (1985) Common Tern foraging: seasonal trends in prey fish densities and competition with bluefish. *Ecology*, **66**, 1457–63.

Sale, P.F. (1978) Coexistence of coral reef fishes – a lottery for living space. *Environmental Biology of Fishes*, **3**, 85–102.

Sale, P.F. & Dybdahl, R. (1975) Determinants of community structure for coral reef fishes in an experimental habitat. *Ecology*, **56**, 1343–55.

Salo, J. (1987) Pleistocene forest refuges in the Amazon: evaluation of the biostratigraphical, lithostratigraphical and geomorphological data. *Annales Zoologici Fennici*, **24**, 203–11.

Salt, G.W. (1983) Roles: their limits and responsibilities in ecological and evolutionary research. *The American Naturalist*, **122**, 697–705.

Samson, F.B. & Lewis, S.J. (1979) Experiments on population regulation in two North American parids. *The Wilson Bulletin*, **91**, 222–33.

Sasvári, L., Török, J., & Tóth, L. (1987) Density dependent effects between three competitive bird species. *Oecologia*, **72**, 127–30.

Saunders, D.A. & de Rebeira, C.P. (1985) Turnover in breeding bird populations on Rottnest I., Western Australia. *Australian Wildlife Research*, **12**, 467–77.

Saunders, D.A., Arnold, G.W., Burbidge, A.A. & Hopkins, A.J.M., eds. (1987) *Nature Conservation: the role of remnants of native vegetation*. Chipping Norton, NSW: Surrey Beatty & Sons.

Savidge, J.A. (1987) Extinction of an island forest avifauna by an introduced snake. *Ecology*, **68**, 660–8.

Schemske, D.W. & Brokaw, N. (1981) Treefalls and the distribution of understory birds in a tropical forest. *Ecology*, **62**, 938–45.

Schluter, D. (1981) Does the theory of optimal diets apply in complex environments? *The American Naturalist*, **118**, 139–47.

(1982a) Seed and patch selection by Galápagos ground finches: relation to foraging efficiency and food supply. *Ecology*, **63**, 1106–20.

(1982b) Distributions of Galápagos ground finches along an altitudinal gradient: the importance of food supply. *Ecology*, **63**, 1504–17.

(1984) A variance test for detecting species associations with some example applications. *Ecology*, **65**, 998–1005.

(1986) Character displacement between distantly related taxa? Finches and bees in the Galápagos. *The American Naturalist*, **127**, 95–102.

Schluter, D. & Grant, P.R. (1982) The distributions of *Geospiza difficilis* in relation to *G. fuliginosa* in the Galápagos Islands: tests of three hypotheses. *Evolution*, **36**, 1213–26.

(1984) Determinants of morphological patterns in communities of Darwin's finches. *The American Naturalist*, **123**, 175–96.

Schmidt-Nielsen, K. (1984) *Scaling. Why is Animal Size so Important?* Cambridge: Cambridge University Press.

Schneider, D.C. & Duffy, D.C. (1985) Scale-dependent variability in seabird abundance. *Marine Ecology – Progress Series*, **25**, 211–8.

Schneider, D.C. & Piatt, J.F. (1986) Scale-dependent correlation of seabirds with schooling fish in a coastal ecosystem. *Marine Ecology – Progress Series*, **32**, 237–46.

Schoener, T.W. (1974a) Resource partitioning in ecological communities. *Science*, **185**, 27–39.

(1974b) Some methods for calculating competition coefficients from resource-utilization spectra. *The American Naturalist*, **108**, 332–340.

(1982) The controversy over interspecific competition. *The American Scientist*, **70**, 586–95.

(1983a) Reply to John Wiens [Letter to the Editor]. *The American Scientist*, **71**, 235.

(1983b) Field experiments on interspecific competition. *The American Naturalist*, **122**, 240–85.

(1983c) Rate of species turnover decreases from lower to higher organisms: a review of the data. *Oikos*, **41**, 372–77.

(1986a) Mechanistic approaches to community ecology: A new reductionism? *The American Zoologist*, **26**, 81–106.

(1986b) Overview: kinds of ecological communities – ecology becomes pluralistic. In *Community Ecology*, ed. Diamond, J. & Case, T.J., pp. 467–79. New York: Harper & Row.

Schoener, T.W. & Schoener, A. (1978) Inverse relation of survival of lizards with island size and avifaunal richness. *Nature*, **274**, 685–7.

Schreiber, R.W. & Schreiber, E.A. (1984) Central Pacific seabirds and the El Niño Southern Oscillation: 1982 to 1983 perspectives. *Science*, **225**, 713–6.

(1987) Tropical seabirds and the El Niño: 1985 and 1986 perspectives. In *Proceedings of the Jean Delacour Symposium*, pp. 352–9. North Hollywood, California: International Foundation for the Conservation of Birds.

Schroeder, M.H. & Sturges, D.L. (1975) The effect on the Brewer's Sparrow of spraying big sagebrush. *Journal of Range Management*, **28**, 294–7.

Scott, J.M., Mountainspring, S., Ramsey, F.L. & Kepler, C.B. (1986) Forest bird communities of the Hawaiian Islands: their dynamics, ecology, and conservation. *Studies in Avian Biology*, **9**, 1–431.

Seagle, S.W. (1986) Generation of species-area curves by a model of animal-habitat dynamics. In *Wildlife 2000. Modeling Habitat Relationships of Terrestrial Vertebrates*, ed. Verner, J., Morrison, M.L., & Ralph, C.J., pp. 281–5. Madison: University of Wisconsin Press.

Sharpe, D.M., Stearns, F.W., Burgess, R.L. & Johnson, W.C. (1981) Spatio-temporal patterns of forest ecosystems in man-dominated landscapes. In *Perspectives in Landscape Ecology*, ed. Tjallingii, S.P. & de Veer, A.A., pp. 109–16. Wageningen, The Netherlands: Pudoc.

Sherry, T.W. (1979) Competitive interactions and adaptive strategies of American Redstarts and Least Flycatchers in a northern hardwoods forest. *The Auk*, **96**, 265–83.

(1984) Comparative dietary ecology of sympatric, insectivorous neotropical flycatchers (Tyrannidae). *Ecological Monographs*, **54**, 313–38.

(1985) Adaptation to a novel environment: food, foraging, and morphology of the Cocos Island Flycatcher. *Ornithological Monographs,* **36**, 908–20.

Sherry, T.W. & Holmes, R.T. (1985) Dispersion patterns and habitat responses of birds in northern hardwoods forests. In *Habitat Selection in Birds*, ed. Cody, M.L., pp. 283–309. New York: Academic Press.

(1988) Habitat selection by breeding American Redstarts in response to a dominant competitor, the Least Flycatcher. *The Auk*, **105**, 350–64.

Shields, W.M. & Bildstein, K.L. (1979) Birds versus bats: behavioral interactions at a localized food source. *Ecology,* **60**, 468–74.

Short, L.L. (1979) Burdens of the picid hole-excavating habit. *The Wilson Bulletin*, **91**, 16–28.

Shugart, H.H. Jr. (1984) *A Theory of Forest Dynamics.* New York: Springer-Verlag.

Simberloff, D. (1976) Species turnover and equilibrium island biogeography. *Science,* **194**, 572–8.

(1978) Using island biogeographic distributions to determine if colonization is stochastic. *The American Naturalist,* **112**, 713–26.

(1982a) A succession of paradigms in ecology: essentialism to materialism and probabilism. In *Conceptual Issues in Ecology*, ed. Saarinen, E., pp. 63–99. Boston: D. Reidel Publishing Company.

(1982b) Reply. In *Conceptual Issues in Ecology*, ed. Saarinen, E., pp. 139–53. Boston: D Reidel Publishing Company.

(1982c) The status of competition theory in ecology. *Annales Zoologici Fennici,* **19**, 241–53.

(1983a) Competition theory, hypothesis-testing, and other community ecological buzzwords. *The American Naturalist,* **122**, 626–35.

(1983b) Sizes of coexisting species. In *Coevolution*, ed. Futuyma, D.J. & Slatkin, M., pp. 404–30. Sunderland, Massachusetts: Sinauer Associates.

(1983c) When is an island community in equilibrium? *Science,* **220**, 1275–6.

Simberloff, D. & Abele, L.G. (1976) Island biogeography theory and conservation practice. *Science,* **191**, 285–6.

(1982) Refuge design and island biogeographic theory: effects of fragmentation. *The American Naturalist,* **120**, 41–50.

(1984) Conservation and obfuscation: subdivision of reserves. *Oikos,* **42**, 399–401.

Simberloff, D. & Cox, J. (1987) Consequences and costs of conservation corridors. *Conservation Biology,* **1**, 63–71.

Skutch, A.F. (1985) Clutch size, nesting success, and predation on nests of Neotropical birds, reviewed. *Ornithological Monographs,* **36**, 575–94.

Slagsvold, T. (1975) Competition between the Great Tit *Parus major* and the Pied Flycatcher *Ficedula hypoleuca* in the breeding season. *Ornis Scandinavica,* **6**, 179–90.

(1978) Competition between the Great Tit *Parus major* and the Pied Flycatcher *Ficedula hypoleuca*: an experiment. *Ornis Scandinavica,* **9**, 46–50.

(1980) Habitat selection in birds: On the presence of other bird species with special regard to *Turdus pilaris*. *Journal of Animal Ecology,* **49**, 523–36.

Slobodkin, L.B. (1975) Comments from a biologist to a mathematician. In *Ecosystem Analysis and Prediction*, ed. Levin, S.A., pp. 318–29. Philadelphia: Society for Industrial and Applied Mathematics.

Smith, C.C. & Balda, R.P. (1979) Competition among insects, birds and mammals for conifer seeds. *The American Zoologist,* **19**, 1065–83.

Smith, F.E. (1975) Comments Revised – or, what I wish I had said. In *New Directions in the Analysis of Ecological Systems*, ed. Innis, G.S., pp. 231–5. La Jolla, California: The Society for Computer Simulation.

Smith, J.N.M., Grant, P.R., Grant, B.R., Abbott, I.J. & Abbott, L.K. (1978) Seasonal variation in feeding habits of Darwin's ground finches. *Ecology*, **59**, 1137–50.
Smith, K.G. (1982) Drought-induced changes in avian community structure along a montane sere. *Ecology*, **63**, 952–61.
(1986a) Winter population dynamics of three species of mast-eating birds in the eastern United States. *The Wilson Bulletin*, **98**, 407–18.
(1986b) Winter population dynamics of Blue Jays, Red-headed Woodpeckers, and Northern Mockingbirds in the Ozarks. *The American Midland Naturalist*, **115**, 52–62.
Smith, N.G. (1968) The advantage of being parasitized. *Nature*, **219**, 690–4.
Snyder, N.F.R. & Snyder, H.A. (1975) Raptors in range habitat. In *Proceedings of a Symposium on Management of Forest and Range Habitats for Nongame Birds. General Technical Report WO-1*, ed. Smith, D.R., pp. 190–209. Washington, DC: USDA Forest Service.
Solomon, M.E., Glen, D.M., Kendall, D.A. & Milson, N.F. (1976) Predation of overwintering larvae of codling moth (*Cydia pomonella* (L.)) by birds. *Journal of Applied Ecology*, **13**, 341–52.
Sousa, W.P. (1979) Disturbance in marine intertidal boulder fields: the nonequilibrium maintenance of species diversity. *Ecology*, **60**, 1225–39.
(1984) The role of disurbance in natural communities. *Annual Review of Ecology and Systematics*, **15**, 353–91.
Springer, A.M., Roseneau, D.G., Lloyd, D.S., McRoy, C.P. & Murphy, E.C. (1986) Seabird responses to fluctuating prey availability in the eastern Bering Sea. *Marine Ecology – Progress Series*, **32**, 1–12.
Stairs, G.R. (1985) Predation on overwintering codling moth populations by birds. *Ornis Scandinavica*, **16**, 323–4.
Steadman, D.W. & Olson, S.L. (1985) Bird remains from an archeological site on Henderson Island, South Pacific: Man-caused extinctions on an 'uninhabited' island. *Proceedings of the National Academy of Sciences, USA*, **82**, 6191.
Stearns, S.C. & Schmid-Hempel, P. (1987) Evolutionary insights should not be wasted. *Oikos*, **49**, 118–25.
Steenhof, K. & Kochert, M.N. (1985) Dietary shifts of sympatric buteos during a prey decline. *Oecologia*, **66**, 6–16.
Stewart, A.M. & Craig, J.L. (1985) Movements, status, access to nectar, and spatial organisation of the Tui. *New Zealand Journal of Zoology*, **12**, 649–66.
Stiles, F.G. (1978) Ecological and evolutionary implications of bird pollination. *The American Zoologist*, **18**, 715–27.
(1985) Seasonal patterns and coevolution in the hummingbird-flower community of a Costa Rican subtropical forest. *Ornithological Monographs*, **36**, 757–87.
Storass, T. (1988) A comparison of losses in artificial and naturally occurring Capercaillie nests. *Journal of Wildlife Management*, **52**, 123–6.
Strong, D.R. Jr. (1983) Natural variability and the manifold mechanisms of ecological communities. *The American Naturalist*, **122**, 636–60.
(1984a) Exorcising the ghost of competition past: phytophagous insects. In *Ecological Communities. Conceptual Issues and the Evidence*, ed. Strong, D.R. Jr, Simberloff, D., Abele, L.G. & Thistle, A.B., pp. 28–41. Princeton: Princeton University Press.
(1984b) Density-vague ecology and liberal population regulation in insects. In *A New Ecology, Novel Approaches to Interactive Systems*, ed. Price, P.W. & Slobodchikoff, C.N. Gaud W.S., pp. 313–27. New York: John Wiley & Sons.
(1986) Density vagueness: Abiding the variance in the demography of real populations. In *Community Ecology*, ed. Diamond, J. & Case, T.J., pp. 257–68. New York: Harper & Row.

Strong, D.R. Jr, Szyska, L.A. & Simberloff, D. (1979) Tests of community-wide character displacement against null hypotheses. *Evolution*, **33**, 897–913.

Sugden, G.L. & Beyersbergen, G.W. (1986) Effect of density and concealment on American Crow predation of simulated nests. *Journal of Wildlife Management*, **50**, 9–14.

Svärdson, G. (1949) Competition and habitat selection in birds. *Oikos*, **1**, 157–74.

Svensson, S., Carlsson, U.T. & Liljedahl, G. (1984) Structure and dynamics of an alpine bird community, a 20-year study. *Annales Zoologici Fennici*, **21**, 339–50.

Szaro, R.C. & Balda, R.P. (1979) Bird community dynamics in a ponderosa pine forest. *Studies in Avian Biology*, **3**, 1–66.

Szaro, R.C. & Jakle, M.D. (1985) Avian use of a desert riparian island and its adjacent scrub habitat. *The Condor*, **87**, 511–9.

Taylor, L.R. & Woiwod, I.P. (1980) Temporal stability as a density-dependent species characteristic. *Journal of Animal Ecology*, **49**, 209–24.

(1982) Comparative synoptic dynamics. I. Relationships between inter- and intra-specific spatial and temporal variance/mean population parameters. *Journal of Animal Ecology*, **51**, 879–906.

Taylor, L.R., Woiwod, I.P. & Perry, J.N. (1978) The density-dependence of spatial behaviour and the rarity of randomness. *Journal of Animal Ecology*, **47**, 383–406.

Taylor, R.J. (1984) *Predation*. London: Chapman and Hall.

Temple, S.A. (1986) Predicting impacts of habitat fragmentation on forest birds: a comparison of two models. In *Wildlife 2000. Modeling Habitat Relationships of Terrestrial Vertebrates*, ed. Verner, J., Morrison, M.L. & Ralph, C.J., pp. 301–4. Madison: University of Wisconsin Press.

Terborgh, J. (1980) The conservation status of neotropical migrants: present and future. In *Migrant Birds in the Neotropics. Ecology, Behavior, Distribution, and Conservation*, ed. Keast, A. & Morton, E.S., pp. 21–30. Washington, DC: Smithsonian Institution Press.

Terborgh, J. & Faaborg, J. (1973) Turnover and ecological release in the avifauna of Mona Island, Puerto Rico. *The Auk*, **90**, 759–79.

(1980) Factors affecting the distribution and abundance of North American migrants in the eastern Caribbean region. In *Migrant Birds in the Neotropics. Ecology, Behavior, Distribution, and Conservation*, ed. Keast, A. & Morton, E.S., pp. 145–55. Washington, DC: Smithsonian Institution Press.

Terborgh, J. & Weske, J.S. (1975) The role of competition in the distribution of Andean birds. *Ecology*, **56**, 562–76.

Terborgh, J. & Winter, B. (1980) Some causes of extinction. In *Conservation Biology: An Evolutionary-Ecological Perspective*, ed. Soulé, M.E. & Wilcox, B.A., pp. 119–33. Sunderland, Massachusetts: Sinauer Associates.

Terborgh, J., Faaborg, J.W. & Brockmann, H.J. (1978) Island colonization by Lesser Antillean birds. *The Auk*, **95**, 59–72.

Thompson, C.F. & Nolan, V. Jr. (1973) Population biology of the Yellow-breasted Chat (*Icteria virens* L.) in southern Indiana. *Ecological Monographs*, **43**, 145–71.

Thompson, P.M. & Lawton, J.H. (1983) Seed size diversity, bird species diversity and interspecific competition. *Ornis Scandinavica*, **14**, 327–36.

Thomson, J.D. (1980) Implications of different sorts of evidence for competition. *The American Naturalist*, **116**, 719–26.

Tiainen, J. (1983) Dynamics of a local population of the Willow Warbler *Phylloscopus trochillus* in southern Finland. *Ornis Scandinavica*, **14**, 1–15.

Tilman, D. (1982) *Resource Competition and Community Structure*. Princeton: Princeton University Press.

(1986) A consumer-resource approach to community structure. *The American Zoologist*, **26**, 5–22.
(1987) The importance of the mechanisms of interspecific competition. *The American Naturalist*, **129**, 769–74.
Toft, C.A., Trauger, D.L. & Murdy, H.W. (1982) Tests for species interactions: breeding phenology and habitat use in subarctic ducks. *The American Naturalist*, **120**, 586–613.
Tomiałojć, L. (1979) The impact of predation on urban and rural Woodpigeon [*Columba palumbus*: (L.)] populations. *Polish Ecological Studies*, **5**, 141–220.
Tomiałojć, L. & Profus, P. (1977) Comparative analysis of the breeding bird communities in two parks of Wroclaw and adjacent Querco-Carpinetum forest. *Acta Ornithologica*, **16**, 117–77.
Török, J. (1987) Competition for food between Great Tit *Parus major* and Blue Tit *P. caeruleus* during the breeding season. *Acta Reg. Soc. Sci. Litt. Gothoburgensis. Zoologica*, **14**, 149–52.
Ulfstrand, S. (1977) Foraging niche dynamics and overlap in a guild of passerine birds in a south Swedish coniferous woodland. *Oecologia*, **27**, 23–45.
Underwood, A.J. (1986a) What is a community? In *Patterns and Processes in the History of Life*, ed. Raup, D.M. & Jablonski, D., pp. 351–67. Berlin: Springer-Verlag.
(1986b) The analysis of competition by field experiments. In *Community Ecology. Pattern and Process*, ed. Kikkawa, J. & Anderson, D.J., pp. 240–68. Oxford: Blackwell Scientific Publications.
Underwood, A.J. & Denley, E.J. (1984) Paradigms, explanations, and generalizations in models for the structure of intertidal communities on rocky shores. In *Ecological Communities. Conceptual Issues and the Evidence*, ed. Strong, D.R. Jr, Simberloff, D., Abele, L.G. & Thistle, A.B., pp. 151–80. Princeton: Princeton University Press.
Urban, D.L. (1986) 'Forest Bird Demography in a Landscape Mosaic.' Ph.D. Dissertation. Knoxville: University of Tennessee.
Urban, D.L. & Shugart, H.H. Jr. (1986) Avian demography in mosaic landscapes: modeling paradigm and preliminary results. In *Wildlife 2000. Modeling Habitat Relationships of Terrestrial Vertebrates*, ed. Verner, J., Morrison, M.L. & Ralph, C.J., pp. 273–9. Madison: University of Wisconsin Press.
Urban, D.L., O'Neill, R.V. & Shugart, H.H.Jr. (1987) Landscape ecology. *BioScience*, **37**, 119–27.
Väisänen, R.A. & Järvinen, O. (1977a) Structure and fluctuation of the breeding bird fauna of a north Finnish peatland area. *Ornis Fennica*, **54**, 143–53.
(1977b) Dynamics of protected bird communities in a Finnish archipelago. *Journal of Animal Ecology*, **46**, 891–908.
Väisänen, R.A., Järvinen, O. & Rauhala, P. (1986) How are extensive, human-caused habitat alterations expressed on the scale of local bird populations in boreal forests? *Ornis Scandinavica*, **17**, 282–92.
Valle, C.A. (1985) Aleracion en las boblaciones del cormoran no volador, el pinguino y otras aves marinas en Galápagos por efecto de El Niño 1982–83 y su subsequente recuperacion. In *El Niño in the Galápagos Islands: the 1982–1983 Event*, ed. Robinson, G. & del Pino, E.M., pp. 245–58. Quito, Ecuador: Fundacion Darwin.
Vandermeer, J. (1981) *Elementary Mathematical Ecology*. New York: John Wiley & Sons.
Van Devender, T.R. (1986) Climatic cadences and the composition of Chihuahuan desert communities: the Late Pleistocene packrat midden record. In *Community ecology*, ed. Diamond, J. & Case, T.J., pp. 285–99. New York: Harper & Row.
Van Riper C., III, Van Riper, S.G., Goff, M.L. & Laird, M. (1986) The epizootiology and ecological significance of malaria in Hawaiian land birds. *Ecological Monographs*, **56**, 327–44.

Van Valen, L. & Pitelka, F.A. (1974) Commentary – intellectual censorship in ecology. *Ecology*, **55**, 925–6.

Verner, J. (1986) Summary: predicting effects of habitat patchiness and fragmentation – the researcher's viewpoint. In *Wildlife 2000, Modeling Habitat Relationships of Terrestrial Vertebrates*, ed. Verner, J., Morrison, M.L., & Ralph, C.J., pp. 327–9. Madison: University of Wisconsin Press.

Verner, J. & Ritter, L.V. (1983) Current status of the Brown-headed Cowbird in the Sierra National Forest. *The Auk*, **100**, 355–68.

von Haartman, L. (1971) Population dynamics. In *Avian Biology*, ed. Farner, D.S. & King, J.R., pp. 391–459. New York: Academic Press.

(1973) Changes in the breeding bird fauna of North Europe. In *Breeding Biology of Birds*, ed. Farner, D.S., pp. 448–81. Washington, DC: National Academy of Sciences.

Vuilleumier, F. & Simberloff, D. (1980) Ecology versus history as determinants of patchy and insular distributions in high Andean birds. *Evolutionary Biology*, **12**, 235–379.

Wagner, J.L. (1981) Seasonal change in guild structure: oak woodland insectivorous birds. *Ecology*, **62**, 973–81.

Waide, R.B. & Reagan, D.P. (1983) Competition between West Indian anoles and birds. *The American Naturalist*, **121**, 133–8.

Waite, R.K. (1984) Sympatric corvids: Effects of social behaviour, aggression and avoidance on feeding. *Behavioral Ecology and Sociobiology*, **15**, 55–9.

Walkinshaw, L.H. (1941) The Prothonotary Warbler, a comparison of nesting conditions in Tennessee and Michigan. *The Wilson Bulletin*, **53**, 3–21.

Walter, G.H., Hulley, P.E. & Craig, A.J.F.K. (1984) Speciation, adaptation and interspecific competition. *Oikos*, **43**, 246–8.

Ward, P. (1965) Feeding ecology of the Black-faced Dioch *Quelea quelea* in Nigeria. *Ibis*, **107**, 173–214.

Warner, R.E. (1968) The role of introduced diseases in the extinction of the endemic Hawaiian avifauna. *The Condor*, **70**, 101–20.

Waser, P.M. & Wiley, R.H. (1979) Mechanisms and evolution of spacing in animals. In *Handbook of Behavioral Neurobiology. Volume 3. Social Behavior and Communication*, ed. Marler, P. & Vandenbergh, J.G., pp. 159–223. New York: Plenum Press.

Watson, A. & Moss, R. (1972) A current model of population dynamics in Red Grouse. *Proceedings 15th International Ornithological Congress*, 134–49.

Watt, K.E.F. (1971) Dynamics of populations: a synthesis. In *Dynamics of Populations 1970*, ed. den Boer, P.J. & Gradwell, G.R., pp. 568–80. Waageningen: Centre for Agricultural Publication and Documentation.

Weatherhead, P.J. (1986) How unusual are unusual events? *The American Naturalist*, **128**, 150–4.

Weathers, W.W. & van Riper, C. III. (1982) Temperature regulation in two endangered Hawaiian honeycreepers: the Palila (*Psittirostra bailleui*) and the Laysan Finch (*Psittirostra cantans*). *The Auk*, **99**, 667–74.

Wegner, J.F. & Merriam, G. (1979) Movements by birds and small mammals between a wood and adjoining farmland habitats. *Journal of Applied Ecology*, **16**, 349–57.

Welden, C.W. & Slauson, W.L. (1986) The intensity of competition versus its importance: an overlooked distinction and some implications. *The Quarterly Review of Biology*, **61**, 23–44.

Welty, J.C. (1982) *The Life of Birds*, 3rd edn. Philadelphia: Saunders College Publishing.

Wendland, V. (1984) The influence of prey fluctuations on the breeding success of the tawny owl *Strix aluco*. *Ibis*, **126**, 284–95.

Wetmore, A. (1925) Bird life among lava rock and coral sand. *National Geographic Magazine*, **48** (7), 76–108.

Whitcomb, B.L., Whitcomb, R.F. & Bystrak, D. (1977) Island biogeography and 'habitat islands' of eastern forest. Part III. Long-term turnover and effects of selective logging on the avifauna of forest fragments. *American Birds*, **31**, 17–23.

Whitcomb, R.F., Lynch, J.F., Klimkiewicz, M.K., Robbins, C.S., Whitcomb, B.L. & Bystrak, D. (1981) Effects of forest fragmentation on avifauna of the eastern deciduous forest. In *Forest Island Dynamics in Man-dominated Landscapes*, ed. Burgess, R.L. & Sharpe, D.M., pp. 125–205. New York: Springer-Verlag.

Whitcomb, R.F., Lynch, J.F., Opler, P.A. & Robbins, C.S. (1976) Island biogeography and conservation: strategies and limitations. *Science*, **193**, 1030–2.

White, P.S. (1979) Pattern, process, and natural disturbance in vegetation. *Botanical Review*, **45**, 229–99.

White, P.S. & Pickett, S.T.A. (1985) Natural disturbance and patch dynamics: an introduction. In *The Ecology of Natural Disturbance and Patch Dynamics*, ed. Pickett, S.T.A. & White, P.S., pp. 3–13. New York: Academic Press.

Whittaker, R.H. (1975) *Communities and Ecosystems*, 2nd edn. New York: Macmillan.

Widén, P., Andrén, H., Angelstam, P. & Lindström, E. (1987) The effect of prey vulnerability: Goshawk predation and population fluctuations of small game. *Oikos*, **49**, 233–5.

Wiens, J.A. (1974) Climatic instability and the 'ecological saturation' of bird communities in North American grasslands. *The Condor*, **76**, 385–400.

(1976a) Review of 'Competition and the Structure of Bird Communities' by M. Cody. *The Auk*, **93**, 396–400.

(1977b) Population responses to patchy environments. *Annual Review of Ecology and Systematics*, **7**, 81–120.

(1977) On competition and variable environments. *The American Scientist*, **65**, 590–7.

(1981a) Scale problems in avian censusing. *Studies in Avian Biology*, **6**, 513–21.

(1981b) Single-sample surveys of communities: Are the revealed patterns real? *The American Naturalist*, **117**, 90–8.

(1981c) Modelling the energy requirements of seabird populations. In *Seabird Energetics*, ed. Whittow, G.C. & Rahn, H., pp. 255–84. New York: Plenum Press.

(1983a) Avian community ecology: an iconoclastic view. In *Perspectives in Ornithology*, ed. Brush, A.H. & Clark, G.A. Jr, pp. 355–403. Cambridge: Cambridge University Press.

(1983b) Interspecific competition [Letter to the editor]. *The American Scientist*, **71**, 234–5.

(1984a) On understanding a non-equilibrium world: myth and reality in community patterns and processes. In *Ecological Communities. Conceptual Issues and the Evidence*, ed. Strong, D.R. Jr, Simberloff, D., Abele, L.G. & Thistle, A.B., pp. 439–57. Princeton: Princeton University Press.

(1984b) Resource systems, populations, and communities. In *A New Ecology. Novel Approaches to Interactive Systems*, ed. Price, P.W., Slobodchikoff, C.N. Gaud, W.S., pp. 397–436. New York: John Wiley & Sons.

(1984c) The place of long-term studies in ornithology. *The Auk*, **101**, 202–03.

(1985a) Vertebrate responses to environmental patchiness in arid and semiarid ecosystems. In *The Ecology of Natural Disturbance and Patch Dynamics*, ed. Pickett, S.T.A. & White, P.S., pp. 169–93. New York: Academic Press.

(1985b) Habitat selection in variable environments: shrub-steppe birds. In *Habitat Selection in Birds*, ed. Cody, M.L., pp. 227–51. New York: Academic Press.

(1986) Spatial scale and temporal variation in studies of shrubsteppe birds. In *Community Ecology*, ed. Diamond, J. & Case, T.J., pp. 154–72. New York: Harper & Row.

Wiens, J.A. & Dyer, M.I. (1975a) Simulation modelling of Red-winged Blackbird impact on grain crops. *Journal of Applied Ecology*, **12**, 63–82.
(1975b) Rangeland avifaunas: Their composition, energetics, and role in the ecosystem. In *Symposium on Management of Forest and Range Habitats for Nongame Birds*, ed. Smith, D.R., pp. 176–82. *USDA Forest Service General Technical Report WO-1*, Washington, DC: USDA Forest Service.
(1977) Assessing the potential impact of granivorous birds in ecosystems. In *Granivorous Birds in Ecosystems*, ed. Pinowski, J. & Kendeigh, S.C., pp. 205–66. Cambridge: Cambridge University Press.
Wiens, J.A. & Johnston, R.F. (1977) Adaptive correlates of granivory in birds. In *Granivorous Birds in Ecosystems*, ed. Pinowski, J. & Kendeigh, S.C., pp. 301–40. Cambridge: Cambridge University Press.
Wiens, J.A. & Rotenberry, J.T. (1979) Diet niche relationships among North American grassland and shrubsteppe birds. *Oecologia*, **42**, 253–92.
(1980) Patterns of morphology and ecology in grassland and shrubsteppe bird populations. *Ecological Monographs*, **50**, 287–308.
(1981a) Habitat associations and community structure of birds in shrubsteppe environments. *Ecological Monographs*, **51**, 21–41.
(1981b) Censusing and the evaluation of avian habitat occupancy. *Studies in Avian Biology*, **6**, 522–32.
(1985) Response of breeding passerine birds to rangeland alteration in a North American shrubsteppe locality. *Journal of Applied Ecology*, **22**, 655–68.
(1987) Shrub-steppe birds and the generality of community models: a response to Dunning. *The American Naturalist*, **129**, 920–7.
Wiens, J.A., Addicott, J.F., Case, T.J. & Diamond, J. (1986) Overview: The importance of spatial and temporal scale in ecological investigations. In *Community Ecology*, ed. Diamond, J. & Case, T.J., pp. 145–53. New York: Harper & Row.
Wiens, J.A., Crawford, C.S., & Gosz, J.R. (1985) Boundary dynamics: a conceptual framework for studying landscape ecosystems. *Oikos*, **45**, 421–7.
Wiens, J.A., Rotenberry, J.T. & Van Horne, B. (1986) A lesson in the limitations of field experiments: shrubsteppe birds and habitat alteration. *Ecology*, **67**, 365–76.
(1987) Habitat occupancy patterns of North American shrubsteppe birds: the effects of spatial scale. *Oikos*, **48**, 132–47.
Wilcove, D.S. (1985) Nest predation in forest tracts and the decline of migratory songbirds. *Ecology*, **85**, 1211–4.
Wilcove, D.S. & Terborgh, J.W. (1984) Patterns of population decline in birds. *American Birds*, **38**, 10–13.
Wilcove, D.S. & Whitcomb, R.F. (1983) Gone with the trees. *Natural History Magazine (September)*, 82–91.
Wilcove, D.S., McLellan, C.H. & Dobson, A.P. (1986) Habitat fragmentation in the temperate zone. In *Conservation Biology. The Science of Scarcity and Diversity*, ed. Soulé, M.E., pp. 237–56. Sunderland, Massachusetts: Sinauer Associates.
Wilcox, B.A. & Murphy, D.D. (1985) Conservation strategy: the effects of fragmentation on extinction. *The American Naturalist*, **125**, 879–87.
Wiley, J.W. (1985) Shiny cowbird parasitism in two avian communities in Puerto Rico. *The Condor*, **87**, 165–76.
Williams, E.E. (1969) The ecology of colonization as seen in the zoogeography of anoline lizards on small islands. *Quarterly Review of Biology*, **44**, 345–89.
Williams, G.R. (1984) Has island biogeography theory any relevance to the design of biological reserves in New Zealand? *Journal of the Royal Society of New Zealand*, **14**, 7–10.

Williams, J.B. & Batzli, G.O. (1979) Interference competition and niche shifts in the bark-foraging guild in central Illinois. *The Wilson Bulletin,* **91**, 400–11.
Williamson, G.B. (1978) A comment on equilibrium turnover rates for islands. *The American Naturalist,* **112**, 241–3.
Williamson, K. (1975) Birds and climatic change. *Bird Study,* **22**, 143–64.
Williamson, M. (1981) *Island Populations.* Oxford: Oxford University Press.
(1983) The land-bird community of Skokholm: ordination and turnover. *Oikos,* **41**, 378–84.
(1984) The measurement of population variability. *Ecological Entomology,* **9**, 239–41.
Willis, E.O. (1974) Populations and local extinctions of birds on Barro Colorado Island, Panama. *Ecological Monographs,* **44**, 153–69.
Wilson, E.O. (1975) *Sociobiology. The New Synthesis.* Cambridge: Harvard University Press.
Wilson, E.O. & Willis, E.O. (1975) Applied biogeography. In *Ecology and Evolution of Communities*, ed. Cody, M.L. & Diamond, J.M., pp. 522–34. Cambridge: Harvard University Press.
Winstanley, D., Spencer, R. & Williamson, K. (1974) Where have all the Whitethroats gone? *Bird Study,* **21**, 1–14.
Winternitz, B.L. (1976) Temporal change and habitat preference of some montane breeding birds. *The Condor,* **78**, 383–93.
Woinarski, J.C.Z. (1984) Small birds, lerp-feeding and the problem of honeyeaters. *The Emu,* **84**, 137–41.
Wolda, H. (1983) Seasonality of Homoptera on Barro Colorado Island. In *The Ecology of a Tropical Forest*, ed. Leigh, E.G., Rand, A.S. & Windsor, D.M., pp. 319–30. Oxford: Oxford University Press.
Wolf, L.L., Hainsworth, F.R. & Gill, F.B. (1975) Foraging efficiencies and time budgets of nectar feeding birds. *Ecology,* **56**, 117–28.
Wolf, L.L., Stiles, F.G. & Hainsworth, F.R. (1976) Ecological organization of a tropical, highland hummingbird community. *Journal of Animal Ecology,* **45**, 349–79.
Wright, S.J. (1979) Competition between insectivorous lizards and birds in central Panama. *American Zoologist,* **19**, 1145–56.
(1981) Extinction-mediated competition: the *Anolis* lizards and insectivorous birds of the West Indies. *The American Naturalist,* **117**, 181–92.
(1985) How isolation affects rates of turnover of species on islands. *Oikos,* **44**, 331–40.
Wright, S.J. & Hubbell, S.P. (1983) Stochastic extinction and reserve size: a focal species approach. *Oikos,* **41**, 466–76.
Wright, S.J., Faaborg, J. & Campbell, C.J. (1985) Birds form tightly structured communities in the Pearl Archipelago, Panama. *Ornithological Monographs,* **36**, 798–812.
Wunderle, J.M. Jr. & Pollock, K.H. (1985) The bananaquit-wasp nesting association and a random choice model. *Ornithological Monographs,* **36**, 595–603.
Wunderle, J.M. Jr, Diaz, A., Velazquez, I. & Scharrón, R. (1987) Forest openings and the distribution of understory birds in a Puerto Rican rainforest. *The Wilson Bulletin,* **99**, 22–37.
Yahner, R.H. (1983) Seasonal dynamics, habitat relationships, and management of avifauna in farmstead shelterbelts. *Journal of Wildlife Management,* **47**, 85–104.
(1984) Effects of habitat patchiness created by a Ruffed Grouse management plan on breeding bird communities. *The American Midland Naturalist,* **111**, 409–13.
(1987) Short-term avifaunal turnover in small even-aged forest habitats. *Biological Conservation,* **39**, 39–47.
Yahner, R.H. & Wright, A.L. (1985) Depredation on artificial ground nests: effects of edge and plot age. *Journal of Wildlife Management*, **49**, 508–13.

Zeleny, L. (1978) Nesting box programs for bluebirds and other passerines. In *Endangered Birds. Management Techniques for Preserving Threatened Species*, ed. Temple, S.A., pp. 55–60. Madison: University of Wisconsin Press.

Zimmerman, B.L. & Bierregaard, R.O. (1986) Relevance of the equilibrium theory of island biogeography and species-area relations to conservation with a case from Amazonia. *Journal of Biogeography*, **13**, 133–43.

Zimmerman, J.L. (1971) The territory and its density dependent effect in *Spiza americana*. *The Auk*, **88**, 591–612.

(1984) Nest predation and its relationship to habitat and nest density in Dickcissels. *The Condor*, **86**, 68–72.

Zumeta, D.C. & Holmes, R.T. (1978) Habitat shift and roadside mortality of Scarlet Tanagers during a cold wet New England spring. *The Wilson Bulletin*, **90**, 575–86.

# Author index

# Subject index